'An invaluable and comprehensive overview  historic Paris Agreement that brings togeth who evaluate its effectiveness from a varie., national circumstances – a major contribution, illuminating hopes while voicing concerns.'

Richard Falk, Professor Emeritus of International Law,
Princeton University, USA

'In this exciting new book, Vesselin Popovski and colleagues take on the pressing, but as yet-understudied, question of "will the Paris Agreement work?" Over 20 expert scholars provide an impressively broad set of perspectives into whether the Paris Agreement's adoption of a new bottom-up set of obligations and a much better-defined structure for monitoring and implementation will pay off. The book is particularly valuable because its chapters adopt both legal as well as ethical perspectives, derive valuable insights from the Kyoto Protocol experience, provide important details about all aspects of Paris' implementation mechanisms, and clarify the many ways local conditions (in India, Latin America, Europe, Asia and the US) will shape the prospects for the Paris Agreement's successful implementation. This book should be on the shelf of anyone interested in the fate of the Paris Agreement and of the planet.'

Ronald Mitchell, Professor of Political Science and
Environmental Studies, University of Oregon, USA

'This timely and valuable work recognizes the irony in the international climate change regime's tendency to enforce strict compliance with procedural obligations in such matters as transparency in state reporting on mitigation actions while being utterly unable to create agreement on aggregate emission reductions sufficient to avoid dangerous anthropogenic interference with the climate system. Among the many important insights presented by the book's contributing authors is the suggestion that the Paris Agreement will not succeed in its aims unless the parties to the treaty develop a broadly conceived and strong compliance system that keeps them faithful to the Agreement's target of limiting warming to well below 2 degrees Celsius.'

Alexander Zahar, Asst Professor, Macquerie University, Australia, and
Founder and editor-in-chief of the peer-reviewed journal Climate Law

'States' implementation of international law depends on a mixture of interests and goals (political, economic, social, and reputational), expectations, knowledge, and capability – as well as external pressures or incentives – that move governments towards making and upholding international agreements. The contributors to this volume offer rich, detailed and varied analyses of the problems and prospects for efforts to implement the Paris climate agreement, addressing what is surely one of the key global problems of our time.'

Prof. Alistair Edgar Professor, Department of Political
Science, Wilfrid Laurier University, Canada

'Climate change surpasses the ability of any one country to tackle on their own and requires global collective action. This book offers a comprehensive analysis of the architectural framework for collective action on climate change and incisive insights into the mechanisms for facilitation of implementation of the Paris Agreement with compelling examples across levels of governance around the world.'

Maria Ivanova, Associate Professor of Global Governance,
University of Massachusetts Boston, USA

The Implementation of the Paris Agreement on Climate Change

In December 2015, 196 parties to the United Nations Framework Convention on Climate Change (UNFCCC) adopted the Paris Agreement, seen as a decisive landmark for global action to stop human-induced climate change. The Paris Agreement will replace the 1997 Kyoto Protocol which expires in 2020, and it creates legally binding obligations on the parties, based on their own bottom-up voluntary commitments to implement Nationally Determined Contributions (NDCs). The codification of the climate change regime has advanced well, but the implementation of it remains uncertain.

This book focuses on the implementation prospects of the Agreement, which is a challenge for all and will require a fully comprehensive burden-sharing framework. Parties need to meet their own NDCs, but also to finance and transfer technology to others who do not have enough. How equity-based and facilitative the process will be, is of crucial importance. The volume examines a broad range of issues including the lessons that can be learnt from the implementation of previous environmental legal regimes, climate policies at national and sub-national levels and whether the implementation mechanisms in the Paris Agreement are likely to be sufficient.

Written by leading experts and practitioners, the book diagnoses the gaps and lays the ground for future exploration of implementation options. This collection will be of interest to policy-makers, academics, practitioners, students and researchers focusing on climate change governance.

Vesselin Popovski is Professor and Vice Dean of the Law School, Executive Director of the Centre for the Study of the UN, Jindal Global University, India. Until 2014 he was Senior Academic Officer at the United Nations University, Tokyo. He has published numerous articles in peer-reviewed journals and has authored and edited over 20 books.

Law, Ethics and Governance Series

Recent history has emphasized the potentially devastating effects of governance failures in governments, government agencies, corporations and the institutions of civil society. 'Good governance' is seen as necessary, if not crucial, for economic success and human development.

Although the disciplines of law, ethics, politics, economics and management theory can provide insights into the governance of organizations, governance issues can only be dealt with by interdisciplinary studies, combining several (and sometimes all) of those disciplines. This series aims to provide such interdisciplinary studies for students, researchers and relevant practitioners.

Series Editor: Charles Sampford, Director, Institute for Ethics, Governance and Law, Griffith University, Australia

Recent titles in this series

Positive Social Identity
The Quantitative Analysis of Ethics
Nick Duncan

Global Governance and Regulation
Order and Disorder in the 21st Century
Edited by Leon Wolff and Danielle Ireland-Piper

For more information about this series, please visit:
www.routledge.com/Law-Ethics-and-Governance/book-series/LEAG

The Implementation of the Paris Agreement on Climate Change

Edited by Vesselin Popovski

LONDON AND NEW YORK

First published 2019 by Routledge

2 Park Square, Milton Park, Abingdon, Oxfordshire OX14 4RN
52 Vanderbilt Avenue, New York, NY 10017

Routledge is an imprint of the Taylor & Francis Group, an informa business

First issued in paperback 2019

Copyright © 2019 selection and editorial matter, Vesselin Popovski; individual chapters, the contributors

The right of Vesselin Popovski to be identified as the author of the editorial material, and of the authors for their individual chapters, has been asserted in accordance with sections 77 and 78 of the Copyright, Designs and Patents Act 1988.

All rights reserved. No part of this book may be reprinted or reproduced or utilised in any form or by any electronic, mechanical, or other means, now known or hereafter invented, including photocopying and recording, or in any information storage or retrieval system, without permission in writing from the publishers.

Notice:
Product or corporate names may be trademarks or registered trademarks, and are used only for identification and explanation without intent to infringe.

British Library Cataloguing-in-Publication Data
A catalogue record for this book is available from the British Library

Library of Congress Cataloging-in-Publication Data
Names: Popovski, Vesselin, author.
Title: The implementation of the Paris Agreement on
Climate Change / Vesselin Popovski.
Description: New York, NY : Routledge, 2018. |
Series: Law, ethics and governance |
Includes bibliographical references and index.
Identifiers: LCCN 2018017035 | ISBN 9780415791236 (hardback)
Subjects: LCSH: United Nations Framework Convention on
Climate Change (1992 May 9). Protocols, etc. (2015 December 12) |
Climatic changes–Government policy. | Climate change mitigation.
Classification: LCC K3585.5.A42015 P67 2018 | DDC 344.04/6342–dc23
LC record available at https://lccn.loc.gov/2018017035

ISBN: 978-0-415-79123-6 (hbk)
ISBN: 978-0-367-48148-3 (pbk)

Typeset in Galliard
by Out of House Publishing

Contents

List of figures	x
List of tables	xi
Notes on contributors	xii
Series editor's preface	xvii
Foreword	xix
Acknowledgements	xxii

1 Implementation of international environmental agreements 1
   VESSELIN POPOVSKI

2 'Hard' and 'soft' law on climate change: comparing the 1997 Kyoto Protocol with the 2015 Paris Agreement 19
   VESSELIN POPOVSKI

3 A comparative architectural analysis of the 1997 Kyoto Protocol and the 2015 Paris Agreement and other ways to counter environmental 'ratification fatigue' 42
   TRUDY FRASER

4 Promoting the implementation of international environmental law: mechanisms, obligations and indicators 57
   NATALIA ESCOBAR-PEMBERTHY

5 Strengthening compliance under the Convention on Biological Diversity: comparing follow-up and review systems with the global climate regime 79
   ANA MARÍA ULLOA AND SYLVIA I. KARLSSON-VINKHUYZEN

6 Five short words and a moral reckoning: the Paris regime's CMA-APA equity stocktake process 104
   HUGH BREAKEY

7 Equity in the global stocktake   126
   SWAPNA PATHAK AND SIDDHARTH PATHAK

8 Stakeholder perceptions of the implementation capacity
   of the climate change regime   138
   TIM CADMAN AND TEK MARASENI

9 Technological ethics, faith and climate control: the
   misleading rhetoric surrounding the Paris Agreement   151
   HAROLD P. SJURSEN

10 The implementation of the principle of common but
   differentiated responsibilities within the Paris Agreement:
   a governance values analysis   164
   ANNA HUGGINS AND ROWENA MAGUIRE

11 After Paris: do we need an international agreement on green
   compulsory licensing?   179
   DONG QIN

12 Low-carbon market opportunities and a brief discussion on
   lessons learned from the Adaptation Fund   194
   ANDREA FERRAZ YOUNG

13 Understanding the relationship between global and national
   climate regimes and local realities in India   212
   ARNAB BOSE AND SEEMA SHARMA

14 Paris Agreement and climate change in India: to be
   or not to be?   222
   ADITYA RAMJI

15 Comparing the US and India on climate change: how
   the tables turned   232
   ARMIN ROSENCRANZ AND RAJNISH WADEHRA

16 Cities and the Paris Agreement   263
   KELSEY COOLIDGE

17 Beyond COP 21: what does the Paris Agreement mean for
European climate and energy policy? 283
ANNIKA BOSE STYCZYNSKI

18 Strengths, weaknesses, opportunities and threats to the
implementation of the Paris Agreement in the Latin
American region 297
TRISHNA MOHAN KRIPALANI AND GARGI KATIKITHALA

Index 309

Figures

4.1	Reporting rate for global environmental conventions	75
4.2	Facilitation implementation mechanisms for the Ramsar Convention	76
4.3	Facilitation implementation mechanisms for CITES	76
4.4	Facilitation implementation mechanisms for the Stockholm Convention	77
6.1	Official actors in the stocktake process	107
6.2	Potential actors in the CMA-APA equity stocktake process	124
13.1	Causal mapping of global climate policy at community level	213
13.2	The original eight missions and nodal ministries and departments under the NAPCC	216
13.3	Stage 1: information collection and dissipation	217
13.4	Stage 2: creation of solutions/options inventory	218
13.5	Stage 3: project implementation and maintenance stage	219
14.1	India's sector-wise emissions (CO_2 eq), 2010	226
15.1	US greenhouse gas emissions per capita and per dollar of GDP, 1990–2014	245

Tables

4.1	Facilitation implementation mechanisms in global environmental conventions	70
4.2	Implementation mechanisms included in the analysis by convention	72
7.1	Elements of implementation of the Paris Agreement and their institutional homes	131
8.1	Hierarchical framework for the assessment of governance quality	141
8.2	Summary of survey questions	142
8.3	Total number of survey respondents by sector	143
8.4	UNFCCC – quality of governance by region (February 2015)	144
8.5	UNFCCC – quality of governance by sector (February 2015)	146
13.1	Overview of priority result areas for the Green Climate Fund	215
15.1	An analysis of capacities for power generation in India	256
15.2	Actual generation of power in India	258
17.1	German federal energy concept, 2010	290

Contributors

Arnab Bose is a Fellow of the Jindal Institute of Behavioral Sciences (JIBS) where he pursues a doctoral degree in smart urbanization. He is Assistant Professor of Management Practice at Jindal Global Law School. Arnab was Assistant Vice President (Responsible Banking) at YES Bank and Associate Fellow in the Green Growth Division of TERI working on various consultancy projects. Arnab is a sustainability professional with a Masters in Economics from JNU and an MBA in Finance from the University of Amsterdam.

Hugh Breakey[1] is a Research Fellow at Griffith University's Institute for Ethics, Governance and Law, Australia. Currently a Chief Investigator in two federally funded projects, Hugh's research stretches across the philosophical sub-disciplines of political theory, legal philosophy, normative ethics and applied philosophy. The author of Intellectual Liberty: Natural Rights and Intellectual Property (Ashgate), Hugh's work explores ethical issues spanning myriad practical fields, including peacekeeping, safety industries, institutional governance and personal integrity, climate change and sustainable tourism.

Tim Cadman graduated from Girton College, Cambridge and holds a Doctorate from the School of Government in the University of Tasmania. He is a Research Fellow in the Institute of Ethics, Governance and Law of Griffith University, and was previously Sustainable Business Fellow at the University of Southern Queensland. He is an Earth Systems Governance Fellow and lectures in politics, environmental policy and sustainability. He is a Member of the Australian Centre for Sustainable Business and Development, and is on the Editorial Board of the *Journal of Sustainable Finance and Investment*.

Kelsey Coolidge[2] is a social science researcher with strong qualitative and quantitative research skills and experience in non-profit think-tanks and international organizations. She has a strong interest in using these skills and experiences to strengthen the international non-profit sector. She believes that the wider

---

1 https://griffith.academia.edu/HughBreakey
2 www.linkedin.com/in/kelsey-coolidge-25448923/

public sector can be strengthened by applying research and analytic tools to its strategic mission and ensuring that beneficiaries are treated as primary stakeholders.

Qin Dong, Ph.D of law, is Associate Professor at the School of Law and Politics of Nanjing University of Information and Technology, China. His research field is international environment law and international environment politics. In recent years he has focused on the study of global climate governance and China's climate policy.

Natalia Escobar-Pemberthy[3] is currently Assistant Professor of International Relations and Global Governance in the Department of International Business at Universidad EAFIT in Medellín (Colombia), and research associate for the Center for Governance and Sustainability at the University of Massachusetts, Boston. Her research focuses on environmental governance, particularly on the implementation of global environmental conventions and their linkages to the Sustainable Development Goals. She holds a Ph.D. in Global Governance and Human Security from the University of Massachusetts, Boston, an MSc in International Relations from the London School of Economics and Political Science, and a BA in International Business from Universidad EAFIT.

Trudy Fraser[4] is a Postdoctoral Research Fellow with joint affiliation at the United Nations University in Tokyo, Japan, and the University of Tokyo, Japan.

Anna Huggins[5] is a Senior Lecturer in the Queensland Institute of Technology. Her primary research interests lie in the areas of environmental law and administrative law at the international and domestic levels. A key theme in her research is enhancing verification and compliance processes in international environmental law. Anna's research was selected for presentation at the peer-reviewed NYU/Nottingham/Melbourne Junior Faculty Forum for International Law in 2016.

Sylvia I. Karlsson-Vinkhuyzen is Assistant Professor with the Public Administration and Policy Group of Wageningen University, the Netherlands. Her research is focused on understanding the key determinants of what makes global governance processes and the international norms coming out of these exert influence and build legitimacy.

Gargi Katikithala obtained BA LLB from O.P. Jindal Global University and Masters in International Law from the Graduate Institute in Geneva, where she is currently pursuing a Ph.D.

---

3 https://co.linkedin.com/in/natalia-escobar-pemberthy-5402b238
4 www.amazon.com/Security-Council-Global-Legislator-Institutions/dp/0415743370/ref=la_B0073PHFW6_1_1?s=books&ie=UTF8&qid=1522040618&sr=1-1
5 http://staff.qut.edu.au/staff/hugginsa/

Notes on contributors

Trishna Mohan Kripalani[6] is Assistant Professor, Jindal Global Law School.

Rowena Maguire[7] is Senior Lecturer in the School of Law at the Queensland University of Technology and a visiting fellow at Strathmore Law School, Nairobi, Kenya. She is the theme leader of the Climate and Environmental Governance research hub within the International Law and Global Governance research program within the Faculty of Law. Rowena's research is primarily concerned with equitable design and implementation of climate and environmental law and her publications can be viewed via QUT eprints.

Tek Maraseni[8] has over 20 years' work experience in carbon- and forest-related areas in different countries including Nepal, Thailand and Australia. Whilst working with the Ministry of Forests and Soil Conservation in Nepal, he worked closely with several forest stakeholders and delivered training on several issues including forest inventory.

Siddharth Pathak works as the Director of Partnerships for the 2050 Pathways Platform at the European Climate Foundation. He has been involved with the UN negotiations on climate change for a decade. Siddharth led the political advocacy for Climate Action Network (the largest NGO network working on climate change) during the run-up to the Paris Agreement. He has a Master's Degree in Politics from Jawaharlal Nehru University, India.

Swapna Pathak is Assistant Professor of Environmental Studies at Oberlin College, Ohio. She works on issues related to international environmental politics, environmental policy, and conflict and environment. Swapna holds a Ph.D. in Politics from the University of South Carolina, USA, and a Master's Degree in International Studies from Jawaharlal Nehru University, India.

Vesselin Popovski is Professor and Vice Dean of the Law School, Executive Director of the Centre for the Study of the UN, Jindal Global University, India. Until 2014 he was Senior Academic Officer at the United Nations University, Tokyo. He has published numerous articles in peer-reviewed journals and has authored and edited over 20 books.

Aditya Ramji is a Programme Lead with the Council on Energy, Environment and Water (CEEW), India. He is an energy and development economist by training with a specialization in environmental and resource economics. His key areas of research have been development policy, energy access and energy policy.

Armin Rosencranz[9] is a lawyer and political scientist (A.B., Princeton; J.D., M.A, Ph.D. Stanford). He has taught at Berkeley (1987–94) and Stanford

---

6 http://jgls.edu.in/users/trishna-mohan-kripalani
7 http://staff.qut.edu.au/staff/maguirer/
8 https://researchdata.ands.org.au/dr-tek-maraseni/17423
9 http://jgls.edu.in/users/armin-rosencranz

(1994–2012). He founded and led an international environmental NGO, Pacific Environment (1987–96). At Stanford, his courses were listed in ten departments, from history to biology, and he received three teaching awards. He is the co-author of *Environmental Law and Policy in India* (3d ed. forthcoming) and *Climate Change Science and Policy* (2010). He has had five Fulbright grants, a program record.

Seema Sharma has redefined the concept of resilience (social and ecological) at local level by engaging with communities using non-invasive learning methods and creating resilience centers as an interface between academia-industry-policy-community to deal with issues at local level. She has a Ph.D. in industrial waste management and has more than 12 years of experience in the area of environment and sustainability. She is an Assistant Professor with the University of Delhi and Chief Executive Officer of Resilience Relations.

Harold P. Sjursen[10] is Professor of philosophy and global ethics emeritus at the NYU-Tandon School of Engineering. He is an affiliated faculty member at NYU Abu Dhabi as well as NYU Shanghai. He is a guest lecturer at Shanghai Jiao Tong University. He has previously held teaching or research appointments at Augustana College, the University of Chicago, the University of Iowa, Pace University, and the New School for Social Research. Sjursen's research and teaching focuses on problems of technology and ethics in the context of comparative philosophy and global studies.

Annika Bose Styczynski[11] is Assistant Professor and Assistant Dean (Research and International Collaboration) of German nationality at the Jindal School of Government and Public Policy (JSGP). In her research, she is focusing on the governance of socio-technical transitions and the market formation for low-GHG emission technologies in both advanced industrial as well as developing economies. She has more than seven years of experience in renewable energy and electric vehicles policy development in pioneering markets (Japan, USA, China, at EU-level and Germany, France, Norway) and beyond.

Ana Maria Ulloa is a junior researcher interested in understanding the multi-level, multi-layered and multi-factorial processes determining the influence of environmental policies and environmental governance processes. She recently obtained a Master's Degree at the Technical University of Munich and currently is working towards consolidating a Ph.D. project.

Rajnish Wadehra researches energy policy and governance. He handled relationships with the government as an Advisor with India's leading corporates and researched as a Visiting Fellow with the Observer Research Foundation at New Delhi. In his younger years he designed and manufactured rug collections for USA, European and Japanese markets.

---

10 https://shanghai.nyu.edu/academics/faculty/directory/harold-sjursen
11 http://jgu.edu.in/users/annika-bose-styczynski

Andrea Ferraz Young[12] is a Young Researcher (FAPESP) at the National Center for Monitoring and Alert of Natural Disasters (CEMADEN). Her graduation and research background is in Architecture and Urban Planning with Geographic Information System specialization for Social Science, Landscape Architecture and Environmental Planning.

12 www.researchgate.net/profile/Andrea_Young10

Series editor's preface

This is the third of three edited collections arising from workshops funded by an Australian Research Council funded 'Discovery' Project: "Towards Global Carbon Integrity: Applying integrity systems methodology to the 'Global Carbon Crisis'".

The project of which this workshop based collection is a part of has exposed the team to a wide range of views and perspectives. Those who write and seek to act on climate change are a mixed lot – even more diverse than those writing in this and the previous two edited collections. We vary in age, origin, discipline and optimism.

The diversity in our ages is much greater than the period during which we have been writing and taking action. A 40 year old is as likely to have become engaged in the 1990s as a 60 or 70 year old. And a 70 year old is as likely to share the passion of a 30 year old and vice versa.

The diverse origins of the contributors reinforces that this is the quintessential issue without borders. While environmental problems are not confined to any one country and pollution does not stop on borders, it is the one form of pollution that does not diminish with the distance from its source. This can bring us all together but it also creates an even greater tragedy of the commons and even greater incentives to free-ride and externalize costs on to others. The differential contribution to the problem has long raised critical issues over the differential responsibility in response – though mitigated by the increasing efficiency of low carbon technologies.

Optimism suffers as the atmospheric carbon rises slowly - and the predictions rise rapidly. Some have hope, some despair and many think it is too late but we just have to keep trying. One of the difficult issues is how to balance mitigation and adaptation. They are, in theory, mutually supportive – the more we mitigate the less we need to adapt but recognizing the scale of mitigation required and the risks that it will not be achieved means that we must, as risk calculators, recognize and prepare for adaptation. These difficulties point to the poignancy of the intersection of optimism, pessimism and realism.

The variety disciplines reflects the different but complementary and necessary contributions to understanding and addressing the Global Carbon Crisis.

Science predicts; engineers provide practical responses. But the Governance disciplines (law, ethics, political science and economics) are needed to understand how, and whether, particular responses can be effected.

One of the key ideas was to apply the integrity systems approach (previously used for analysing 'domestic' governance systems) to the governance of climate change responses. The integrity systems approach arose out of the recognition by Queensland governance reformers in the early 1990s that the best way of combating corruption and promoting integrity is not through a single law and anti-corruption agency (the Hong Kong ICAC model). What is needed is a *combination* of state institutions and agencies (courts, parliament, police, prosecutors, DPP), state watchdog agencies (ombudsman, auditor general, parliamentary committees), non-governmental organisations (NGOs) and the norms (including values and laws) and incentive mechanisms by which relevant groups live. I called it an 'ethics regime'. The OECD renamed it an 'ethics infrastructure' and Transparency International called it a 'National Integrity System' – a name that stuck. The application to global problems was suggested by global governance expert Professor Ramesh Thakur when he was UNU Senior Vice Rector and UN Assistant Secretary General.

In this project we sought to adapt *National Integrity Systems Assessment* (NISA) methodology to the global carbon integrity system (GCIS). The number of international norms, institutions and mechanisms was so extensive and complex that they could not be represented on a single plane but only in a 3-D representation. The complexity is daunting and points to some of the difficulties of co-ordinating climate action and the opportunities for spoilers. However, most integrity systems are not planned as a whole but grow as new problems are encountered, a wide variety of new allies found and new norms, mechanisms and institutions are instituted and developed.

The span of generations, states, traditions and disciplines that are recognizing the need for climate action and are becoming a part of that Global Climate integrity system tends to add to that strength. Or at least that is the optimists hope and the reason for continuing this work.

Professor Charles Sampford.
Foundation Dean of Law, Research Professor in Ethics and
Director of the Institute for Ethics, Governance and Law

Foreword

The Paris Agreement has captured the attention of both academia and mainstream public as an inspiring outcome of positive multilateralism. Its rapid entry into force, at an unprecedented speed, echoed the unanimous sense of urgency with which our international community assesses Climate Change, its impacts and its opportunities.

As Special Advisor to the Presidency of COP22/CMP12/CMA1, I had the privilege of playing a small role in helping our community move from Negotiation to Implementation, from Agreement to Action, in a consensual manner that aims at enabling, supporting, coordinating and accelerating impactful ambition everywhere and at all levels, with a specific focus on the most vulnerable among us.

In December 2018, the world is set to agree on a Paris Agreement "rulebook", also referred to as Implementation Guidelines. The granularity, robustness and balance of the outcome that will emerge from COP 24 will be a key test of our global ability to continue to move forward, together, on the important issue of Climate Change.

However, the Paris Agreement is not the only instrument to play an important role in our global response to the threats of Climate Change. Indeed, the Kyoto Protocol and its Doha Amendment remain instrumental in our efforts to meet the objectives of both the Paris Agreement and the 2030 Agenda for Sustainable Development. I am encouraged to see that there has been a marked acceleration as of late in terms of ratification by Countries and other Parties of the Doha Amendment to the Kyoto Protocol. Pre-2020 is a tremendously important period for our ability to succeed, or fail, to meet the objectives of the Paris Agreement.

This book is a very welcome addition to the reflection around the central issue of implementation. The Implementation of the Paris Agreement is a matter that is at the same time global, regional, national, subnational and local. This book provides both backward-looking and forward-looking elements on all these dimensions.

Reading the various chapters from leading researchers that you will soon discover, I was struck by the many insights they share. Three elements in particular come to mind.

First, how important it is not to fall into silo conceptualization and silo implementation. Several chapters of this volume look at how we can learn across disciplines and from other international agreements, both within the scope of climate change and the broader environmental context.

Second, several chapters of this book aim at taking stock, both as a finite process as well as a key element of raising ambition. The Talanoa Dialogue that is to take place during COP24 in Katowice is structured around three questions: Where are we, where do we need to be and how do we get there? Taking stock is a very pertinent exercise to answer these three central questions. In addition to the Talanoa dialogue process, there will also be stocktakes on pre-2020 ambition and implementation during both COP24 and COP 25.

Third, the focus on implementation at the national and regional levels is another strength of this book. Indeed, it is my conviction that broad and deep ownership of the Paris Agreement at the National level will unlock channels to bolster our global capacity to meet the objectives of the Paris Agreement. To that effect, the Nationally Determined Contributions (NDCs) provide pathways that can benefit from a host of enhancements including comprehensive legal and governance tools. Furthermore, National Adaptation Plans (NAPs) are also strategic tools for implementation. Regional cooperation is a tremendous asset for both design and implementation of NDCs and NAPs. Indeed, Climate Change knows no border and regional cooperation provide significant opportunities to leverage proximity and the similarity of Climate Contexts to learn and implement at scale.

The Marrakech Action Proclamation for Our Climate and Sustainable Development, a document that was obtained unanimously during the high-level segment of COP22, strikes me as a concise and clear roadmap for our global mobilization to rise to the challenge of Climate Change. Indeed, the Proclamation notably states "As we now turn towards implementation and action, we reiterate our resolve to inspire solidarity, hope and opportunity for current and future generations." Solidarity, hope and opportunity are central elements of the road ahead.

Solidarity as Climate Change Injustice is an undeniable and very unfortunate reality of our world. Solidarity is indispensable to support the most vulnerable among us, that have contributed the least to the emergence of the issue of Climate Change.

Hope, as one must inspire to federate, and it is a scientific fact that, with resolute implementation, we can meet the objectives of the Paris Agreement and the 2030 Agenda for Sustainable Development.

Opportunity as well designed and implemented Climate Change Action, in the context of sustainable development, can provide an opportunity to leapfrog and accelerate our ability to meet the needs of current and future generations. It is often said that there is no shortcut in development. And that is certainly an important aspect to keep in mind as there is plenty of hard work involved. However, there is also no obligation whatsoever to remain prisoners of a pathway that has already proven its limits and to only focus on developing solutions to problems that are becoming irrelevant. There is no shortcut to development but there are several paths to sustainable development. One must seize the opportunities of tomorrow and not provide solutions for yesterday.

To conclude this humble preface, I would like to share with you another excerpt from the Marrakech Action Proclamation:

> "Our task now is to rapidly build on that momentum, together, moving forward purposefully to reduce greenhouse gas emissions and to foster adaptation efforts, thereby benefiting and supporting the 2030 Agenda for Sustainable Development and its Sustainable Development Goals."

As you will note from reading the many fascinating articles from this volume, implementing the Paris Agreement requires a holistic and interdisciplinary approach, built on the deep interconnections and interlinkages between the Climate Agenda and the 2030 Agenda for Sustainable Development. There is an irreversible global momentum for Climate Action and Implementation is already happening on the ground, often not labelled as Climate Action. I commend the authors of this book for providing timely and useful considerations to facilitate further implementation, at the scale we need it to be in order to meet the objectives of the Paris Agreement.

Ayman Cherkaoui
Executive Director of the Global Compact Network Morocco
Lead Counsel for Climate Change at the Center for
International Sustainable Development Law
Former Special Advisor to the Presidency of COP22/CMP12/CMA1

Acknowledgements

This book is the third edited collection arising from an Australian Research Council (ARC) "Discovery Project" 2014-2017. My first sincere thanks go primarily to Professor Charles Sampford, Griffith University's Foundation Dean of Law and the Foundation Director of the Institute of Ethics, Law and Governance, Griffith University, Brisbane, Australia, for drafting the research idea and raising the grant, and subsequently for leading the research efforts to completion of the team comprising also of Hugh Breakey, Rowena Maquire, Tim Cadman and myself.

The workshop at which these papers were first discussed was held in the O.P. Jindal Global University, Sonipat, Haryana, India and sincere thanks also go to its Founding Vice-Chancellor Professor C. Raj Kumar for his unrivalled leadership of this world-class private university, and of many of my colleagues who not only eagerly participated in the workshop, but also committed writing chapters afterwards. I would like to specially thank my interns in the Centre for the Study of United Nations Atish Ghoshal, Juhi Rana, Madhavi Goplalakrishnan, Shalini Sachdeva, Namon Deep Jain, Dushyant Kishan Kaul and Joysheel Shrivastava for their prompt and continuous assistance in formatting the chapters and taking care of the footnotes, commas and all other details to bring the book to the final publishable style.

Not least, Alexandra Buckley at Routledge Publishers, was extremely helpful with guidance, understanding and patient to see this book into fruition.

1 Implementation of international environmental agreements

Vesselin Popovski

The law is developed through three processes – codification, interpretation and implementation – and these are undertaken by three branches of power: the codification is done by the legislative branch, the parliaments; the interpretation by the judicial branch; and the implementation by the executive branch, the governments. These three processes and powers are clearly established in domestic constitutions, but in international law the picture is different. States negotiate and adopt international treaties, and they also are those who implement these treaties.

Louis Henkin began his book *How Nations Behave*[1] by asserting that almost all nations observe almost all laws almost all of the times, and he further explained that they do so not simply because of threat of sanctions, but also because they consider implementation to be in their national interests, because they think this is the moral thing to do, because they would like to maintain friendly relations with other states, because they don't like to be the subject of criticisms, etc. Even when states do not implement international law, this might not necessarily be because of ignorance, malign intentions or bad faith, but because of lack of information or lack of capacity.

The codification of international law has significantly developed over the last century, but such progress has not been paralleled with similar progress in implementation. Various conventions have been adopted to protect human rights, for example, but still millions of people suffer from violations due to disregard of human rights or poor implementation in many states. In another example, the Non-Proliferation Treaty was solemnly adopted in 1968, but its implementation dramatically failed and as a result the number of nuclear powers has doubled, but also the existing five nuclear powers at the time did nothing to reduce and abolish their nuclear weapons. The codification of international humanitarian law has been admirable after World War II with the adoption of the 1948 Genocide Convention, 1949 Geneva Conventions, and its 1977 Additional Protocols, but the implementation has lagged behind and only the establishment of two *ad hoc* international criminal tribunals and of the International Criminal Court in the last two decades made progress in implementing the international humanitarian law.

---

1 Henkin, L. (1979), *How Nations Behave*, Columbia University Press

One reason for insufficient implementation of international law is the lack of global government and the limited global law enforcement. In domestic law, when crimes happen, the police, the investigation, the prosecution and the courts would normally be capacitated to deal with these violations. But if states violate international law, the only hope is that the UN Security Council will impose sanctions and punish those states. Occasionally the Security Council indeed imposed sanctions, but because of the veto it will never act against one of its five permanent members, nor against a close friend-state of a permanent member. Even when the Security Council is united and imposes sanctions against a state, those sanctions might not be effective to exercise the necessary pressure on that state to co-operate.

Implementation mechanisms are usually discussed in the process of codification and inserted in the text of treaties. These could be either 'hard' or 'soft' provisions, depending on the mandatory nature of the obligations and on the sanctions envisaged. 'Hard' law has developed when states adopted rules and put clear enforcement mechanisms in place to sanction those who will disregard the law. 'Soft' law has developed when states felt urgency to adopt rules but were unprepared to put in place sanctioning mechanisms, or when instead of binding commitments they opted for voluntary commitments.

Whether the international law is 'hard' or 'soft' depends also on which organ imposes it. The resolutions of the General Assembly are considered 'soft' and non-binding, because the Assembly does not have enforcement powers. The resolutions of the Security Council are 'hard' and binding, as the Council can punish states that do not carry out its decisions.

This book discusses how 'hard' or 'soft' are the implementation provisions of the 2015 Paris Agreement under the United Nations Framework Convention for Climate Change (UNFCCC). It argues that the Paris Agreement is a result of a pioneering bottom-up approach in international law that allows states to define their nationally determined commitments (NDCs) voluntarily, but once they do, the commitments become legally binding. Instead of 'Sanctions Committee' the Agreement creates a much friendlier mechanism – 'Facilitation Committee'. The hope is that the synergy of 'hard' obligations towards long-term goals, being made in a voluntary facilitative manner with the opportunity to readjust goals and targets over periods of time, will bring international consensus and support to address one of the biggest global challenges that humanity has ever faced in history. The goals in the Paris Agreement are clear – to hold the increase in the global average temperature to "well below 2°C above pre-industrial levels"[2], and the implementation will depend on ensuring transparency, accountability, technological transfers, finance and long-term commitments.

The book follows from a previous co-edited book on ethical values and integrity of the climate regime, edited together with Hugh Breakey and Rowena

---

2 United Nations (2015), *The Paris Agreement*: http://unfccc.int/files/essential_background/convention/application/pdf/english_paris_agreement.pdf

Maguire,[3] who are also authors of chapters in this book. This time the purpose is to analyze the implementation of the 2015 Paris Agreement, an essential task as the targets encapsulated in the Agreement will require a fully comprehensive burden-sharing framework both in legal and ethical obligations and in practical facilitation, unrivalled in importance, and incomparable in challenges to all previous agreements. The Paris Agreement is flexible as it provides for reviews of the commitments every five years, aiming at long-term goals without temporal limits.

Questions and challenges

The questions that this book addresses are: How will Parties be held accountable for their NDCs and how will these be reviewed? Will the implementation facilitation mechanisms in the Paris Agreement be sufficient? What role can the facilitation mechanisms play at national and sub-national levels?

The book diagnoses challenges that the implementation may face, such as lack of commitment from large polluters, lack of capacity in least developed countries, unwillingness to share technology, political instability, inadequate finance, etc. Parties may submit their NDCs, but a question remains as to how will they sustain and implement these commitments. Finance, technical assistance, capacity-building and other support would be of a paramount importance to meet the NDCs. One big challenge is how to satisfy the growing energy needs in developing countries by introducing renewable energy at strategic level. One in five people primarily in rural areas of Africa and South Asia lacks access to electricity. The dominant model of electricity service delivery in these regions remains centralized power generation connected to extensive national grids for transmission and delivery. While this model has worked well for more than a century in developed countries, it has drawbacks which penalize developing countries that are yet to provide access to a large part of their population through extending the grid. The high investment cost involved, combined with the need to deliver the service in a commercially viable way, means economics dictates coverage. With the reduction in the cost of renewable energy technologies and with more efficient end-use appliances, the decentralized renewable energy-based distributed power generation is becoming an increasingly viable option. There are estimates that by 2030, 70% cent of rural areas will be connected either to mini-grid (65%) or stand-alone off-grid solutions (35%).[4]

The challenges to implementation are not only financial, technological or lack-of-political-will. Even when there is sufficient awareness and commitment,

---

3 Breakey, H., Popovski, V. and Maguire, R. (eds.) (2015), *Ethical Values and Integrity of the Climate Change Regime*, Ashgate
4 International Energy Agency (2010), *Energy Poverty – How to Make Modern Energy Access Universal?*, Paris; www.se4all.org/wp-content/uploads/2013/09/Special_Excerpt_of_WEO_2010.pdf

governments may lack the capacity to ensure that the public sector will create an enabling environment for investments. The knowledge and political clout to create and enact appropriate regulations and tariffs, that allow bottom-up initiatives to unfold and grow, might be missing. Potential entrepreneurs might be discouraged by bureaucratic processes or lack of resources to provide timely public administration. Public utilities might be heavily indebted, or suffer from mismanagement and corruption.

Literature review

The literature on international environmental law generally, and on climate agreements in particular, has expanded recently. Although much has been written on the negotiations and adoption of the agreements, much less has been written on compliance and implementation of these agreements.

Alexander Zahar in *International Climate Change and State Compliance* (Routledge, 2014) attempted to fill the gap, probing the inconsistent compliance with the procedural and substantive obligations under the UNFCCC and the Kyoto Protocol. He showed how the international climate regime for only 20 years in existence has developed normative rules, binding on states, but he also explored the feeble consequences of non-compliance. Zahar demonstrated that the state conduct under the climate change law is characterized by generally high compliance in areas where equity is not a major concern, and by contrast, there is low compliance in matters requiring a burden-sharing agreement among states to reduce emissions. In a sober analysis he argued that the substantive climate law presently in place must be further developed through normative rules that bind states individually to top-down mitigation commitments. While a solution to the problem of climate change must take this form, the law development in this direction is likely to be hesitant and slow, predicted Zahar. Another contribution of his book was that it looks not only at individual emission reduction commitments and reporting obligations, but also delved into a deeper range of individual and collective commitments and their interplay under the climate regime to better understand the compliance challenge.

Peter H. Sand and Jonathan B. Wiener questioned whether we are moving "Towards a New International Law of the Atmosphere?", *Goettingen Journal of International Law* (2016), Vol. 7, pp. 197–223, reflecting on the inclusion of the item 'protection of the atmosphere' in the codification agenda of the International Law Commission (ILC). This is a long overdue recognition that the scope of contemporary international law for the atmosphere extends beyond the traditional disciplines 'space law' or 'air navigation law'. The authors discussed the atmospheric commons, regulated by a 'regime complex' comprising a multitude of economic uses, global communications, pollutant emissions and diffusion, in different geographical sectors and vertical zones, in the face of different categories of risks. They assessed the 'highly restrictive ILC initial understanding' in 2013 and the reports and debates in 2014–15, and also addressed earlier attempts at identifying crosscutting legal rules and principles by the UN Environment Programme, the International Law Association and the Institut de Droit International.

On the implementation of the 1997 Kyoto Protocol three Norwegian scholars, Olav Stokke, Jon Hovi and Geir Ulfstein, edited the book *Implementing the Climate Regime: International Compliance* (Routledge, 2005) and described in detail the negotiation leading to the compliance structure, adopted with the 2001 Marrakesh Accords: the Compliance Committee, composed of 20 members, with its Facilitative Branch and Enforcement Branch with ten members each. Because the Kyoto Protocol effectively had binding effect only on industrialized (Annex I) countries, the authors focused mostly on domestic hard enforcement compliance. Interestingly, Russia was the most regular client of the Enforcement Branch, even if it enjoyed large emissions allowances and entered into tonnes of 'hot air' trading, because of its weak institutions and different, even conflicting, promises made by the government and by the semi-private company *Gazprom* (p. 26).

Hans Blix, in an article "Developing International Law and Inducing Compliance" in *Columbia Journal of Transnational Law* (2002), Vol. 41, no. 1, pp. 12–38, defined that treaties serve as contracts establishing reciprocal interstate obligations, as constitutions of intergovernmental organizations, or as instruments for codification of global or regional legislation. An interesting practice, creating international legal norms without parallels in domestic law, is the adoption of declarations made by states, bilaterally, regionally or globally within the framework of international organizations. The predilection of governments for making declarations as treaty precursors is understandable – they may want to test their way and stake out broader guidelines in various matters before they bind themselves formally. Supporting a norm as a policy and guiding principle is one thing, committing oneself to every detail is quite another, argued Blix.

Susan Subak wrote "Verifying Compliance with an Unmonitorable Climate Convention" (Centre for Social and Economic Research on the Global Environment, Working Paper, 1995) examining various implementation arrangements, ranging from secretariats collecting reports from parties, to intrusive measures such as surprise on-site inspections carried out by multilateral teams. In the case of the UNFCCC, Subak demonstrated how significant was the investment in development of methodology for reporting, training and participation in reporting, compared to all previous international environmental agreements.

Elizabeth Barratt–Brown in "Building a Monitoring and Compliance Regime of the Montreal Protocol" in *Yale Journal of International Law* (1991), Vol. 16, pp. 519–552, examined already existing multilateral environmental treaties. The Montreal Protocol on Substances that Deplete the Ozone Layer is one of the first to address a serious global environmental issue, when scientific findings of the reality of ozone depletion shocked the world and set the stage for collective action. Other multilateral treaties referred to by Barratt-Brown are the UN human rights regime, the ILO, and the nuclear non-proliferation regime, the International Atomic Energy Agency. She listed the following factors constructing an effective compliance regime: formation of a governing body, incorporation of NGOs into the compliance, creation of a mechanism to ensure compliance, placing experts on compliance committees and full disclosure and transparency of all reports made by these committees.

The success of the Montreal Protocol is also analyzed by Duncan Brack in "International Trade and the Montreal Protocol" (Royal Institute of International Affairs, 1996). He pointed out that among other non-encouraging stories of international environmental co-operation, the Montreal Protocol stands as a shining light, because of its effective set of procedures and institutions centred around an implementation committee, a well-funded financial mechanism to assist with compliance, and a credible threat of trade sanctions in case of persistent non-compliance. The Montreal Protocol had a successful record in dealing with non-compliance of the transition economies, and, although it faced a major challenge with regard to developing countries, there is reason to believe that it can cope successfully.

Henry Lee edited *Shaping National Responses to Global Climate Change: A Post-Rio Guide* (Island Press, 1995), revealing methods of designing, implementing and gaining political support – both domestically and internationally – for strategies and policies to reduce emissions of greenhouse gases. The book framed the economic, policy and management trade-offs involved in designing strategies for international agreements, developing and implementing the means to enforce international agreements, comparing alternative policy responses, transferring technology from developed to developing countries, and transitioning to domestic agenda to implement the agreement. The book presented a strategic framework from which specific alternatives can be assessed and compared. One of the chapters in the book, written by Ronald B. Mitchell and Abram Chayes, "Improving Compliance with the Climate Change Treaty" (pp. 115–145), emphasized the need to facilitate compliance, reporting, verification and responses to non-compliances by those actors already predisposed to perform these tasks. Efforts to alter incentives and capacities, the authors argued, are less likely to succeed than efforts to elicit possible co-operation from existing incentives and abilities. Several other processes can and should be set in motion to address the underlying factors inhibiting compliance.

Ronald B. Mitchell together with Edward A. Parson wrote "Implementing the Climate Change Regime's Clean Development Mechanism", *Journal of Environment and Development* (June 2001), Vol. 10, Issue 2, pp. 125–146, and also the chapter "Institutional Aspects of Implementation, Compliance, and Effectiveness" (pp. 221–244) in the book *International Relations and Global Climate Change*, edited by Urs Luterbacher and Detlef Sprinz (MIT Press, 2001). He evaluated the UNFCCC concerns with effectiveness and raised two institutional design questions: first, how should international institutions be designed to maximize the chances that the regime will achieve agreed-on goals. Second, how should they be designed to allow the regime to assess its progress towards those goals. The nature of the UNFCCC regime, according to Mitchell, highlights several obstacles, common to other international regimes, but also poses several novel institutional challenges.

Clare Briedenich and Daniel Bodansky in the report "Measurement, Reporting and Verification in Post-2012 Climate Agreements" (Pew Center,

2009) addressed the full range of mitigation, adaptation, technology and finance, focusing on the 2012 Bali plan which introduced the measurement, reporting and verification (MRV) of three categories of action: developed country mitigation commitments, developing country mitigation actions, and the provision of support for developing country mitigation actions. Placing MRV as a core element of the climate agreement, the authors argued, depends on parties' confidence that commitments can be reliably measured, reported and verified. Parties' experiences to date with reporting and review under the Kyoto Protocol offer insights into the future design of MRV. Credible MRV rests on clearly defined commitments – the more specific the commitment, the more readily appropriate metrics and processes can be established for measuring and verifying the implementation.

David Victor, Kal Raustiala and Eugene Skolnikoff edited *Implementation and Effectiveness of International Environmental Commitments* (MIT Press, 1998). They defined that implementation translates intent into action, and is vital to effective public policy, especially to international agreements that regulate complex behaviour. The book has two parts, one on international mechanisms for monitoring and systems for implementation review, and another on national implementation. The authors presented their three-year research project with 14 case studies, examining how international commitments are implemented at the international and national levels, analyzing both national and international mechanisms for monitoring and reviewing implementation. They ended with a sceptical conclusion that, given the many complexities of implementing international environmental commitments, it is impossible to draw any systematic conclusions about the implementation process and the ways to enhance implementation.

Oliver Meier and Clare Tenner in "Non-Governmental Monitoring of International Agreements", *Verification Yearbook* (2001), VERTIC, pp. 209–227, described the explicit verification provisions in the multilateral environmental agreements, such as the Montreal Protocol and the Kyoto Protocol. They showed new verification technologies becoming involved in monitoring compliance with these agreements. Much has been written about the role of NGOs in initiating and influencing negotiations on multilateral agreements, in monitoring the activities of governments and non-state actors to detect breaches. In some cases NGOs assisted states in bringing themselves back into compliance. The authors argued that NGOs can make a unique contribution to the monitoring of international agreements as their strengths enable them to identify and highlight treaty violations in ways that established verification mechanisms can't. More serious from a verification point of view is the political dilemma that NGOs must tackle if they want to move beyond unofficial monitoring roles and become involved with official political mechanisms. There is a trade-off between political independence and involvement with official bodies. From an NGO perspective, there are benefits to be gained from such involvement, such as better access to information, enabling them to assess problems more accurately and potentially improving the quality of their work. NGO monitoring activities can also become too dependent on official data, impairing their judgement. NGOs, just like governments, have to weigh the benefits of publicly accusing states of non-compliance against the benefits of working quietly behind the scenes. Exposing non-compliant behaviour through press releases and other media activity can have great political impact and satisfy the demands of journalists or organization members, however NGOs may remain vulnerable and the legitimacy of their role in treaty implementation may not be widely accepted.

Implementation of the 2015 Paris Agreement

The implementation of the 2015 Paris Agreement will depend on states' serious take on the goals, their strategic long-term approach reflected in the NDCs, their compliance with the reporting and verification mechanisms and the effectiveness of the implementation facilitation mechanisms. Many industrialized and developing countries have taken part in the development of emissions inventory guidelines and methodological research, but the challenge remains as to the degree to which parties tolerate uncertainty in many areas of planning, committing and reporting. Despite the high degree of compliance with the reporting requirements, evidence of progress towards compliance with the objective of the convention, the soft target of aiming to stabilize emissions is much weaker. A different and fundamental problem in achieving compliance with the Paris Agreement, and in verifying that it has been accomplished, is that the precision of the baseline for many emissions sources is very low and will probably remain so indefinitely. If convincing verification does not develop through these or other approaches, a failure of confidence in the climate regime is not inevitable, if a good balance is established between intrusiveness and trust. Ideally the solution would be a verification system that is intrusive enough to allow the regime's strongest supports from other parties and the public to be reassured of progress in reducing the emissions. At the same time, the system should avoid the type of over-watching and intrusiveness that can enhance distrust.

The role of non-binding policy instruments, such as for example the Stockholm Declaration on the Human Environment, is also important. While most of the principles contained in the Stockholm Declaration were of a nature that governments did not consider as being suitable for direct transformation into binding treaty commitments, there is little doubt that several of these non-binding principles had a significant conceptual and policy impact on various instruments adopted subsequently in the area of the environment. Furthermore, declarations are not the only intergovernmental instruments that provide non-binding 'soft' law for state conduct. Most of the analytical and negotiating energy surrounding the development of the climate change regime focused on substantive limitations on net emissions, whether through targets and timetables, emission permits, taxes or technological standards. But no matter how stringent these commitments are, the regime will not succeed unless the parties comply with it. The compliance must be addresses from the outset and a system must be designed from the beginning. In the climate change regime the costs and magnitude of the required behavioural changes, and the regulatory breadth and complexity, pose especially difficult compliance problems.

What is unique with the Paris Agreement is that effectively it must alter not only the behaviour of states, but also the behaviour of corporate bodies and ordinary citizens whose activities account for the emissions. The Agreement must encourage national governments not only to comply with its provisions by adopting legislation and other appropriate policies, but also to take action to facilitate compliance and condemn violations of corporate and individual actors.

Numerous international and national factors influence whether a nation complies with a given rule of a given agreement at a given time – for example, the distribution of international power or the administrative efficiency and constitutional structure of the parties. These factors bound the aggregate level of compliance that can be expected under the agreement. With respect to compliance, nations and sub-national actors can be divided into three categories. First, some nations will have incentives and abilities that lead them to comply with a given agreement independent of systems established to identify and respond to non-compliance. Second, for some nations compliance will be contingent on the type and likelihood of response to non-compliance institutionalized in the international regime. Third, some nations will have incentives and abilities that will lead them either to not sign an agreement or to sign and violate the agreement independent of the systems established to identify and respond to compliance.

It is important to emphasize the need to facilitate compliance, reporting, verification and responses to non-compliances by those actors already predisposed to perform these tasks. Efforts to alter incentives and capacities are less likely to succeed than are efforts to elicit the highest possible co-operation from existing incentives and abilities, which may be making a virtue of necessity.

Processes should be set in motion to address the underlying factors inhibiting compliance. It is important to assess which actors are liable for non-compliance and how the facilitation system must respond most effectively. The goal of the system should be to respond in ways that target the source of non-compliance and promote future compliance. As a first approximation, this could involve providing the financial, administrative or technical resources deemed lacking in cases of incapacity; providing technical advice and new, extended, but specific compliance deadlines in cases of inadvertent policy or programme failure; and adopting sanctions in cases of intentional violation. Effectiveness is also likely to be fostered by rewarding compliance. Providing positive incentives for compliance, and for positive behaviours that produce emissions reductions larger or sooner than required, could help the regime achieve aggregate environmental improvements that exceed rather than merely meet the goals.

The response system must be able to differentiate compliance from non-compliance; also to differentiate non-compliance due to inadvertence, incapacities and intentionality; and induce differentiated responses to behaviours and outcomes that make goal-promoting behaviours more likely. Sanctioning those assessed as having intentionally violated their commitments provides those actors with incentives to bring themselves into compliance while simultaneously deterring others who might be tempted to intentionally violate in the future. Unfortunately, governments often prove reluctant to impose trade sanctions or other penalties on other states, because of collective action problems and the costs of sanctioning.

The Paris Agreement could facilitate sanctioning by removing legal barriers that inhibit those predisposed to enforce the agreement – for example, altering World Trade Organization (WTO) rules to permit trade sanctions in response to non-compliance. And governments may engage in various forms of diplomatic shaming that may induce compliance. Unfortunately, experience suggests that sanctioning is unlikely to be sufficiently frequent or severe to alter the non-compliance behaviour in many cases.

These obstacles to an effective sanction-based system and the recognition that sanctions are not appropriate when non-compliance is not intentional have prompted interest in alternative approaches. The best response to non-compliance that stems from incapacity is to provide the financial, administrative or technical resources needed to remedy the incapacity. Financial and technology transfers and training may prove most helpful when capacity rather than will is the source of the problem. The international wetlands convention has sought to prevent wetlands degradation by providing technical advisors to countries experiencing difficulty doing so on their own while also publishing a list of wetlands at risk that provides a basis for mobilizing either assistance or shaming. Unfortunately, such programmes require funding from governments and experience with the Global Environment Facility (GEF). Technology transfer programmes demonstrate that governments often prove as reluctant to fund such programmes as they do to impose sanctions. Indeed, governments have yet to develop mechanisms to induce developed countries to provide the funds needed by developing countries to contribute to the climate goals. When non-compliance stems from inadvertence, the best approach for the regime may be to provide various avenues for the non-compliant party to bring itself into compliance. These avenues could include a specified but extended deadline for compliance, allowing the *post hoc* purchase of emissions credits, or contributing to the financial mechanism in an amount sufficient to fund the quantity of reductions needed to bring it into compliance.

Finally, provisions should be made to reward over-compliance and innovation. Precisely because current emissions reduction targets fall far short of what most scientists consider necessary to avert climate change, significant progress requires incentives for going beyond what is required and for undertaking risky projects that provide uncertain, but potentially large, reductions at low cost. Countries, corporations and NGOs that exceed their required emissions reductions should be rewarded by creating awards and a 'white list', by providing access to the financial mechanism if appropriate, by reducing the verification requirements imposed, or by other similar incentives.

A walk through the book

The next chapter by Vesselin Popovski compares the implementation clauses of the Paris Agreement with those of the Kyoto Protocol and analyzes the bottom-up approach and the creation of facilitation mechanisms helping states to achieve the targets. Such bottom-up approach in international law, the author argues, bears less sovereignty costs, it is more practical for larger package deals, specifically when states face uncertainty and hesitance to commit to strictly binding legal obligations. The Paris Agreement is legally binding, but it relies not on sanctions, rather on capacity-building, transparency, technological and financial assistance and other forms of facilitation. International regimes that combined sanctions with facilitating implementation mechanisms proved to work better than those relying only on sanctions. Certainly the approach towards facilitation should neither diminish, nor jeopardize efforts to sanction states for deliberate non-implementation. Being 'soft' to co-operating states and offering facilitation should not reduce the demand to be 'hard' with unwilling and non-cooperating states. Sanctions remain as important to have in the toolkit, as are the facilitating mechanisms. The Paris Agreement encourages flexible bottom-up commitments by parties that can be reviewed over time and achieved through mechanisms of facilitating the implementation of these commitments in both mitigation and adaptation, and regular assessments and reviews. States can commit ambitiously first and then revise their commitments after five years. Or they can commit modestly, and increase their commitments, if the methodologies work well. The traditional 'hard' international law, based on sanctions, would not be the best way forward to guide the global climate change regime, because states experience a natural reluctance to accept uncertain obligations, failing which they will be sanctioned. Therefore the efforts have been focused on facilitation of implementation.

The chapter by Trudy Fraser continues to evaluate and compare the architectural framework of the Kyoto Protocol and the Paris Agreement, comparing the top-down, bottom-up and hybrid approaches, and how these approaches produce varied compliance and obligations. The chapter contrasts the relative merits of the Kyoto Protocol's top-down approach with the development of a more hybrid bottom-up approach in the Paris Agreement and introduces some alternative legal options that may be viable as to make states recover from substantive 'ratification fatigue' which has plagued the legal instruments developed to support the UNFCCC. Fraser explains the top-down approach, which claims that the nature of obligations and responsibilities stems quantitatively from the Convention itself that sets definite terms of reference by a general international consensus. By contrast, the bottom-up approach offers a relative understanding of the state's own self-imposing mandates in the NDCs. The chapter succinctly analyses instances where alternative methods of dispute resolutions and other international legal mandates may produce environmental measures and implementation mechanisms to ensure state compliance, ranging from national prosecution, international arbitration, Security Council involvement and the possibility of emergence of new forms of customary international law on climate change. The chapter ends by evaluating the above methods and arriving at a conclusion that there is no certainty with respect to how the environmental obligations may be enforced and whether the Paris Agreement would also attract 'ratification fatigue' within NDCs, as it did in internationally determined targets. Fraser concludes that it is a matter of time as to how these obligations crystallize, the catalyst for which has to be a legal and a normative tipping point which is yet to arrive.

Natalia Escobar-Pemberthy in her chapter goes beyond the Paris Agreement and presents comprehensively the implementation – both in achievements and shortages – of several environmental conventions. She draws a conceptual picture and categorization of facilitation implementation mechanisms, such as reporting, institutional arrangements, finance, capacity-building and technology transfer. She argues that the study of the facilitation mechanisms is fundamental for three reasons: First, because a better analytical framework about its operations would contribute to expand its scope and coverage. Facilitation mechanisms can point to specific obligations in order to develop targeted approaches that guarantee effective and efficient solutions. Second, national reports – as implementation mechanisms themselves – need to collect data on how the mechanisms work and their results in terms of effectiveness. And third, mapping the existing facilitation implementation mechanisms opens the door to the identification of clusters and synergies that address the challenges presented by the proliferation of instruments.

Ana María Ulloa and Sylvia Karlsson-Vinkhuyzen compare the implementation of the UNFCCC and the Convention on Biological Diversity (CBD). While the former has achieved decent success over the past 25 years, the latter has been criticized for failing in its objectives of conserving the biodiversity and providing for its sustainable use. Biodiversity is declining at unprecedented rates. The authors observe that the CBD has very little impact on states' practice, reflecting a noteworthy degree of non-compliance. This can be largely attributed to the dilution of the essence of the CBD, by the use of words such as 'subject to national legislation', 'subject to patent law', 'as far as possible' and 'as appropriate'. This results in the CBD being viewed as 'softer' law. The authors seek to assess the effectiveness of the CBD, but making such an assessment is challenging since biological systems are complex, detrimental activities do not have immediate consequences and the data is often outdated or unavailable. To carry out the task, the chapter reviews the compliance and review system under the CBD, and does a comparative analysis with the climate change regime under the UNFCCC.

Hugh Breakey analyzes the global stocktake as a major part of the pledge-and-review system contained in the Paris Agreement, which is expected to monitor the extent to which each country's commitments are in accordance with the Convention's principles, particularly *in the light of equity* as written in Article 14(1) of the Paris Agreement. Breakey argues that this is the strongest indication so far that some equity-based consideration of exploration of the NDCs will occur for three reasons: first, the purpose of the Agreement itself includes reference to equity and common but differentiated responsibilities (CBDR), which have distributive implications. Second, while equity itself may have an array of meanings, they all have clear consequences for distributive ethics and the allocation of burdens and entitlements between parties, and a review *in the light of equity* has no alternative but to consider them. And third, the stocktake will focus on the NDCs not collectively, but individually. Breakey considers both the benefits and pitfalls that this process may encounter and how it can harness the potential of moral language and argument, and avoid divisiveness and acrimony. The global stocktake was termed as a task, rather than a process, because of the focus on deliberative procedural values, rather than on substantive questions of distributive justice. Breakey analyses in more detail the notions of 'equity' and 'facilitative manner', comparing the top-down model of responsibility used in the Kyoto Protocol with the bottom-up model in the Paris Agreement, arguing persuasively in favour of the latter as a means to achieve substantive objectives, although it is conceded that this may affect equity outcomes. It is necessary calling out low-performers on the basis of an agreed-upon criteria, and of highlighting efficient and targeted efforts. Breakey notes the importance of practical encouragement as a means of ensuring that targets are met, for constructive engagement with nations that do not meet commitments, and for nations to do a 'moral stocktake' prior to assessments becoming public, giving them time to improve and reflect.

The chapter by Swapna and Siddharth Pathak also deals with the global stocktake and argues that this process will be crucial in enhancing ambition when the implementation of the Paris Agreement begins to take shape. In order to be equitable, the global stocktake should look beyond mitigation to other critical elements of the agreement, such as adaptation and finance. Moreover, the global stocktake should pay attention to disaggregated data and not just focus on collective progress. This will be the key to ensure equity based on common but differentiated responsibilities and respective capabilities, in the light of different national circumstances, the authors argue. Collective stocktake will also have to look into other elements that potentially address equity, like human rights, economic diversification as well as sustainable development goals. Even if these elements are not directly implemented in the global stocktake, they can provide a good assessment of the quality of implementation of climate action at national level and create synergies for a more comprehensive global climate regime.

The chapter by Tim Cadman and Tek Maraseni is a qualitative and quantitative analysis of stakeholder perceptions of the UNFCCC, based on a framework of Principles, Criteria and Indicators for evaluating governance quality. The evaluation reveals that the climate regime performed relatively well, but there were distinct differences between respondents from the developed countries and developing countries. Although respondents generally found UNFCCC to be inclusive of their interests, they did not consider that there was the necessary capacity, or resources, for meaningful participation. Developing countries tended to criticize the power imbalance between them and developed states, while there was universal doubt about the UNFCCC's capability to actually solve the issues that it addresses. The authors conclude with the observation that while the UNFCCC is substantively superior to the Kyoto Protocol, an increase in funding is essential to ensure the inclusion of developing countries in the international discourse around climate change, and to ensure that climate change efforts are not co-opted by northern nations solely due to their resources.

The chapter by Harold Sjursen compares two views on implementation – one based on moral commitment, and the second on economic costs. He cites the UN Secretary-General António Gutteres' affirmative view that all questions on climate science are settled, the devastating consequences are foreseen, and the nations of the world now have a moral imperative to act. However there is also a sceptical view, expressed by Bjorn Lomborg, that unless a solution is economically viable its implementation would be unlikely because of unintended consequences. Lomborg situates climate problems within the constraints of contemporary capitalism and social/political expectations and does not acknowledge ethical imperatives that might question either the efficacy or the justice of these constraints or an approach that demands their revision. Whereas Guterres sees a clear moral imperative to arrest an impending global crisis, Lomborg sees such an approach as unwarranted, ineffective and ultimately damaging. Sjursen offers critique to both, claiming, first, that ethical imperative may compel sacrifice or deprivation; and second, that what from an economic point of view may not be sustainable, may as an exception be compelled by a moral or ethical imperative. The forgiving of loans provides an example of this principle – in general it is not economically sustainable for a financial institution to forgive debts, but under specific circumstances this can be done on ethical grounds. A question arises over whether the scientifically calculated prospect of climate change is ethically required, even if doing so may weaken global finances in a way that undermines the economic well-being of many. As a platform on which to debate the ethical issue the general proposition of Lomborg will be granted, not because it is evidently correct, but because it requires that the ethical questions surrounding global climate change be given full consideration, argues Sjursen. The criticism of Lomborg is considered from an ethical perspective that is not limited to the conditions proscribed by axiomatic economic values. Indeed, the Paris Agreement does not limit itself to a singular set of economic principles in so far as it explicitly honours that different countries will, according to their own various governance models, implement the principles in various ways. This is a different approach from Kyoto or Copenhagen, that makes all signatories responsible for the achievement of the overall goal.

The chapter by Anna Huggins and Rowena Maguire deals with the principle of common but differentiated responsibilities (CBDR) which suggests that all states have international environmental obligations, but these obligations and the manner to meet them vary, based on the level of economic development and contribution to the environmental degradation. Essentially, states have own obligations towards environmental protection based on their economic status, as well as their contribution to environmental pollution, while recognising the difference in their capabilities when it comes to taking responsibility.

The chapter by Dong Qin discusses the compulsory licensing of green technologies. The Paris Agreement ensures that the parties stay true to their commitments, but while the developed countries have the requisite environment-sound or green technology, the developing countries do not. It is observed that there has been a minimal level of international transfer of technology between developed and developing countries. Even though developing countries have massive potential to reduce greenhouse gases, the goals of the Agreement may never be achieved if the developed countries hesitate to share their green technology with developing countries. The compulsory licensing is a legal system that allows courts or patent administrations to permit someone to use the patents without the permission of their owners. This is an important legal tool to protect public interests by preventing patent right holders from abusing their rights, which is a reality in the context of green technology. In light of this, Dong Qin assesses the need for an international agreement on green compulsory licensing that can facilitate the green technology transfer to those who need it most.

The chapter by Andrea Ferraz Young stresses that debates on environmental justice have occurred on an international scale, but the majority of them focused on human rights declarations. The environmental justice has been mostly pursued on a national scale, which brings attention to issues such as the recognition that low-carbon market opportunities can be associated with social environmental responsibilities in order to change global perception. Young proposes to understand environmental justice as a foundation, not as a simple income distribution, which includes the recognition of social responsibilities, massive education, legal and administrative procedures, effective rights for all in an unequal society, hierarchically organized by the principles of justice. Young then applies this conceptual reference to examine adaptation actions, financing and renewable energy opportunities in four countries: India, Indonesia, Colombia and Zambia. She argues that adaptation practices should be linked to the idea of environmental justice, examining how renewable source opportunities can be articulated with the Adaptation Fund. She concludes that it is possible to shape and implement adaptation projects with interventions that seek to ensure environmental justice.

The chapter by Arnab Bose and Seema Sharma deals with the differences in understanding, policy-making and implementation at global, national and local levels, in the context of climate change regimes. The authors collaborate with a number of institutions, most importantly the Resilience Center Global Network of Delhi University and the Resilience Center Vivekananda College Chapter, to identify problems and propose how to overcome these differences. They look at the mechanisms under the global (UNFCCC) and national regimes; and observe that while there is an understanding of sustainability at the higher level, this does not translate into an understanding at the local community level. Even if individuals are aware of what must be done, the real problem lies in the implementation of plans adopted at a higher level without giving the local community a voice. Therefore, even though the intentions behind policies and plans might be good, they may fail at multiple levels.

The chapter by Aditya Ramji focuses on the actions, taken by India, towards the Paris Agreement. Considering that 2016 and 2017 saw several instances of extreme weather, both heat and unprecedented rains and flooding, it is of paramount importance that the country turns its focus to invest in an economy that, while built on principles of equity and justice, will strengthen India's stand at the UNFCCC. At the Durban Platform for Enhanced Action (COP 17) in 2010, developing countries pressed upon the need to ensure that equity be made a part of any agreement. COP 18 at Doha was of even more significance to India, as it ensured that the principle of CBDR continued to be a part of the negotiations. This is because the Indian government made its pledge of focusing on climate change contingent on the inclusion of CBDR. Between COP 19 and COP 21 there were various changes in terminology and nomenclature. Fortunately, the final Agreement in Paris retained the much-sought principles of equity and CBDR, with India being a key player in negotiations and other smaller developing countries looking at India to lead.

The chapter by Armin Rosencranz and Rajnish Wadehra presents 'how the tables turned' – an interesting and unique comparative study between the dramatic decline in the USA climate change commitments from President Obama to President Trump, and – on the opposite side – the growing engagement with climate change initiatives of the initially reluctant-to-do-so Indian government.

The chapter by Kelsey Coolidge examines the implementation of the Paris Agreement with respect to cities. After President Trump pulled the United States out of the Paris Agreement, many US cities defied that decision and asserted that they would continue to abide by the Agreement. Regardless of the US government's stance, and perhaps directly in spite of it, many US cities have actively engaged in mitigating climate change. This presents an interesting question for international law and treaty-making: whether sub-state actors can comply with international agreements even if the central state defect from existing treaties. Can sub-state involvement in international treaties shame governments and reduce their ability to govern across multiple issues? The literature on cities and climate change mitigation under-emphasizes legal or political implications relating to international relations. It analyzes the effectiveness of efforts at mitigating climate change, partnerships between cities and other stakeholders, and public policy limitations of operating at city level. Coolidge reaffirms that sub-state actors can play a role in supporting international treaties and carrying out promises. She explores the ways in which major cities have participated in international mitigation networks, such as C40 Cities Climate Leadership Group or Local Governments for Sustainability. She reminds us that the Paris Agreement in fact calls on parties to recognize how local actors are making progress towards their reduction targets. Given their participation over time, the US cities have proven to be far more reliable partners in mitigating climate change than the US federal government. It is undeniable that cities have a role to play, but that role has been limited to informal activities with support from traditional international organizations, like the UN and the Compact of Mayors programme.

The chapter by Annika Bose Styczynski looks at how the implementation of the Paris Agreement is viewed in the European Union in general and in some European countries – Germany, Norway and France – in particular. It analyses the Storting Policy, the Triple 20 and other initiatives that form part of low-carbon technology strategy. Despite all adversities, the country cases show positive and promising signs of alignment to the climate protection movement, preserving overall integrity of the endeavour. Domestic political-economic factors are probably best in explaining the variation in compliance among states to national policy targets and international agreements. The research on climate sensitivity and vast uncertainties about timing and intensity of climate responsiveness to emissions reductions remain two cases in point, concludes Bose Styczynski.

Trishna Mohan Kripalani and Gargi Katikithala focus on Latin America through the lenses of strengths, weakness, opportunities and threats. The chapter identifies overarching issues such as the contribution of indigenous communities and state policies on economic development. The authors show how the implementation suffers due to two factors: a lack of integrity and of political will. This is evident from unambitious and insufficient NDCs; for instance, Brazil's NDCs, considered among the best in the region, are very minimalistic. Brazil has put in place laws and frameworks that seek to combat climate change, but these are not supported by the mechanisms that would lead to successful implementation. There is a clash of the right to development with the rights of indigenous people and the protection of the environment. For example, Ecuador introduced a national development plan based on the indigenous concept *sumak kawsay* (or good living), but this posed a threat to climate action. Latin America is yet to achieve a balance between the right to development and policies of climate change. The key is to not only include the indigenous people in the decision-making, but also rely on them for possible solutions.

Conclusion

Will the parties achieve the goals they set in the Paris Agreement? Many years need to pass before a serious assessment can be made and this question answered. Indeed, the Agreement may never solve the challenge of climate change. It will, at best, find ways to manage the problem over time. Successfully accomplishing even the more limited goal requires the regime and its parties to establish rules, information systems, facilitation mechanisms and an evaluation system that provide clear expectations about what is required, and to encourage compliance while discouraging non-compliance. Even with the best-designed implementation system imaginable, the effectiveness of the regime at inducing the economic, social and political changes necessary to avert climate change will depend on nations, corporations, NGOs and individuals dedicating significantly greater resources to the task of preventing climate change than they have dedicated to any previous environmental problem. Having discussed the political implications of the climate change regime and the challenges of compliance and implementation of the climate policies it is important to assess the place of the climate change regime within the general framework of international environmental agreements. What are the common aspects and what are the differences? Can one draw some inferences for the climate change regime from the experiences of other environmental accords?

The chapters that follow will tackle these questions and will recommend views on the implementation approaches, such as reporting and verification. A fundamental problem in achieving the implementation of the Paris Agreement is that the precision of the baseline for many emissions sources is very low and will probably remain so indefinitely. If approaches do not convince and if verification fails, the confidence in the climate regime will fade away. There should be a balance of intrusiveness and trust – the solution would be a system that is robust enough to allow the regimes' strongest supports to be reassured of progress in reducing the emissions. At the same time, the system should avoid over-watching and intrusiveness that can enhance distrust.

2 'Hard' and 'soft' law on climate change

Comparing the 1997 Kyoto Protocol with the 2015 Paris Agreement

Vesselin Popovski

The development of international law in the 20th century introduced various norms and rules – prohibition of the use of force, protection of civilians, prohibition of chemical, biological, nuclear weapons, landmines, protection of human rights, gender equality, rights of the child, law of the sea, etc. – and imposed those rules on states. This was 'hard law' development, top-down imposition of rules in treaties and conventions creating binding obligations. If states did not comply with these laws, they were warned, sanctioned and punished. States joined such international treaties, sometimes reluctantly, but even if they signed and ratified these treaties, they did not always properly implement them.

'Soft law' refers to norms and rules that are less binding and enforceable[1]. The soft law can be expressed in treaties consisting only soft legal obligations, or non-binding resolutions or codes of conduct of international and regional organizations[2]. Soft law is a complex phenomenon as it serves various purposes. Alan Boyle and Christine Chinkin distinguish the following purposes of the soft law: (1) alternative to treaty law; (2) authoritative interpretation of treaties; (3) guidance on the implementation of a treaty; (4) step in the development of international legal principles; (5) evidence of *opinio juris* in the formation of international customary law[3].

The UN General Assembly resolutions, for example, can be considered as 'soft law', as far as the Assembly cannot punish those states that may disregard these resolutions. The 'soft law' is viewed differently by different legal theories. The legal purists reject the notion of 'soft law', insisting that international law should always be 'hard', based on binding obligations. The legal realists believe exactly the opposite, claiming that every international law is in fact 'soft', because of the absence of global government to enforce it. The legal pragmatists argue that 'soft

---

1 Boyle, A. (2014), "Soft Law in International Law-Making", in M.D. Evans, *International Law*, 4th edition, Oxford, pp. 118-136; Abbott, K. and Snidal, D. (2000), "Hard and Soft Law in International Governance", *International Governance*, 54, pp. 421–456
2 Chinkin, C. (1989), "The Challenge of Soft Law: Development and Change in International Law", *International Comparative Law Quarterly*, 38, pp. 850–866
3 Boyle, A. and Chinkin, C. (2007), *The Making of International Law*, Oxford, Ch. 5.2

law' emerges when consensus is difficult or when states are reluctant to accept strict legal obligations, and as a result they adopt vague texts leaving freedom of interpretation. States also may adopt 'hard' law, but 'soften' it by making reservations and opt-outs.

The 1997 Kyoto Protocol of the UN Framework Convention on Climate Change (UNFCCC)[4] was a classical 'hard law' and, not surprisingly, states had difficulties accepting top-down imposition of its targets and the Protocol faced serious implementation problems[5]. The 2015 Paris Agreement on Climate Change[6] emerged as a new type of international law, based on bottom-up commitments. Richard Falk described the Paris Agreement as 'Voluntary International Law', which comes with worrisome concerns as far as governments can ignore such non-binding agreement or simply make insufficient national commitments, but it also leaves the implementation open for civil society activism and often this has worked[7]. Anne-Marie Slaughter went further by referring to a new kind of global governance, manifested by the Paris Agreement, where states make rolling commitments and instead of suing some for non-compliance, they try to outperform others, concluding that the Paris Agreement is a 'sprawling, rolling, overlapping set of national commitments brought about by a broad conglomeration of parties and stakeholders. It is not law. It is a bold move toward public problem solving on a global scale. And it is the only approach that could work'[8].

States enjoy the freedom to determine and enact national commitments, while articulating a legally binding long-term collective agenda. The Agreement establishes a committee facilitating its implementation by sharing information and technologies, which is a radically different body from the 'sanctions committees' known from the 'hard law'. The 'soft law' in the Paris Agreement makes states comfortable; they can make commitments voluntarily and revise these later. Some may plan ambitiously and adjust the targets down, if problems occur. Others, vice versa, can start with modest initial commitments and adjust those up if methods work better than expected, or if domestic priorities shift in favour of climate policy.

This chapter compares the 1997 Kyoto Protocol with the 2015 Paris Agreement and demonstrates the shift from hard top-down law, based on sanctions, to soft bottom-up law, based on facilitation, making the hypothesis that 'soft law' should not be dismissed as it may offer a viable alternative to 'hard law'.

---

4 http://kyotoprotocol.com/
5 Victor, D. (2001), *The Collapse of the Kyoto Protocol and the Struggle to Slow Global Warming*, Princeton University Press
6 http://unfccc.int/paris_agreement/items/9485.php
7 Richard Falk, "Voluntary International Law and the Paris Agreement" at https://richardfalk.wordpress.com/2016/01/16/voluntary-international-law-and-the-paris-agreement/
8 Anne-Marie Slaughter, "Paris Approach to Global Governance" at https://scholar.princeton.edu/sites/default/files/slaughter/files/projectsyndicate12.28.2015.pdf

1997 Kyoto Protocol

The Rio Earth Summit in 1992 lead to the adoption of the UNFCCC, the Convention on Biological Diversity and the Convention to Combat Desertification. A Joint Liaison Group was set up to boost the implementation of these three Conventions. Later the Ramsar Convention on Wetlands was also added.

The UNFCCC aims to prevent dangerous human interference with the climate system and it entered into force on 21 March 1994, achieving since then near-universal membership. The Kyoto Protocol of the UNFCCC was adopted in 1997, and it set internationally binding targets for reducing greenhouse gas emissions. It required industrialized countries (Annex I parties) to cut their greenhouse gas emissions to 1990 levels, to report regularly on their climate policies and to submit an annual inventory of their greenhouse gas emissions, including data for the base year (1990) and all years since. Developing countries (Non-Annex I parties) report in more general terms on their climate actions, and less regularly than Annex I parties do. Their reporting is made contingent on getting funding for the preparation of the reports, particularly in the case of the least developed countries. The Kyoto Protocol accepted that the share of greenhouse gas emissions produced by developing nations may grow in the coming years, as economic development is vital for them. In the interests of fulfilling its ultimate goal, the Protocol sought to help such countries limit greenhouse gas emissions in ways that will not hinder their economic progress. The Kyoto Protocol (Article 2.1) demanded the parties 'enhance the individual and combined effectiveness of their policies and measures adopted under the Convention. To this end, these Parties shall take steps to share their experience and exchange information on such policies and measures, including developing ways of improving their comparability, transparency and effectiveness'[9].

The Kyoto Protocol, recognizing that industrialized countries are principally responsible for the high levels of emissions in the atmosphere as a result of their 150 years of industrial activity, places a heavier burden on them under the principle of common but differentiated responsibilities (CBDR). The detailed rules for the implementation of the Protocol[10] were adopted at COP 7 in Marrakesh, Morocco, in 2001. The first commitment period started in 2008 and ended in 2012. In Doha, Qatar, on 8 December 2012, the Doha Amendment was adopted, which added commitments for Annex I Parties in a second period from 1 January 2013 to 31 December 2020. The Doha Amendment contained also a revised list of greenhouse gas emissions reduction to be reported by parties.

The Kyoto Protocol, despite its broad ratification and initial expectations, has encountered growing scepticism both in the environmental and in the economic

---

9 http://unfccc.int/resource/docs/convkp/kpeng.pdf
10 See analysis in Freestone, D. and Streck, C. (eds.) (2005), *Legal Aspects of Implementing the Kyoto Protocol Mechanisms: Making Kyoto Work*, Oxford University Press

communities. The purpose of the Protocol (Article 2) was admirable – 'stabilization of greenhouse gas concentrations in the atmosphere at a level that would prevent dangerous anthropogenic interference with the climate system' – but the method to achieve this was unrealistic, and industrialized nations were asked to reduce their emissions from 1990 levels by an average of 5% over the period 2008 to 2012 with the expectation that the developing nations would also do the same later[11]. Those who signed on to the Doha Amendment commitments only represent 15% of world emissions.

The Kyoto Protocol set firm objectives for emissions reduction, but its deficiency was that it did not tell countries exactly how to implement these, apart from offering three market-based mechanisms: (1) Trading and Carbon Market (Cap and Trade); (2) Clean Development Mechanism; and (3) Joint Implementation. The first mechanism, 'Cap and Trade', sets limit on the emissions, and companies were allotted permits to emit only specific amounts into the atmosphere. However, some companies based on fossil fuels could still pollute more by purchasing 'credits' from those that pollute less[12]. Each country committed to the Kyoto Protocol had an assigned amount of units (AAUs) to emit as a target towards limiting or reducing emissions. Article 17 of the Protocol allowed countries that have AAUs allotted, but not used, to sell them to countries that have surpassed their limits.

The emissions were made a marketable commodity, but this has done little to solve the problem[13]. Developing countries that collected AAUs ultimately sold them to the highest-bidding industrialized nations, so instead of taking measures towards reducing the emissions, rich economies were capable of purchasing AAUs in an open, unlimited manner. Article 12 of the Kyoto Protocol established a Clean Development Mechanism (CDM) allowing industrialized countries to invest in projects that reduce emissions in developing countries as an alternative to expensive emission reductions in their own countries. In turn, the country implementing the CDM project can attain certified emission reduction credits that can be sold or used to offset that country's limitation targets. The CDM indeed stimulated emission reductions, by giving industrialized countries some flexibility in how they can meet their emission reduction[14]. Yet, as an alternative to domestic emissions reductions, the CDM, even where not misused, did not produce any more or less reductions than without its use.

Article 6 of the Kyoto Protocol defined a Joint Implementation Mechanism, which allowed two or more industrialized nations to join in partnerships to reduce carbon emissions in order to earn emission reduction units (ERUs) from joint projects. Credits from the resulting improvements can be used and counted

---

11 Oberthür, O. and Ott, H.E. (1999), *The Kyoto Protocol: International Climate Policy for the 21st Century*, International and European Environmental Policy Series, Springer
12 Lecocq, F. (2005), *State and Trends of the Carbon Market*, World Bank Working Paper 44, World Bank Publications, May
13 Douma, W., Massai, L. and Montini, M. (eds.) (2007), *The Kyoto Protocol and Beyond*, Asseer Press
14 Olawuyi, D. (2010), *Aspects of Implementing the Clean Development Mechanism: Lessons from India and Nigeria*, LAP Lambert Publishing

towards meeting the Kyoto targets. Unlike the CDM, the Joint Implementation Mechanism has caused less concern regarding spurious emission reductions, as it took place in countries which have emission reduction requirements. Flexible mechanisms intend to facilitate countries to fulfil their commitments to reduce emissions, but they also present a risk of changes in energy and transportation systems. Conversely, while the climate effect will be the same no matter where emissions take place, the cost of reduction will vary considerably from one place to another. Thus, the focus on industrialized nations has been more fixed upon the financial rather than on the environmental impact. The Kyoto Protocol generally could achieve very little in developing countries, like those in Africa, South Asia, Southeast Asia and Latin America.

In terms of compliance, the Kyoto Protocol provided for monitoring of emissions targets through a registry system tracking and recording transactions by parties. The UN Climate Change Secretariat keeps an international transaction log to verify that transactions are consistent with the rules of the Protocol. Reporting is done by parties submitting their annual emission inventories and national reports under the Protocol at regular intervals. A compliance system ensures that parties are meeting their commitments and helps them if they have problems doing so.

One problem with the Kyoto Protocol was the so-called 'hot air trading'. Russia and some Eastern European countries, for example, could easily reduce their emissions by 20% or even 30% below their 1990 levels, but these reductions were the result of economic decline or restructuring, not of climate-relevant policies. The Protocol allowed them to trade the surplus difference to industrialized countries, and essentially they got credits without having taken any actions to de-carbonize their economies. As their surplus permits were bought up by other countries, the result could even be an increase of emissions[15]. The Republican US Administration, coming to power in 2001, withdrew from the Kyoto Protocol immediately, seeing the obligations as arbitrary and ineffective in nature. Many countries were completely exempted from the Protocol, including China and India, two of the top emitters in the world. The Kyoto Protocol failed in several ways: the temperatures continue to rise dangerously and extreme weather events – typhoons, hurricanes and tsunamis, flooding and fires – continue to hit all parts of the Earth. Accordingly, climate change has continued to be a top concern, and the International Panel for Climate Change (IPCC) has kept issuing its reports urging states to come together to reach an agreement.

2015 Paris Agreement

The 2015 Paris Agreement is an 'umbrella agreement', that allows parties to make concrete bottom-up commitments. Its preamble links climate change with

---

15 Korppoo, A., Karas, J. and Grubb, M. (2006), *Russia and the Kyoto Protocol: Opportunities and Challenges*, Chatham House, 27 March

sustainable development and the eradication of poverty, and contains examples of the best-ever human rights language in international environmental law. It stresses that climate change is a common concern of humankind, and urges parties, when taking climate action, to respect and promote human rights, the right to health, the right to development, the rights of indigenous peoples, local communities, migrants, children, persons with disabilities and people in vulnerable situations, as well as gender equality, empowerment of women and intergenerational equity. No other treaty has ever listed in such detail the human rights of all vulnerable groups.

The Paris Agreement builds upon the UNFCCC and Kyoto Protocol and – for the first time – brings states into a common cause to undertake ambitious efforts to combat climate change and adapt to its effects, with enhanced support to assist developing countries to do so. The Paris Agreement's aim is to strengthen the global response to the threat of climate change by keeping global temperature rise well below 2° Celsius above pre-industrial levels and to pursue efforts to limit the temperature increase even further to 1.5° Celsius. Additionally, the agreement aims to strengthen the ability of countries to deal with the impacts of climate change. To reach these ambitious goals, the Agreement provided for appropriate financial flows, established a new technology framework and an enhanced capacity-building framework. These aimed to support action by developing countries and the most vulnerable countries, in line with their national objectives.

The Paris Agreement provides for enhanced transparency of action and support through a robust transparency framework. It requires all parties to put forward their best efforts through 'nationally determined contributions' (NDCs) and to strengthen these efforts in the years ahead. This includes requirements that all parties report regularly on their emissions and on their implementation efforts. During 2018, parties should take stock of the collective efforts in relation to progress towards the goal set in the Agreement and inform how they prepare their NDCs. There will be a global stocktake every five years to assess the collective progress towards achieving the purpose of the Agreement and to inform further individual actions by parties.

The word 'implementation' is repeated dozens of times in the Paris Agreement in a serious effort to ensure that the Agreement remains not simply a wish-list. Article 15 urge parties to establish a special mechanism to facilitate implementation and promote compliance, non-existent in the Kyoto Protocol.

The implementation, therefore, was of paramount concern to all those who drafted and lobbied for the adoption of the Paris Agreement.

Article 2, para. 1 of the Paris Agreement stresses that the agreement is 'enhancing the implementation of the Convention', and, importantly, connects this with sustainable development goals (SDGs) and efforts to eradicate poverty. Therefore, not only do the SDGs (Goal 13) now include climate action, but also the climate actions are closely seen in the context of the SDGs[16]. The Paris Agreement

---

16 Article 4, para. 1 in fact repeats the Article 2 demand to achieve the long-term temperature goal 'in the context of sustainable development and efforts to eradicate poverty'

calls for holding the increase in the global average temperature to well below 2° Celsius above pre-industrial levels (para. 1a). This is short of the more ambitious goal, discussed at the preparation meetings, of phasing out entirely all emissions to net zero in the second half of the 21st century. The Agreement pointed to the need to increase the ability to adapt to the adverse impacts of climate change and to foster climate resilience in a manner that does not threaten food production (para. 1b). The resilience-building of communities facing climate impacts is supposed to be achieved through the collective co-operative actions of all countries. Similarly to the Kyoto Protocol, the Paris Agreement 'will be implemented to reflect equity and the principle of common but differentiated responsibilities and respective capabilities' (para. 2).

Article 3 demands parties undertake Nationally Determined Contributions (NDCs) and communicate ambitious efforts to achieving the purpose of the Agreement. These efforts 'will represent a progression over time, while recognizing the need to support developing country Parties for the effective implementation of this Agreement'. This is a strong message to urge developed countries to think not only about how they will implement the Agreement, but also how are they going to assist the developing countries to implement it.

Article 4, para. 2 asks parties to prepare, communicate and maintain successive NDCs – a clear bottom-up approach, different from Kyoto. Para. 3 urges parties to adopt the highest possible ambitions, reflecting common but differentiated responsibilities in the light of different national circumstances. Para. 4 demands developed countries continue to take the lead by undertaking economy-wide absolute emission reduction targets, but also invites developing countries also to move over time to such economy-wide emission reduction or limitation targets. Para. 5 urges developed countries to support developing countries in the implementation of the Agreement, which should allow developing countries to reach higher ambitions in their own actions. Even the least-developed and small islands may prepare and communicate strategies for low emissions reflecting their special circumstances (para. 6). This is another example how the Paris Agreement, in contrast to the Kyoto Protocol, brings together all states in a voluntary manner, and combines 'hard' with 'soft' law.

Initially, the draft of the Paris Agreement mentioned the three market mechanisms established by Kyoto Protocol – cap and trade, CDM and joint implementation – in three different articles, acknowledging their role in enhancing and promoting the cost-effectiveness of the mitigation actions. However, in another shift from top-down to bottom-up approach, these texts were dropped from the final text, leaving more flexibility as to how parties would like to achieve their mitigation targets. There is a requirement to communicate NDCs every five years and to be informed by the outcomes of the global stocktake (para. 9). Crucially, a party may at any time adjust its existing NDC with a view to enhancing its level of ambition (para. 11). This is possibly the most flexible ever arrangement in international law – states can decide their own NDC and after five years adjust it. They can start ambitiously and adjust the NDC down, if problems occur. Or they can start modestly and adjust up, if the work goes better than expected, or if new opportunities to mitigate emissions arise. Para. 13 asks parties to promote integrity, transparency, accuracy, completeness, comparability and consistency, and to ensure avoidance of double counting, in a text that clearly and strongly emphasizes the ethical values and integrity embedded in the climate change regime[17]. Parties may decide to undertake joint actions or act as part of regional economic integration agreements (paras. 16–18) bringing yet another bottom-up modality. They should strive to formulate long-term strategies (para. 19) in the light of their different national circumstances, a clear invitation to act both bottom-up and long-term. Article 4 on NDCs is the longest in the Paris Agreement, giving parties a lot of choice and flexibility and making them feel comfortable with the goals.

Article 5 of the Paris Agreement aims at sustainable management of forests. It asks parties to conserve and enhance sinks and reservoirs to reduce emissions resulting from deforestation.

Article 6 deals with mitigation and manifests the innovative way of making international law – the word 'voluntary' is repeated in every one of its paragraphs and emphasizes very well the 'soft law' bottom-up character of the Agreement. This departs significantly not only from the Kyoto Protocol, but also from early drafts of the Paris Agreement itself. Some parties (para. 1) can choose to pursue voluntary co-operation in the implementation of their NDCs to allow for higher ambition in their mitigation and adaptation actions. This is a manifestation of how creating a co-operative 'soft' spirit of the law may in fact produce more ambitious outcomes. Parties voluntarily ensure environmental integrity and transparency, including in governance, but also apply robust accounting to ensure the avoidance of double counting (para. 2). We see an excellent equilibrium of voluntary commitments, but accompanied with strict applicability of principles such as integrity, transparency, accountability. In the same way, the use of internationally transferred mitigation outcomes to achieve NDCs (para. 3) is both voluntary, and authorized by participating parties. States can volunteer, but are obliged to follow the principles and procedures. They are not told how much to reduce emissions, but the letter and spirit of the Paris Agreement creates a competion for higher ambitions as well as facilitation and co-operation to achieve these ambitions. The Agreement establishes a (para. 4) Sustainable Development Mechanism (SDM) to contribute to the mitigation of the emissions on a voluntary basis, supervised by a body designated by the COP. This mechanism initially created tensions between pro-market liberals and anti-market socialists. The first group argued that the parties should focus on carbon trading, while the second believed that governments should be more interventionistist and ambitious. The COP 23 in Bonn in November 2017 clarified that it is not all about market and non-market mechanisms, it is also about true process, as opposed to stalling tactics. The COP 23 also discussed what rules, modalities and procedures (para. 7)

---

17 In a previous book the author of this chapter addressed these issues: Breakey, H., Popovski, V. and Maguire, R. (eds.) (2015), *Ethical Values and Integrity of the Climate Change Regime*, Ashgate

are needed. Still, a lot has been left for future deliberations and agreements. The process of continuous legislation – as started in Paris and continuing since – is something which classical international lawyers would have strongly criticized, but actually such flexibility both empowers and incentivizes parties, and facilitates their participation in reaching higher ambitions. In comparison with the Kyoto Protocol, heavily focused on market mechanisms, the Paris Agreement (para. 8) talks about integrated, holistic and balanced non-market approaches to assist in the implementation of NDCs in the context of sustainable development and poverty eradication.

Article 7 of the Paris Agreement deals with adaption. It recognizes that adaptation is a global challenge faced by all, and it has local, sub-national, national, regional and international dimensions (para. 2). The parties recognize (para. 5) that adaptation should follow a country-driven, gender-responsive, participatory and fully transparent approach, taking into account vulnerable groups, communities and ecosystems, guided by the best available science, and, as appropriate, by traditional knowledge, knowledge of indigenous people and local knowledge systems, in order to achieve the long-term goals. This is, indeed, a revolutionary text, unseen not only in the Kyoto Protocol but also in any previous international legal documents. The parties strengthen their co-operation on adaptation, taking into account the Cancun Adaptation Framework (para. 7).

Article 8 addresses the importance of averting and minimizing loss and damage, a concept already established by the Kyoto Protocol. Para. 4 lists areas of co-operation and facilitation to enhance understanding and action, such as early warning systems, emergency preparedness, risk insurance facilities and others.

Article 9 deals with finance, and urges developed countries to provide financial resources to assist developing countries in both mitigation and adaptation (para. 1). Other parties – with reference mostly to China – are encouraged to provide such assistance voluntarily (para. 2). Developed countries should take the lead (para. 3) in mobilizing climate finance from a wide variety of sources, noting the significant role of public funds, and supporting country-driven strategies. The aim is on public finance because, while it may be necessary to increase private sector involvement in adaptation finance over time, the majority of adaptation finance currently comes from public sources. Developed countries shall biennially communicate (para. 5) indicative quantitative and qualitative information. Here we see a shortened version from a much more ambitious text at the preparatory stage, where parties were urged to scale up and shift all investments, public and private, to be low-carbon and climate-resilient, and to ensure that they are aligned with national sustainable development objectives. The initial draft (Article 6: Finance) asked the Meeting of the Parties to clarify which other parties will mobilize finance in the form of contributions between developing countries, seeking to broaden the pool of contributors over time, guided by national circumstances. Another passage was dropped from the draft: 'In order to shift all financial flows to be low-carbon and climate-resilient mandates parties to improve national institutional frameworks for climate finance and ensure that their legal and policy frameworks promote this aim.' The initial draft also asked that, every five years, parties would submit strategies regarding finance, aligned to the adaptation and mitigation cycles. The first strategies are due in 2018 to cover the period 2025–30. All parties are required to report on national efforts to shift and scale up their investments. Parties receiving climate finance had to include information on their national investment plans and any gaps and needs; and contributor parties will include information on their plans and channels for scaling up climate finance. A Standing Committee on Finance (SCF) will produce a synthesis and aggregate analysis of parties' strategies regarding finance, no later than nine months prior to the date when the Meeting of Parties is due to consider the adaptation efforts. Regular assessments of the investment portfolios of national and international financial institutions will identify climate-related risks that arise from climate change impacts and from stranded investments in carbon-intensive assets. Assessments could include climate change stress-testing of cross-country financial flows and benchmarking to measure progress towards the long-term mitigation and adaptation goals of the Agreement, including the goals of phasing out emissions to net zero as early as possible, and building the climate resilience of communities.

Finance became a central theme at COP 23. The dilemma that developed countries face is for what exactly will they pay? Should the funds go to vulnerable countries for sustainable development and poverty alleviation, or for emissions removals as a new form of mitigation? Or for adaptation to the current climate crisis? Or for all of these? This is no longer a dialogue about fair share, or historical obligations for the emission sins of the past. How to prioritize funding remains a crucial issue to be resolved.

Article 10 of the Paris Agreement welcomes parties to share a long-term vision on the importance of fully realizing technology development and transfer. At this juncture, it is clear that serious measures to ensure innovation and technology transfer must be adopted, but the final text is once again much shorter than Article 7 of the initial draft, which dealt with technology transfer. Research shows that the global diffusion rate of climate technology must be doubled by 2025 to have any chance of achieving the objectives. In the initial draft, Article 7, para. 1 aimed to operationalize the attempt to respond to this need. External institutions and the private sector should be engaged to direct more expertise toward the end of achieving the long-term mitigation and adaptation goals. Parties were called upon in para. 2 to strengthen global co-operation, not only to achieve an overall increase in the global diffusion rate of climate technology, but also to provide support to developing countries with funding for research and development, enhancement of national enabling environments, and achievement of appropriate cost reductions in the access to climate technologies. This is imperative given the recognition that new models for resilience and methods for managing shocks are needed to adapt to escalating climate change impacts.

Article 9 (Finance) and Article 10 (Technology transfer) aimed largely to provide the necessary support to developing countries, in particular least-developed countries and small island developing states, to be assisted in compliance with this process. From the initial draft – disappointingly – Article 7, para. 6 was dropped, which authorized collaboration with relevant international organizations, such as WTO, WIPO and expert bodies, to address the current limitations in the absorption, development and transfer of climate technology by and for developing countries. Also, Article 8 of the Kyoto Protocol envisaged expert review undergoing thorough and comprehensive technical assessment of implementation, circulated to all parties.

Article 11 of the Paris Agreement defines yet another arrangement to give proper support to the parties by providing a sustainable and substantial capacity-building mechanism so that developing states can take effective climate change mitigation and adaptation action (para. 1). Because not all parties have the ability to contribute equally to the fulfilment of the objectives of the Agreement, capacity-building should be (para. 2) country-driven, based on and responsive to national needs, and foster country ownership of parties. This mechanism aims to address gaps in capacity and consolidate existing provisions into a framework ensuring that capacity-building is undertaken in a co-ordinated, integrated and sustained manner. In the initial draft, Article 8, para. 1 was designed effectively to make the long-term interventions necessary to assist all parties to achieve the objective of the Convention. Under the draft Article 8 para. 3, all parties must be represented within this mechanism in an equitable and balanced manner within a transparent system of governance. It is critical to ensure that the mechanism has the trust and confidence of all parties and the necessary support to carry out its functions effectively. The capacity-building mechanism will focus on providing support to states that need it most. It recognizes that only through global co-operation will it be possible to achieve the long-term goals. Under Article 8, para. 4, the mechanism ensures adequate capacity-building for effective implementation of the developing countries' commitments, and undertakes regular assessment of the financial and technical assistance being received for capacity-building and to ensure that such assistance is being utilized in the most effective way possible. The mechanism also sought to widen the scope and impact of capacity-building programmes by collaborating with expert institutions under and outside the Convention, facilitating capacity-building at the national and regional levels, and providing guidance and counsel to the Meeting of the Parties. The final version of Article 11 in the Paris Agreement only mentioned regular communication between parties on all capacity-building actions (para. 4) and enhancing the capacity-building with appropriate institutional arrangements (para. 5).

Article 12 encourages parties to co-operate in climate change education, training, public awareness, participation and access to information, recognizing the importance of these steps with respect to enhancing actions.

Article 13 of the Paris Agreement creates a very detailed transparency framework for action and support (para. 1) with built-in flexibility (para. 2) in the implementation of the provisions, based on parties' different capacities. It generally followed the ambition of the draft, although it limited it mostly to transparency, whereas the draft talked about improving the transparency, accountability, consistency, completeness, comparability and accuracy of data over time. The transparency framework replaces the monitoring, reporting and verification (MRV) language of the Kyoto Protocol. Availability of support for developing countries (finance, capacity-building, technology development and transfer) in fulfilling their requirements is envisaged, in particular for the least-developed countries and small island states (para. 3). The transparency and accountability in the pursuit of long-term adaptation and mitigation goals will be achieved by regularly providing information such as national inventory reports (para. 7). In line with the overall approach developed for the Agreement, parties would work toward continuous improvement of the quality of information provided to enhance understanding, track progress against the long-term goals, and build trust among the wider international community. They should also provide information on climate change impacts and adaptation (para. 8). Developed and other countries that provide support should provide information on finance, technology transfer and capacity-building support (para. 9) and the developing countries should provide information once these are received (para. 10). All information received under paras. 7, 8 and 9 will undergo a technical expert review (para. 11), which can identify areas of improvement and include a review of the consistency of the information with the modalities, procedures and guidelines (para. 12). Article 13, para. 13 recommends that the COP adopts common modalities, procedures and guidelines. The idea is to have a common type of national reports and assessment processes in place after 2020, acknowledging that the quality of the information would not be expected to be the same for all countries, taking into account differing capabilities and circumstances. Still, reporting guidance could be crafted after Paris in a way that enables parties to aspire to more robust and credible data over time.

Article 13 in the final text of the Agreement is the second-longest after Article 4, but it avoids naming the periods of time, as in the draft, where we see differentiation in the content and frequency (every two years and every four years) of the reports, based on the relevance of the information required – for example, the reported information could be different if a country provides support, receives support, or does both. The transparency framework reiterates the need to acknowledge the different capabilities and circumstances of parties, recognizing that some states may require flexibility in the implementation of their commitments, while at the same time promoting a cycle of improvement. It therefore establishes the direction for all countries, building on the progress made over the past few years and on lessons learned, with a view to ensuring that post-2020 requirements should not be less than those that currently exist. The more robust transparency framework requires commensurate financial and technical support to remove current capacity barriers that prevent developing countries from fulfilling their transparency requirements. It also emphasizes the special needs of countries with fewer capabilities.

Article 14 suggests that the COP shall periodically take stock of the implementation of the Agreement, assessing the collective progress towards achieving its purposes and long-term goals (para. 1). It shall undertake such a 'global stocktake' in a comprehensive and facilitative manner starting in 2023, and do this every five years thereafter (para. 2). This is indeed an excellent innovative equity framework to guide the implementation of the Agreement, specifically through integration in five-year cycles for adaptation, mitigation and support. The parallel and mutually supportive five-year cycles for adaptation, mitigation, and the means of implementation and support, would help actualize the long-term goals, and, in effect, all the objectives of the Convention. They will allow for durability and continuous strengthening of the regime, while rendering unnecessary the return to prolonged negotiations at the end of every commitment period. Putting five-year improvement cycles in place would also enable the identification of, and support for, action and leadership, while maintaining predictability and confidence in the process. The Kyoto Protocol was imposing less discipline, saying only that the COP shall initiate the consideration of such commitments at least seven years before the end of the first commitment. Kyoto was essentially procedural, encouraging co-operation among parties to follow the framework along with regional and economic integration. The outcome of the global stocktake will help parties in 'updating and enhancing their actions and support... as well as in enhancing international co-operation for climate action' (para. 3).

Facilitating implementation

Article 15 of the Paris Agreement establishes a mechanism to facilitate the implementation and to promote the compliance with the regime. It recognizes that transparency and accountability are not enough, and that there is a need to establish a mechanism to facilitate implementation and promote compliance (para. 1) to engage all parties in an atmosphere of trust and confidence. The mechanism consists of an expert-based committee functioning in a transparent, non-adversarial and non-punitive manner, paying attention to the respective national capabilities and circumstances (para. 2). A combination of facilitation and non-adversarial operation is seen as the most politically feasible manner of promoting compliance. Again, the final text is short, but if we look at the draft of the Paris Agreement, we will see moves in establishing governance arrangements for the mechanism, requiring it to have an equitable and balanced representation of parties and effective decision-making procedures. The draft also clarified how proceedings before the facilitating committee can be triggered, and stipulated that it shall function independently and have effective measures available. The mechanism for facilitating and promoting implementation should mobilize support from governments as well as from business and economic actors.

The Article 15 mechanism demonstrates well the difference between the 'sanctions committees' established by the United Nations to monitor various disarmament[18], counter-terrorist[19] or non-proliferation[20] regimes, and the Paris

---

18 Imposed on Iraq (1991–2003), Iran or North Korea
19 Security Council Resolution 1373
20 Security Council Resolution 1540

Agreement. The compliance is exercised not under threat of sanctions and punishment, but rather is facilitated. Therefore, the mechanism to facilitate implementation and promote compliance in the Paris Agreement (compared with Article 11 of the Kyoto Protocol) can be seen as a new development in international law. The 'hard law' and 'sanctions committees' step aside and give way to 'soft law' and 'facilitation committees'.

The Paris Agreement is a 'soft law' that gives parties freedom to determine and enact NDCs, although these NDCs clearly articulate a legally binding long-term collective agenda. The facilitating committee provides for information and technology sharing. States can adjust to what they initially committed. Some may plan ambitiously and adjust down, if problems occur. Others can start modestly and adjust up, if new opportunities arise. Creating the mechanism to facilitate implementation, the Paris Agreement signals that domestic political interests cannot be the primary determinant of action and parties must work together to raise their climate change ambitions.

The Paris Agreement sends a message that parties should engage in an atmosphere of trust and confidence. They should keep domestic and political interests at bay and focus on tackling the effects of climate change[21]. The mechanism for the facilitation of implementation and the promotion of compliance mobilizes support from both governments and economic actors[22]. The mechanism should have an equitable representation and function in a goal-orientated non-adversarial manner. It describes the nature of proceedings before the committee and how parties may be able to avail of effective measures[23]. The Paris Agreement ensures that the three cycles of mitigation, adaptation and support are applied universally to all countries. However, it is important to note that each cycle would be applied in intrinsically different ways, unique to the circumstances of the country it is sought to be applied in. The focus on national commitments seems to be balanced by a sustainable process to determine an equity framework, which would be able to effectively guide future negotiations and commitments.

Novelty in the Paris Agreement, compared with the Kyoto Protocol

The main difference in the Paris Agreement is that it replaces the temporary commitments in emissions reduction from the Kyoto Protocol with long-term phase-out goals by the second half of the 21st century. We see definitions of new key terms such as 'mitigation commitments', 'adaptation efforts statements', 'long-term adaptation goal', 'long-term mitigation goal' and 'support'. If the

---

21 Gallagher, L. (2014), "Political Economy of the Paris Climate Agreement", *Agreement for Climate Transformation 2015 (ACT 2015)*, p. 6; www.e3g.org/docs/ACT_2015_FINAL_Political_Framing%281%29.pdf
22 Ibid.
23 Oberthür, S. (2014), "Options for a Compliance Mechanism in a 2015 Climate Agreement", *Agreement for Climate Transformation 2015 (ACT 2015)*, p. 6

Kyoto Protocol focused solely on mitigation and briefly mentioned adaptation, the Paris Agreement sets out long-term goals that evolve a recognition of the interdependence of mitigation, adaptation and support strategies, not only of the governments, but also of the private sector and the public. This is significantly different from the Kyoto Protocol which sought to devise a top-down approach wherein governments use regulatory mechanisms to rein in the levels of carbon emission in the atmosphere. The Kyoto Protocol seemed to have alienated the private sector and the public, who began to question the need to sacrifice economic growth for the unsubstantiated (at the time so-believed) effects of climate change. The Paris Agreement purports a more inclusive approach, attempting to gather suggestions and solutions from all factions of society and a more socially conscious private sector. It realizes the irreversible effects of climate change and suggests not only a long-term phase-out goal, but also a long-term adaption strategy, which seeks to strengthen the capabilities of communities facing irreversible climate effects. It requires immediate collective action and co-operation of the international community and the support of the private sector.

Apart from the phase-out goals on emissions by the second half of the 21st century, an emphasis is made on the building of a durable mechanism of resilience against climate impacts. These long-term commitments are integral to the creation of an environment of compliance and an acknowledgement of a globally recognized mandate. The Agreement recognizes the impending need to adapt to the adverse climate effects that have already begun to affect populations across the world[24]. The Kyoto Protocol failed to include provisions for adapting to a rapidly changing environment and fell short of encapsulating the almost symbiotically reinforcing nature of mitigation and adaptation. Both the Protocol and the Agreement recognized the common but differentiated responsibility (CBDR), based on the respective capabilities of countries; however, in the Paris Agreement no country has been excluded from CBDR obligations, only different timeframes have been provided to industrialized and developing countries.

A major governance innovation in the Paris Agreement are the five-year cycles to transpose an element of predictability to the outcome of the climate actions taken by countries in a just, fair and equitable manner. Every five years countries commit to strengthen their NDCs to achieve in the long-term the complete phase-out. The Paris Agreement also proposes the setting-up of an independent expert panel to promote mechanisms by which obstructions to enhanced action of economic instruments, market-based incentives and regulatory policies can be overcome, and the mitigation potential can be leveraged. The Kyoto Protocol failed to provide for such an assessment mechanism for allowing countries to introspect on the reasons for their failure to achieve the proposed mitigation targets and to assist them in overcoming such obstructions, to whatever extent possible.

---

24 Morgan, J., Dagnet, Y., Höhne, N., Oberthür, S. and Li., L. (2014), "Race to the Top: Driving Ambition in the Post-2020 International Climate Agreement", *Agreement for Climate Transformation 2015 (ACT 2015)*, p. 2

While for the least-developed countries adaptation is a priority above mitigation, the Paris Agreement requires parties to submit statements with regard to their adaptation efforts, and a specialized committee to be set up to review the progress in adaptation, in addition to assisting with the facilitation of the mitigation. The five-year cycles are instrumental in the effectuation of long-term goals and simplify the negotiation process between countries with regards to climate commitments, thus maintaining a predictable outcome of the process. Perhaps the biggest governance shortfall of the Kyoto Protocol was the failure to include short-term review processes before the commencement of the 2008–12 commitment period, allowing countries to deviate from the commitments they had made. As a result, a sense of apathy emerged in both developed and developing countries towards the impending need to reverse climate effects.

The Paris Agreement sets a review mechanism strengthened with data provided by the IPCC as a baseline for assessing the status of climate change and making appropriate decisions in accordance with the same. The Agreement provides for the undertaking of quantified emission limitations, objectives of reduction, long-term nationally determined goals and other policies to enhance the viability of the carbon-mitigation mechanisms. These obligations would have to be undertaken in a transparent manner that is both comprehensive and reflective of a responsible initiative, without backsliding on commitments[25]. The commitments of the parties go to 2025, with the option of extending further. The intent of the Paris Agreement is to focus on actions taken towards achieving a complete phase-out, not simply individual results. This is a positive shift away from the goal-orientated outlook of the Kyoto Protocol that was theoretically sound but often impractical to implement. One problem with the Kyoto Protocol was that the Clean Development Mechanism provided for projects that brought too much in the way of carbon trade benefits to countries and companies, such that at some point some of them lost interest to pursue more green projects. The Kyoto Protocol was too much market-orientated and less integrity-orientated[26].

The Paris Agreement calls for the establishment of an expert panel to assess mitigation efforts on an individual and collective level. It stipulates the institution of a continuous mitigation cycle, whose rules and modalities would *de facto* be in place to guide parties towards the first round. This cycle would operate in collaboration with the adaptation, the support and the finance cycles. All parties put forth their strengthened mitigation commitments followed by a technical review of the information provided and a multilateral assessment by an expert panel which would provide its technical expertise to individual countries by monitoring the 'aggregate emission gap' between the combined efforts of the parties' commitments and the long-term goals of the Agreement. The commitments can be submitted at any time (to allow for parties who seek to make additionally faster

---

25 Morgan, J. *et al.* (2014), *Agreement for Climate Transformation 2015 (ACT 2015)*, p. 4, see note 24

26 Breakey, H., Popovski, V. and Maguire, R. (2015), see note 17

commitments to do so) and will be considered on a collective basis to reflect the progress that had been made towards the long-term mitigation goals.

The compliance mechanisms under the Kyoto Protocol faced difficulties. The penalties for non-compliant countries remained a contentious issue and this was a central discussion during the Marrakech deliberations at COP 7 in 2001, with agreement that if a particular party fell short of a particular emission target during the first period of 2008–2012, it would have to make up for this difference in the second period. An additional penalty of 30% was imposed on countries which failed to comply. The faulty party would be barred from taking part in emissions trading schemes and would have to develop a 'comprehensive plan of action'. It was also decided to establish a Compliance Committee, composed of an operational bureau, a plenary and two separate branches, a facilitative branch which advised the parties and provided with requisite assistance, including early warning to be given to those in danger of not complying, and an enforcement branch to act against non-compliance. A specific rulebook was formulated to provide particular procedures in cases of non-compliance, including stipulations of various review mechanisms to assess a country's eligibility to participate in and benefit from the financial mechanisms included in the Protocol. The compliance with the Protocol was not restricted to the achievement of emission reduction commitments but includes procedural requirements and the preparation of adequate inventories. The penalties instituted for failure to meet the requirements of emission reduction only takes shape during the commitment period. This has resulted in significant uncertainties with regard to the functioning of such penalties.

The Kyoto Protocol attempted to achieve equity through differentiation in the commitments instead of through differentiation in the measures implemented to achieve these commitments. The emphasis of the Paris Agreement is on strengthening commitments through adaptation, support and mitigation strategies, attempting to bridge the gap between the commitments made by developed and developing countries. This would in effect result in a more durable mechanism of sustainable growth than the framework of emission reduction mechanisms in the Kyoto Protocol.

The provisions for the inclusion of support mechanisms, including funding and technology transfers, have been drafted with caution. The historical distrust within the developing community towards such commitments has been further enhanced due to allegations of unbalanced distribution of funds. Developed countries, however, contend that the fiscal and political environments within developing nations often obstruct the actualization of numerous efforts, and the transfer of technology and funds seems futile due to non-transparent and unaccountable domestic fiscal environments. Additionally, countries with higher levels of GDP are not required to make any such financial contributions. The support package in the Paris Agreement requires a stipulated level of clarity and accountability, which gives developing countries the capacity to meet the mitigation and adaptation standards to which they commit. The support provisions provided thus includes an elaborate financial cycle which would attempt to interlink the financial needs with financial mobilization strategies, with a view to aiding developing countries in adapting their domestic legal and institutional mechanisms to enable a scaling-up of investment. The specificity and nature of this financial support system is significantly different from the support mechanisms provided for within the Kyoto Protocol, which were more suggestive than mandatory in nature.

The Paris Agreement elaborated the financial provisions in Article 9. It required parties to ensure that investments are concentrated on low-carbon and climate-resilient initiatives, in line with the sustainable development objectives[27]. The requirement of mobilizing finances from identified sources, especially by developed countries, has also been extensively covered[28]. The funds will be allocated in continuation of the finance cycle in collaboration with the mitigation and adaptation cycles. Parties would have to make consolidated reports of their national efforts and increase the scale of their investments in accordance with their internally determined national investment plans. This applies both for developed and developing countries. The Climate Fund would carry out an extensive analysis of the progress of financial mechanisms within a particular five-year cycle. This ensures that the support is being managed and implemented effectively, in a transparent and accountable fashion. The Climate Fund would be the main operating financial mechanism and would evolve to become the operative global fund for climate change finance[29].

The Paris Agreement is reflective of the realization that, even with concentrated efforts and enhanced mitigation strategies, the adverse impacts of climate change have already begun to cause long-lasting loss and damage within certain regions and this is likely to be felt across an even larger area in the near future. The Kyoto Protocol did not take into account the damage that had already occurred and how countries could possibly effectively manage and adapt to the changing environment. The Paris Agreement has proposed the use of the Warsaw International Mechanism for Loss and Damage associated with the Impacts of Climate Change under the Convention as a platform for enhanced co-operation on loss and damage. Parties can conceive, implement and regulate durable and achievable national adaptation plans that identify their medium- and long-term needs along with the development and implementation of strategies to address the requirements of such adaptation[30]. The continuous adaptation cycle is to be collaboratively adhered to along with the mitigation and finance cycles. The duration of the assessment cycle would allow parties to review the efforts taken towards the international adaptation and bridge gaps in the support structure to aid in effective implementation[31]. Each party has a further obligation to submit a statement of their adaptation efforts every five years. The primary task of the

27 Morgan, J. et al. (2014), *Agreement for Climate Transformation 2015 (ACT 2015)*, pp. 23–24, see note 24
28 UNFCCC (2010), Decision 1/CP.16, p. 16
29 UNFCCC (2011), Decision 3/CP.17, "Launching the Green Climate Fund", p. 62
30 UNFCCC (2010), Decision 1/CP.16, "The Cancun Agreements: Outcome of the work of the Ad Hoc Working Group on Long-term Cooperative Action under the Convention", p. 5
31 Morgan, J. et al. (2014), *Agreement for Climate Transformation 2015 (ACT 2015)*, p. 16, see note 24

Adaptation Committee is to review the progress that has been made towards long-term adaptation goals and enhancing the effort statements of parties towards their individual adaptation efforts. This Committee will also take into account the analysis of the standing Finance Committee and its policies, reports, reviews, recommendations and findings. The implementation will be guided by the equity framework drafted under the Agreement.

In terms of implementation of the goals, the Paris Agreement attempts to further the ideal of CBDR and respective capabilities in light of different national circumstances, as outlined in COP 21 in Copenhagen in 2009. This would mean a differentiating time-frame for the phase-out, similar to the distinction made in the Kyoto Protocol between emission cuts for developed and developing countries. The Paris Agreement takes forward the distinction made in the Kyoto Protocol between the different resources and capabilities of countries at different stages of their development, and proposes concrete mechanisms to bridge the development deficit, consistent with the overarching principle of a common but differentiated responsibility and the respective capabilities of countries in light of their national circumstances. These include a concentrated emphasis on NDCs, significantly different from the more global centric approach envisioned by the Kyoto Protocol. This has been balanced by the creation of a process to determine a framework of equity that would be able to inform future climate negotiation processes in a more systematic manner. The Paris Agreement also purports to incorporate especially flexible and less stringent commitments for the least-developed countries and small island developing countries, especially with regards to the scope and frequency of their commitments. Small island developing states would also be required to concentrate funds on adaption and effective response mechanisms. The Agreement also includes within its ambit the role played by international and non-governmental organizations, which are invited to provide their inputs to the five-year cycles as well. This would encourage an element of procedural equity.

The unprecedented inclusion in the Paris Agreement of a well-defined support structure is remarkable – it purports to incorporate finance, capacity-building measures, technology development and transfer of funds to countries in need. The Kyoto Protocol sought to incorporate such means of assistance but failed to define the extent of the same. Each element within this support mechanism will be reviewed every five years. An additional means of review includes the submission of a five-year strategy by each of the parties, which would reflect the measures taken by them at a national and institutional level. Recipient countries would be required to propound possible plans for national investment and an estimate of future financial needs, to which the countries offering such financial solutions would respond with appropriate mechanisms of affording such financial options. The SCF would then review these recommendations. The Paris Agreement draws a fine balance between a nationally driven approach and multilateral norms in order to inculcate specific long-term goals of adaptation. These goals aid in guiding specific national actions that would be undertaken by parties both on an individual and collective level. The shift of emphasis from multilateral norms to region-specific rules has been consciously made to facilitate greater motivation to uphold individual commitments within the five-year review periods. The text proposes three continuous cycles of progressively strengthening commitments to prevent the rollback on these commitments, as was seen with the Kyoto Protocol. The Agreement attempts to impose legally binding obligations on all parties with the greater objective of creating a balance between the Agreement and the country's internal policies. The mitigation commitments would be inscribed within a specific list, kept with the Secretariat and made available publicly.

The Paris Agreement provides an opportunity to focus on specific areas and concentrate efforts in whatever manner they deem appropriate. This is significantly different from the strictly defined ambit of action envisaged by the Kyoto Protocol. The additional leeway would encourage countries to amalgamate around specific focus areas through joint agreements, involving non-state actors as well. This also enlarges the scope for relevant international organizations to take appropriate action. The Paris Agreement focuses on encouraging the participation of non-state actors in the mitigation. This would require additional co-operation between the parties[32].

The controversial aspects of the Kyoto Protocol are centred upon two provisions. First, the provision of international tradability of the permits does not specify how much of a particular country's obligations to cut down on its national aggregate emissions can be achieved through the purchase of credits from other countries, and the extent to which this needs to be met through solely domestic action. Second, there is no specification of the amount of carbon that has been captured by the soil and forests or through agricultural practices that can be included within the country's efforts to reduce emissions. The most contentious aspect of the Kyoto Protocol was the provision of credits for soils, agricultural practices and forests which absorb the carbon within the atmosphere. The allocation of credits for the subsisting forest cover and land use, land-use change and forestry activities were left largely unanswered within the Protocol. It stipulated that the levels of sequestration that have increased beyond those in 1990 would be counted, but this left a significantly ambiguous realm of determining exactly which activities would be counted as contributions to offset the obligations under the Protocol. The activities that were finally decided upon to be included in the offsetting of obligations were the management of grazing land, re-vegetation, afforestation, deforestation, forest management and the management of croplands. The Protocol did not, however, place any cumulative cap on carbon sinks for individual countries. It incorporated certain limits specific to respective countries. The method by which the carbon has been absorbed within the sinks was the basis for assessing the fulfilment of the country's obligations.

---

32 Garibaldi, J.A., Arias, G. and Szauer, M.T. (2014), "Enhancing Bold Collective Action: A Variable Geometry and Incentives Regime", *Agreement for Climate Transformation 2015 (ACT 2015)*, Working Paper, pp. 4, 6–11; www.wri.org/our-work/project/act-2015/publications.

The Paris Agreement envisages a transparent and accountable environment of application wherein parties, having over time aspired to collate their efforts and report their data, do so in a transparent and accountable manner, in light of their respective circumstances and capabilities. This is especially encouraged of developing countries which would receive financial support, technology transfer and capacity-building assistance in accordance with the Agreement. The Article 15 facilitative mechanism of implementation would allow countries in need to take additional action and enhance their accountability. The Paris Agreement has taken into account the rapid development and the exponential increase in the emission levels of countries that were previously left out of Annex I in the Kyoto Protocol.

The Paris Agreement has reconciled its bottom-up approach with the fact that hazardous and often irreversible effects of climate change can only be effectively dealt with if the obligations upon countries are made mandatory and not just voluntary. The Kyoto Protocol did seek to introduce consequences for non-compliance with the emission reductions, but the Protocol itself did not institute any concrete measures for multilateral collaborative action and left the formulation of the same to the COP. The Paris Agreement is more concrete in this respect, as it seeks to introduce a more accountable process as exemplified within its five-year cycles of support, adaptation and mitigation.

The provision of transparency in the Paris Agreement allows for the permeation of an element of trust within multilateral negotiations. It thus provides for a sense of clarity among investors and private parties about the nature and fulfilment of their investment. This allows for a direct inclusion of the private sector within the suggested legal text itself, and not as a party affected by its provisions. The recognition of the importance of a collaborative understanding between the private sector and governments towards the common goal of climate change mitigation is a conscious shift away from the purely regulatory environment created by the Kyoto Protocol. The accountability and transparency provisions of the Paris Agreement are extensions of similar requirements under the Copenhagen and Cancun frameworks. Further, to ensure the optimum participation of developing countries the suggested legal text has mandated that they will necessarily require additional financial support, capacity-building and technology-transfer mechanisms. The system has been reflected through the continuous improvement of transparency, accuracy, consistency and comparability of data. It also provides for a system of assessment through national reports which are crafted to ensure that all parties provide credible data over time. There is also a provision for the prioritization of data based on the information received and, finally, the availability of adequate support for developing countries in the form of finance, capacity-building, technology development and transfer, especially for least-developed countries and small island developing states.

The Kyoto Protocol, while comprehensive in its structural encapsulation of the outlined goals, failed to include a review mechanism that would have perhaps aided in a better assessment of the implementation mechanisms. The Protocol was not as far-sighted in its approach and focused solely on immediate mitigation

40 *Vesselin Popovski*

strategies through regulatory norms. It however failed to address the needs of countries already facing the adverse effects of climate change, such as rising ocean levels, crop shortages resulting in food crises and the increase of diseases due to warmer, more polluted environments. The Paris Agreement, in contrast, formulated a durable and self-sustaining mechanism by which these concerns can be effectively addressed. It offered tangible goals with the option of frequent review to check the effectiveness of the various measures and assess the deliverables.

Conclusion

The analysis of the texts of the Kyoto Protocol and the Paris Agreement reveals a move towards a well-supported, comprehensive system of facilitating the implementation of the climate change regime goals, unprecedented in the past. It remains still to be seen how this elaborated governance system will work in the years ahead. It would be interesting to examine how international law may break boundaries[33] and incorporate soft law, moving from top-down imposition of rules to bottom-up voluntary commitments. This author has already written on the 'hard' and 'soft' power of the UN Security Council in a book presenting the introduction of thematic debates and adopting thematic resolutions as a fundamental shift in the role of the Security Council from being a global policeman into becoming a global legislator[34]. This chapter presented the 2015 Paris Agreement as an example of synergetic 'hard' and 'soft' law, which encourages bottom-up commitments by states. The implementation is achieved through facilitation mechanisms for both mitigation and adaptation, and regular assessments and reviews.

Some international regimes in the past that combined sanctions with facilitating implementation mechanisms proved to work better than those relying only on sanctions. The counter-terrorism regime established with Security Council Resolution 1373 and the non-proliferation regime established with Security Council Resolution 1540 provided permanent operating committees, which states could approach at any time for advice and facilitation in implementing the scopes of the regimes. Certainly, the approach towards facilitation should neither diminish in scope, nor jeopardize the effort to sanction states for deliberate non-implementation. Being 'soft' towards co-operating states and offering facilitating assistance should not reduce the demand to be 'hard' with unwilling and non-cooperating states; therefore sanctions are as important to have in the toolkit as are the facilitating mechanisms.

Such a synergy between 'soft' and 'hard' implementation of the climate change regime can be compared, for example, with the current emphasis on the

---

33 Baxi, U. (2016), "Some Newly Emerging Geographies: Boundaries and Borders in International Law", *Indiana Journal of Global Legal Studies*, 23 (1), pp. 15–37
34 Popovski, V. and Fraser, T. (eds.) (2014), *The Security Council as Global Legislator*, Routledge, pp. 1–11

implementation of the norm 'Responsibility to Protect' (R2P), where the international community clearly divided the 'soft' Pillar Two from the 'hard' Pillar Three, envisaging different approaches towards states that are unable but willing to protect (Pillar Two assistance), from states that are unwilling to protect or are even complicit in mass crimes (Pillar Three sanctions). Most of the international environmental regimes in the future may follow the Paris Agreement model of a proper balance of 'soft' and 'hard' law implementation. The tendency towards 'soft' law and facilitation should not diminish and jeopardize the traditional 'hard' law-making with firm compliance demands, where sanctions are important to have in the toolkit too.

The 'soft' law bottom-up approach presents states with a larger individual agency to decide and implement self-determined commitments supporting the long-term collective agenda. It creates transparency and accountability and facilitates communication and information-sharing. It enables states to adjust their goals depending on circumstances and provides the flexibility to adjust, upon review and assessment, the methods of achieving the targets. It also allows for states to start with conservative commitments and then build those up if they find their methods are working better than expected.

The Paris Agreement is 'soft' in terms of voluntary commitments, but 'hard' in terms of legally binding open-ended phase-out goals. It sets a precedent for future international law-making, it gives parties freedom to determine and enact NDCs, but these NDCs clearly articulate legally binding long-term collective agenda. One can expect that the bottom-up approach in international law can frame not only the climate change regime, but also global environmental governance in general. This approach offers implicit appropriate practices, justice, equilibrium of rights and duties. The normative and regulative dimensions of the responses to climate change and other environmental challenges should be decisive, combining enhanced integrity with renewed confidence in the technological innovation and holistic approach to sustainability.

3 A comparative architectural analysis of the 1997 Kyoto Protocol and the 2015 Paris Agreement and other ways to counter environmental 'ratification fatigue'

Trudy Fraser

Introduction

Environment regimes such as the 1997 Kyoto Protocol and the 2015 Paris Agreement (both protocols to the 1992 UN Framework Convention on Climate Change (UNFCCC)) are primarily good faith agreements between consenting states, based on voluntary bilateral or multilateral agreements, and are open to the parties' withdrawal if these states deem fit[1]. Such regimes have been criticized as preserving status quo and do not necessarily compel behaviour change post ratification: 'quantitative analysis shows that ratification constraints did not affect bargaining over the Protocol, nor did bargaining outcomes affect ratification[2].'

The efficacy of such regimes largely stems from its fundamental legal foundation, wherein the architecture of such regimes generally conforms to a top-down approach, a bottom-up approach, or a hybrid form of both.

---

1 United Nations (1969), *Vienna Convention on the Law of Treaties*, 23 May 1969, United Nations, Treaty Series, vol. 1155, p. 331; Article 52(2) stipulates that: 'The termination of a treaty, its denunciation or the withdrawal of a party, may take place only as a result of the application of the provisions of the treaty or of the present Convention. The same rule applies to suspension of the operation of a treaty.' Such a legal withdrawal was demonstrated in December 2011 when Canada chose to invoke their legal right to withdraw from the Kyoto Protocol as afforded to them under Article 27 of the Protocol, 1998, which allows that: 1) At any time after three years from the date on which this Protocol has entered into force for a Party, that Party may withdraw from this Protocol by giving written notification to the Depositary; 2) Any such withdrawal shall take effect upon expiry of one year from the date of receipt by the Depositary of the notification of withdrawal, or on such later date as may be specified in the notification of withdrawal. One year later, in December 2012, Canada was officially a non-participant to the treaty

2 McLean, E.V. and Stone, R.W. (2012), "The Kyoto Protocol: Two-Level Bargaining and European Integration", *International Studies Quarterly*, 56 (1), pp. 99–113

1) Top-down approach: where the agreement defines particular policies and measures that parties must undertake[3].

The top-down approach is exemplified in the case of the 1985 Vienna Convention for the Protection of the Ozone Layer (and the subsequent Montreal Protocols to the Convention in 1987). The Protocols established fixed benchmark reductions to be met in parties' production and consumption of ozone-depleting substances and evolved over time via a series of amendments to the point where parties were required to completely phase-out the use of ozone-depleting substances. In 2009, the Vienna Convention became the first Convention of any kind to achieve universal ratification.[4]

2) Bottom-up approach: where parties to the agreement are allowed to define their own unilateral commitments within the scope of a general agreement[5].

The bottom-up approach is exemplified in the case of the 1991 Air Quality Agreement between the United States and Canada, wherein a bilateral Air Quality Committee was established to review progress that had been achieved towards a general agreement and to submit periodic reports between the two countries, but it had no specific targets or deadlines[6].

Negotiations for the 1992 UNFCCC initially looked up to the success of the Vienna Convention's successful top-down approach and sought a regulatory approach in authoring the new climate change agreement. However, several states were hostile to the inclusion of targets and timetables and instead advocated for the development of unilateral strategies within the broader framework of a general consensus. The convention acknowledges common but differentiated responsibilities and respective capabilities (CBDR–RC) of state parties ultimately utilizing both top-down and bottom-up approaches under the UNFCCC:

> Article 4.1, reflects a bottom-up approach, requiring all parties to develop reports on national policies and measures to combat climate change.
>
> Article 4.2, reflects a top-down approach, setting forth a non-binding obligation for developed countries to return their greenhouse emissions to 1990s levels by 2000.[7]

---

3 Bodansky, D. (2011), "A Tale of Two Architectures: The Once and Future UN Climate Change Regime", 1 March, p. 1; available at SSRN: www.ssrn.com/abstract=1773865 or http://dx.doi.org/10.2139/ssrn.1773865
4 *See* United Nations Environment Programme (UNEP) (2006), *Handbook for the Montreal Protocol on Substances that Deplete the Ozone Layer (7th Edition);* www.ozone.unep.org/Publications/Handbooks/MP_Handbook_2006.pdf.
5 Bodansky (2011), see note 3
6 Reitze, A.W. (2001), *Air Pollution Control Law: Compliance and Enforcement,* Washington DC: Environmental Law Institute, p. 247
7 Bodansky (2011), p. 6, see note 3

Since then, the legal foundations for the 1997 Kyoto Protocol and the 2015 Paris Agreement have been built on different architectures in their efforts to compel states' behaviour towards UNFCCC goals – with varying degrees of success and various 'lessons learned' that have shaped subsequent negotiations for further regimes.

This chapter will compare and contrast the relative merits of the 1997 Kyoto Protocol's top-down approach with the development of a more hybrid bottom-up approach for the 2015 Paris Agreement. It will conclude by introducing some alternative legal options that may be viable since the substantive 'ratification fatigue' which has plagued the legal instruments developed to support the UNFCCC.

The road from Kyoto to Paris

Analysis of top-down (categorical) and bottom-up (differentiated) responsibilities

Since its inception, the UNFCCC has struggled to find a formula for its efforts that best motivates states' behaviours towards treaty aims and objectives. Ratification fatigue was a significant contributing factor to the bottom-up structure of the 2015 Paris Agreement which aims to limit global temperature increases to 'well below 2°C above preindustrial levels and pursuing efforts to limit the temperature increase to 1.5°C above preindustrial levels' and to establish binding commitments by all parties to make 'nationally determined contributions' (NDCs), and to pursue domestic measures aimed at achieving them[8].

The 1997 Kyoto Protocol utilized a firm top-down approach by introducing specific quantitative emission limitation and reduction objectives for 38 industrialized countries to reduce their collective greenhouse emissions by approximately 5% below 1990s levels by the end of the Protocol's first five-year commitment period (2008–2012). The Kyoto Protocol has been widely criticized as being highly politicized and legally weak, a criticism that holds up well in the face of empirical analysis of its success. Despite the fact that the signatories to the Kyoto Protocol 'met their target with room to spare, cutting their collective emissions by around 16 per cent',[9] such success did 'not arise from deliberate actions taken by states aiming at compliance but was the consequence of inadvertent developments'[10] such as the collapse of greenhouse gas-producing industries and significant economic contraction in post-communist Europe, and the secondary effects of existing national policies (such as the United Kingdom's

---

8 UNFCCC (2015), Paris Agreement; accessed at http://unfccc.int/files/essential_background/convention/application/pdf/english_paris_agreement.pdf (7 September 2016)
9 Schiermeier, Q. (2012), "The Kyoto Protocol: Hot Air", *Nature*, 28 November; www.nature.com/news/the-kyoto-protocol-hot-air-1.11882.
10 Christoff, P. (2006), "Post-Kyoto? Post-Bush? Towards an Effective 'Climate Coalition of the Willing'", *International Affairs*, 82 (5), p. 834

shift away from coal and towards gas, and France's continued reliance on its established nuclear industry).

> In all, the EU-15 seem unlikely to meet their collective target without buying 'hot air' from Russia... Similar problems abound among the major emitters outside the EU: Japan, with target emissions reduction of six per cent, is likely to increase its total emissions from 1990 by over ten per cent; Canada too will exceed its target.[11]

The Paris Conference resulted in two separate outcome documents: (1) a sessional report of the UNFCCC Conference of the Parties (COP 21); and (2) the Paris Agreement as an annexe to the sessional report. COP 21 does not require ratification or acceptance because it does not constitute a separate legal instrument to the original UNFCCC decision. The Paris Agreement, however, represents a newly conceived legal document pertaining to post-2020 climate regulations. The Paris Agreement was opened for signature in April 2016 and entered into force on 4 November 2016, 30 days after the date on which at least 55 parties (accounting for at least an estimated 55% of total global greenhouse gas emissions) deposited their instruments of ratification, acceptance, approval or accession with the Depositary[12].

The Paris Agreement is based on a hybrid architecture that utilizes both a top-down and a bottom-up approach. It sets a clear target agenda in terms of numbers to be achieved (top-down), but it attempts to help states achieve these target numbers via a negotiated (bottom-up) system that allows states to write their own targets based on their individual abilities to do so. The 29 articles of the Paris Agreement address three key issues in their efforts to achieve its temperature goals:

1) Differentiated strategic goal setting; recognizing that targets will take longer for developing countries.
2) Support; recognizing that some parties will require additional support (including financial support) to mitigate existing climate impact and to build suitable infrastructure to meet climate targets.
3) Stocktaking and scaling up; recognizing that targets are not static but are instead moveable and intended to shift up to continually represent a party's highest possible ambition.

The Paris negotiation, in many ways, 'represents the sum of lessons learned over many years of negotiating to rehabilitate or instill a sense of fairness and trust among parties by striking the necessary balance between the efforts of developed

---

11 Ibid.
12 As of November 2016, 111 of the 197 parties have ratified the Paris Agreement. A current status of ratification is available at UNFCCC, *2015 Paris Agreement – Status of Ratification*: http://unfccc.int/paris_agreement/items/9444.php (accessed 17 November 2016)

and developing countries[13]'. Indeed, 'the fact that any agreement was reached in Paris [or, indeed, any UNFCCC negotiation] is remarkable when one considers the diversity of national circumstances, interests and perspectives that Parties brought to the negotiations[14]'.

The Paris Agreement moves towards a fundamentally different inception of what will work best in meeting UNFCCC aims and objectives. While Kyoto was a classic top-down approach that defined particular policies and measures that parties must undertake, Paris is a classic bottom-up approach; wherein parties to the agreement are generally allowed to define their own unilateral commitments within the scope of a general agreement, and is 'based on the idea that self-imposed, voluntary commitments are more likely to be met than those imposed by the global community, and that demonstrated domestic progress, full transparency and regular review of the collective effort are key to moving Parties beyond no regrets actions[15]'.

The Kyoto Protocol had two lists: one of required emissions reductions by developed countries, another requesting voluntary efforts towards emissions reductions by developing countries. The Paris Agreement is fundamentally different. There are no lists of who has to achieve what by when. Instead, emissions targets are required of all parties – there are no separate lists of developed and developing countries – and they are nationally determined rather than internationally negotiated. Crucially, these nationally determined targets are not intended to be static, legally binding targets, but are instead intended to represent any party's best abilities at any given time – with the hope that parties will be scaling up their target ambitions, but also allowing for the outcome that parties instead have to scale down their target ambitions.

In many ways, the difference between Kyoto's top-down approach and Paris's bottom-up approach lies in states' expectations for what motivates compliance. The top-down approach is based on the assumption that states require strictly imposed rules as a means of changing their behaviour, and that they require clearly established benchmarks to know when this compliance has been met. The bottom-up approach is based on the assumption that states know best how and when they can work towards a self-regulated effort of compliance. Both top-down and bottom-up make room for differentiated responsibilities[16]

---

13 Huang, J. (2015), "The 2015 Climate Agreement: Key Lessons Learned and Legal Issues on the Road to Paris", pp. 33–34. Available at SSRN: https://ssrn.com/abstract=2724109 or http://dx.doi.org/10.2139/ssrn.2724109

14 Doelle, M. (Forthcoming 2016), "The Paris Agreement: Historic Breakthrough or High Stakes Experiment?" (22 December 2015), *Climate Law*, 6 (1–2); http://ssrn.com/abstract+2708148, p. 1

15 Ibid., pp. 2–3

16 The principle of common but differentiated responsibilities and respective capabilities (CBDR-RC) was first articulated in the UNFCCC and has two bases: 1) the different historical responsibilities of parties for causing the climate change problem; and 2) their differing capacities to address it. The Kyoto Protocol additionally allowed for the possibility that states might 'graduate' from one category of states party into another as their responsibilities and capacities changed.

of states, but each has a different conception of how those responsibilities towards a common aim should be articulated and enacted: 'The Kyoto approach is based on the assumption that nation states will always act in self-interest, requiring a global agreement that aligns their self-interest with the global interest through binding commitments and strong compliance. The Paris Outcome is based on the idea that nation states can be moved toward action in the global interest through managerial approaches that build new norms of state behaviour[17].'

The Paris Agreement allows for a more nuanced approach to states' differentiated responsibilities: 'The approach taken in Paris was to move away from these hard lines and agree on more tailored approaches to differentiation depending on the issue, and to allow for more flexibility over time[18].' Indeed, Lavanya Rajamani, an international environmental law professor at the Centre for Policy Research, concurs, suggesting that 'the most feasible solution is a hybrid approach that tailors the manner of differentiation and the use of particular design features to the specific elements of the agreement[19]'.

The Paris Agreement largely completes the move away from the Kyoto Protocol's categorical approach to what Rajamani would describe as a hybrid approach to differentiation. NDCs are established based on states' self-determined differentiated responsibilities that can account for parties' different capacities in different areas, and allow state parties flexibility to determine their needs and abilities to support others on an issue-by-issue basis instead of assigning their abilities based on their 'class' of state (i.e., developed or developing). Top-down versus bottom-up essentially equates to a question of how much latitude is given to states in determining their climate change policies and commitments. The Kyoto Protocol represented a strict compliance system, penalizing those parties that do not make sufficient efforts to bring their emissions below their target[20]. As a result:

> The countries willing to accept Kyoto emission targets represented only about a quarter of global emissions in the first commitment period, and this number has dropped to less than 15 per cent in Kyoto's second commitment period. By contrast, 141 countries put forward emission pledges under the Copenhagen Accord, representing more than 85 per cent of global emissions.[21]

Indeed, Kyoto demonstrated that 'aggressively pushing the process beyond negotiators' comfort zones can negatively impact the sense of fairness and trust[22]'. In many ways, the Paris Agreement was a 'Goldilocks solution that is neither

---

17 Doelle (2016), p. 4, see note 14
18 Ibid.
19 Huang (2015), p. 17, see note 13
20 As already noted, and despite a penalty system for states that did not make sufficient efforts to bring their emissions below their target, there is no international enforcement mechanism to ensure such penalties are delivered or that states are held accountable for their emissions targets.
21 Bodansky (2017), p. 19, see note 3
22 Huang (2015), p. 34, see note 13

too strong (and hence unacceptable to key states) nor too weak (and hence ineffective); it adopts a bottom-up approach, in which the Agreement "reflects rather than drives national policy[23]". The Paris Agreement recognizes that climate change 'implicates virtually every aspect of domestic policy and raises huge domestic sensitivities[24]' and is derived from the assumption that individual states know best how and when they can meet emissions targets. The Paris Agreement gives states the opportunity to develop NDC standards that are borne of, rather than handed down to, domestic policy and domestic constraints.

> Without making too light of a serious endeavor, parties have gone from trying a bed that was "too hard" [Kyoto] to one that was "too soft" [Copenhagen] and now seek to land all parties comfortably in the middle. That middle is often described as a robust, ambitious agreement that is applicable to all, a description that hints at the shortcomings of previous agreements under the UNFCCC.[25]

Of course, there are concerns about the effectiveness of a norms-based bottom-up approach in terms of its capacities to engage with the wider context of climate change. Meinhard Doelle has identified several potential limitations of the Paris outcomes, including the following:

- Gender equity, human rights, intergenerational equity and climate justice are largely limited to the preamble of the Paris Agreement, making their full integration into the implementation of the regime less certain.
- Failure to explicitly signal the phase-out of fossil fuels or the elimination of fossil fuel subsidies risks prolonging the debate over the future of fossil fuels in some countries, and thereby risks diverting attention away from integrated solutions to climate change.
- The failure to address emissions from international shipping and aviation under the UNFCCC will likely continue to plague the regime, as emissions from these sectors are expected to grow and threaten to undermine efforts at the national level, and efforts to deal with the issue outside the UNFCCC have not been successful to date.
- Though not surprising given the pace of negotiations leading up to Paris, there was limited detail provided on the roles of sinks, emissions trading, offsetting and non-market mechanisms.
- There was surprisingly little attention paid in the final agreement to the role of non-state actors and subnational governments, in spite of considerable attention having been given to their potential role in increasing the mitigation ambition of parties.[26]

23 Bodansky (2017), p. 2, see note 3
24 Ibid., p. 19
25 Huang (2015), p. 6, see note 13
26 Doelle (2016), pp. 13–14, see note 14

In terms of supporting states' implementation of international law, the argument can also be made that the bottom-up approach is more conducive to long-term normative change than top-down calculated standards can achieve in premeditated time-frames. Jorge Vinuales writes that the Paris Agreement, because of its bottom-up approach: 'is a realistic instrument and, because of its imperfection, one that is much closer to the human topography than its falsely ambitious predecessor signed in Kyoto. For that reason, it stands a better chance to work. This is one of those times when less is more[27].' The Paris Agreement's managerial approach is certainly more conducive to long-term normative agenda-setting: it allows states a great deal of individual agency to interpret and enact self-determined goals within a clearly articulated long-term collective agenda. It also allows for a great deal of states' interaction, communication and information-sharing, and the abilities of states to adjust their goals if circumstances see fit. This approach allows for states to make and plan for ambitious NDCs with the flexibility to adjust those NDCs down if they find that their methods are not working, or to account for a change in government, domestic priorities, etc. The approach also allows for states to make modest NDCs and then adjust those NDCs up if they find their methods are working better than expected, or if domestic priorities shift in favour of climate mitigation policy enactment.

In attracting and facilitating larger amounts of states' participation, bottom-up approaches also attract greater civil society participation in conference proceedings, which can produce equally important actions by non-state actors in mitigating climate change. Daniel Bodansky notes that 'the Paris process included a number of national, sub-national, and non-state initiatives:

- Pledges by various developed countries to provide $19 billion per year in public finance by 2020, including a pledge of more than $5 billion by France and a doubling by the United States of its support for adaptation, to $800 million.
- Mission Innovation, a joint initiative of the United States, France, and leaders from 18 other countries, who pledged to double their support for clean energy research and development over the next five years.
- The Breakthrough Energy Coalition, a related private initiative spearheaded by Bill Gates to invest in clean energy technologies, financed by 26 investors from ten countries, including Jeff Bezos and Mark Zuckerberg.
- The International Solar Alliance, an initiative involving 120 countries, led by India and France, aimed at promoting solar energy deployment in developing countries.
- The Paris Pledge for Action, which promises the support of non-state stakeholders in implementing the Paris Agreement and meeting or exceeding its 2 degree temperature goal. By mid-January, more than 1,200 non-party

---

27 Vinuales, J.E. (2015), "The Paris Climate Agreement: An Initial Examination", 16 December, C-EENRG Working Papers No. 6; available at http://ssrn.com/abstract=2704670, p. 16

stakeholders had signed the pledge, including more than 600 companies, 180 investors, and 110 cities and regions.
- The Lima to Paris Action Agenda (LPAA) and its associated NAZCA portal, 189, which records actions by sub- and non-state actors. Currently, the NAZCA portal lists approximately 11,000 commitments, more than 2,000 from cities, a roughly equal number from private companies, and more than 230 from civil society organizations.
- The Compact of Mayors, which now involves more than 450 cities.[28]'

Specifically, civil society lobbying succeeded in including a reference to human rights in the Paris Agreement[29]. Civil society engagement should of course be welcomed, but at the same it is necessary to be cautious against complementary civil society actions that duplicate or draw attention away from internationally recognized efforts under UNFCCC regimes and protocols. Such actions should not be seen as an easier alternative that can produce PR victories without the necessary substantive results.

The movement away from top-down to bottom-up institutionalizes a new paradigm of climate regime that is intended to establish a floor beneath further climate damage and to catalyze stronger global action to combat climate change. Although any real outcomes have yet to be measured, and there remains substantive areas in which the Paris Agreement does not engage, 'It is projected to reduce emissions by about 3.5 gigatonnes in 2030 and to reduce expected warming in 2100 by about 1 degree[30]'.

Alternative option for incentivising state compliance?

The differentiation between bottom-up and top-down approaches is important, but neither option fulfils complete compliance nor appears to completely compel state parties to violate their own interests. Clearly, environmental regimes do make a difference[31] but the results of the Kyoto Protocol (Paris Agreement pending) speak for themselves: environmental legal regimes have not yet stopped environmental degradation. For some states, environmental degradation and climate change are acutely felt issues that impact not only long-term sustainability, but also short-term survival. For these states, climate change is an issue of immediate

---

28 Bodansky (2017), pp. 40–41, see note 3
29 'Parties should, when taking action to address climate change, respect, promote, and consider their respective obligations on human rights, the right to health, the rights of indigenous peoples, local communities, migrants, children, persons with disabilities and people in vulnerable situations, and the right to development, as well as gender equality, empowerment of women, and intergenerational equity.' (Paris Agreement, supra note 2, pmbl. para.7); UNFCCC (2015), see note 8
30 Bodansky (2017), p. 25, see note 3
31 See Breitmeier, H., Underdal, A. and Young, O.R. (2011), "The Effectiveness of International Environmental Regimes: Comparing and Contrasting Findings from Quantitative Research", *International Studies Review*, 13 (4) (December), pp. 579–605

international peace and security and whether or not the wider international community takes measures against climate change can be interpreted as a threat against their very survival. Such states (indeed any state that wishes to elevate climate change to an issue of international peace and security, for whatever reason) may look beyond the treaty-based regimes, with their options for top-down (categorical) or bottom-up (differentiated) approaches to compelling international compliance, towards a more enforceable method of international law.

Arbitrating non-compliance

There is existing legal procedure for non-compliance written into the UNFCCC and into the Paris Agreement[32]. Article 14(1) of the UNFCCC states that: 'In the event of a dispute between any two or more Parties concerning the interpretation or application of the Convention, the Parties concerned shall seek a settlement of the dispute through negotiation or any other peaceful means of their own choice[33].' If the dispute still exists after 12 months, Article 14(6) and 14(7) give scope for the creation of a conciliation commission to 'be composed of an equal number of members appointed by each party concerned and a chairman chosen jointly by the members appointed by each party. The commission shall render a recommendatory award, which the parties shall consider in good faith[34].' The crucial line in Article 14 reads *in good faith*. Although Article 14 'opens the possibility for States to accept the compulsory jurisdiction of the International Court of Justice (ICJ) or of an arbitration tribunal, [such a measure] has never been used[35]' and the use of *in good faith* suggests that the movement towards external judicial ruling would be subject to states' acceptance, indeed appearance at, any such process. In the case of the ICJ, the presentation of issues must be done willingly and with the knowledge that judgements apply only to the parties to the dispute; the ICJ is not a criminal court, nor does it have a prosecutor[36].

Prosecuting non-compliance

There is also an emerging legal precedence for domestic court rulings against states' failures to protect their citizens from the effects of climate change. In 2015, a Dutch environmental group, Urgenda, brought forth a case to the Dutch judiciary that their government was failing to protect Dutch citizens from the effects of climate change. On 24 June 2015, a Dutch court ruled that the government of

---

32 Article 24 of the Paris Agreement refers to the dispute settlement clause in Article 14 of the UNFCCC as applicable *mutatis mutandis* to the Agreement
33 UNFCCC, available at: http://unfccc.int/files/essential_background/background_publications_htmlpdf/application/pdf/conveng.pdf (accessed 7 September 2016)
34 Ibid.
35 Vinuales (2015), pp. 14–15, see note 27
36 See Fraser, T. (2015), *Maintaining Peace and Security? The United Nations in a Changing World*, Palgrave Macmillan, pp. 39, 197

the Netherlands had breached its duty of care by taking insufficient measures to prevent dangerous climate change and ordered the government to take measures to increase its efforts to cut emissions by 25% cent below 1990 levels by 2020, up from its current 17% target. The Dutch government has appealed, but the suit is the first in the world to successfully assert human rights as part of the legal basis for protecting citizens against climate change.

> "Before this judgement, the only legal obligations on states were those they agreed among themselves in international treaties," said Dennis van Berkel, legal counsel for Urgenda, the group that brought the suit. "This is the first a time a court has determined that states have an independent legal obligation towards their citizens. That must inform the reduction commitments in Paris because if it doesn't, they can expect pressure from courts in their own jurisdictions."[37]

Similarly, in 2015, 21 children aged from eight to 19 filed a climate-change lawsuit against the United States Federal Government claiming that 'Defendants… knew the harmful impacts of their actions [allowing carbon dioxide pollution] would significantly endanger Plaintiffs, with the damage persisting for a millennium. Despite this knowledge, Defendants continued their policies and practices of allowing the exploitation of fossil fuels[38].' This case and others like it (NPR reports that the governments of Belgium and Norway are facing similar charges from their citizenry[39]) have significant but as yet unknown implications for taking climate commitments under the UNFCCC seriously. Of course, these cases represent domestic, not international, appeals against the effects of climate change and '[t]he best case scenario is one in which [the Dutch ruling] not only works to strengthen individual climate efforts over time, but also reinforces the international environmental law regime[40].'

At the international level, there is scope to consider UN Security Council rulings against states that fail to protect their citizens from the effects of climate change. Some UN member states have already begun considering the Security Council as an alternative route in compelling states' compliance with climate change regimes, where there have been attempts by some members to reposition

---

37 Nelsen, A. (2015), "Dutch government ordered to cut carbon emissions in landmark ruling", *The Guardian*, 24 June; available at www.theguardian.com/environment/2015/jun/24/dutch-government-ordered-cut-carbon-emissions-landmark-ruling (accessed 19 September 2015)

38 See United States District Court, District of Oregon – Eugene Division. Case Number: 6:15-cv-01517-TC; available at: https://static1.squarespace.com/static/571d109b04426270152febe0/t/57a35ac5ebbd1ac03847eece/1470323398409/YouthAmendedComplaintAgainstUS.pdf (accessed 17 November 2016)

39 *NPR Morning Edition* (2015), "The Dutch Ruling On Climate Change That Could Have A Global Impact", 25 June; available at: www.npr.org/2015/06/25/417349227/the-dutch-ruling-on-climate-change-that-could-have-a-global-impact (accessed 19 September 2015)

40 Huang (2015), pp. 36–37, see note 13

climate change as an issue of international peace and security that would therefore fall under the purview of the Council. Such a decisive (and controversial) move would shift the conversation from regime compliance (the burden for which falls to states) to regime implementation (the burden for which falls to the 15-member UN Security Council) and speaks to the most pressing problem of the international legal system: the question of effectiveness. Positioning climate change as a threat to international peace and security brings the issue to the attention of those in the Security Council who could, theoretically, impose judicial consequence against non-compliance and legally prescribe international action against the threat. As I wrote in 2014:

> The inclusion of climate change to the Security Council agenda reflects two separate but overlapping consensuses: first, that the global governance architecture of international agreements do not do enough to incentivize ratification or post-ratification compliance with their own regulatory standards. Second, it is within the experience and scope of practice of the Security Council to adopt a climate change mandate that is tantamount to general legislative practice. The perennial question remains, however: would legislative thematic resolutions be more effective in compelling change than existing treaty-based obligations?[41]

In the event that the Security Council did adopt climate change as an issue of international peace and security, the question remains as to how it would enforce any such decision. To be clear, the Security Council essentially has *carte blanche* to determine threats to international peace and security – but the right to determine a threat is only as effective as the Council's ability to counter that threat, and in the case of climate change, the suitability of the Security Council's toolbox of remedies is a limiting factor. The Council's enforcement measures include 'complete or partial interruption of economic relations and of rail, sea, air, postal, telegraphic, radio, and other means of communication, and the severance of diplomatic relations' (Chapter VII, Article 41 of the United Nations Charter, 1945) or 'action by air, sea, or land forces as may be necessary to maintain or restore international peace and security. Such action may include demonstrations, blockade, and other operations by air, sea, or land forces of Members of the United Nations' (Chapter VII, Article 42, of the UN Charter). However, 'sending in military troops under the UN auspices to prevent trees being cut down or to stop the building of a factory using polluting technology is clearly inappropriate and may itself be a threat to international peace and security…[42]'

---

41 Fraser, T. (2014), "From Environmental Governance to Environmental Legislation: The Case of Climate Change at the Security Council", in Popovski V. and Fraser, T. (eds.), *The Security Council as Global Legislator*, Routledge, p. 237
42 Tinker, C. (1991), "Environmental Security in the United Nations: Not a Matter for the Security Council", *Tennessee Law Review*, 59, p. 794

An alternative theoretical option would be for the Security Council to refer the issue to the International Criminal Court (ICC). Under Article 13 of the ICC's creation treaty (the 1998 Rome Statue)[43] the Security Council can refer situations to the ICC and any such referral by the Security Council is considered a coercive measure under Article 41 of the UN Charter. However, the ICC prosecutes individuals – which may serve as a deterrence mechanism but cannot be relied upon to motivate large-scale or long-term changes in states' behaviours (governments often distance themselves from individual citizens tried in the ICC – relegating their crimes to historical governments or regimes that do not represent current practices[44]).

To date, efforts to motivate Security Council action on the issue of climate change have largely stalled, but that is not to say that viable future options for the UN Security Council to utilize its international peace and security toolbox do not exist.

Exiting formal treaty-based regimes in favour of customary international law

It is worth noting that the concept of the environment as a protected subject already exists in customary international law (ICL) – c.f. the Romans' 'scorched earth' policy, prohibitions against the deliberate destruction of enemy property, and prohibitions against the use of weapons that cause environmental harm. The numerous legal regimes that have sought to formalize environmental protection have relied on such customary law precedent as the guiding 'North Star' to their regimes.

In writing specific treaty language, however, regimes are doing two things at once: they are establishing benchmarks for success (and that is certainly a worthy accomplishment), but they are also potentially establishing goalposts that proscribe a false sense of success when the task is not really, or fully, accomplished. As already discussed, treaty regimes generally allow for states' withdrawal if states decide that it is no longer in their interests to continue to work towards the treaty's established targets for success. The benefit to working outside a treaty-based regime is that there is no proscribed exit mechanism for parties in customary international law: 'the *opinio juris* requirement for [customary international law] provides that even to qualify as customary law in the first place a norm must be perceived as binding. A norm that is intended to be unilaterally revocable would not qualify under this test as it is currently understood[45].'

---

43 1998 Rome Statute of the International Criminal Court; available at: www.icc-cpi.int/nr/rdonlyres/ea9aeff7-5752-4f84-be94-0a655eb30e16/0/rome_statute_english.pdf (accessed 20 November 2016)

44 Of course, there are examples of ICC arrest warrants for sitting heads of state…

45 Brilmayer, L. and Tesfalidet, I.Y. (2011), "Treaty Denunciation and 'Withdrawal' from Customary International Law: An Erroneous Analogy with Dangerous Consequences', *The Yale Law Journal*, 120, 5 January; available at: www.yalelawjournal.org/forum/treaty-denunciation-and-qwithdrawalq-from-customary-international-law-an-erroneous-analogy-with-dangerous-consequences (accessed 17 November 2016)

Therefore, if we understand ICL as an accounting of our norms, we might look to recent normative developments in human protection for a route forward in compelling international climate protection. In a world where we now recognize that to 'do nothing' in the face of states committing crimes against their own citizens is tantamount to a war crime, is it possible to envision environmental protectionism in the same normative framework that has advanced the Responsibility to Protect and the Protection of Civilians as enforceable international policy responsibilities? In other words, is there a universal benchmark that can be determined to gauge if environmental damage is of a magnitude that can warrant that the perpetrators are held to account by other governments?

There are, of course, some immediate obstacles to overcome in such thinking. The primary challenge would be in formulating a basis of responsibility. For example, it would not necessarily suffice to take action against the 'host' country – there are too many ways in which third parties suffer 'downstream' consequences to other states' actions. Similarly, there are issues of environmental negligence that do not bear consequence for many generations, meaning that the responsible governments have long been relegated to the history books. Many of the same issues of differentiated responsibilities that can be found in treaty-based regimes would also exist in a newly developed normative framework of environmental protectionism, i.e., would exceptions be made for developing countries that wish to industrialize/develop even though it would result in long-term environmental consequences?

In developing this line of thought, however, it is worth recalling how normative progress is often hidden and slow, until it is not. Richard Falk writes of the Nuremberg Trials as a 'legislative spasm, which drew upon a common sense of moral outrage to move beyond existing parameters of international law[46]'. Is it worth considering whether such a 'legislative spasm' is conceivable in protecting persons from the effects of climate change? This author would argue that it is – but from where and how such a 'normative spasm' will manifest is yet to be determined.

Whether international environmental protection is born of a treaty-based regime, or if we look beyond treaty-based regimes for environmental solutions, one conclusion is clear: the international community has not yet found the best way to hold its members accountable for international climate change.

The UNFCCC has struggled to find an architecture that can fully support its ambitions. The Kyoto Protocol relied too heavily on a top-down approach that served to alienate state parties from their obligations; the Paris Agreement has adopted a more hybrid bottom-up based approach that has in turn been criticised as too lenient in letting state parties establish targets that serve their own interests before internationally agreed obligations. There is, of course, plenty of room for states to elevate their NDCs in order to work harder towards the internationally

---

46 See Leebaw, B. (2014), "Scorched Earth: Environmental War Crimes and International Justice", *Perspectives on Politics*, 12 (4), p. 777

agreed target of 'well below 2°C above preindustrial levels and pursuing efforts to limit the temperature increase to 1.5°C above preindustrial levels' – but it is far too soon to determine if 'ratification fatigue' will stall nationally determined targets just as it did internationally determined targets.

Alternative legal options that exist beyond treaty-based regimes remain largely hypothetical. However, it is clear is that environmental degradation may yet provoke a universal sense of moral outrage that may compel a normative tipping point, beyond which the rules for environmental protection are yet to conceived.

4 Promoting the implementation of international environmental law
Mechanisms, obligations and indicators

Natalia Escobar-Pemberthy[1]

Environmental issues are central concerns for global governance and human security. For the past four decades, governments and international organizations have worked on the design and implementation of bilateral and multilateral international law instruments – known as environmental conventions – to protect natural resources and promote sustainability. They promote co-ordinated action, define policy frameworks, gather information, and raise awareness[2] in order to address global threats such as climate change, biodiversity loss, pollution and desertification. Only ten of them – out of the estimated 1,100 existing agreements[3] – are considered global because of their membership and scope of their goals.

Academic literature on global environmental conventions addresses multiple issues, including the motivations behind them, treaty formation, the role of the conventions as international law instruments, their institutional performance, and their effectiveness in addressing global environmental problems. Among these issues, implementation is at the core of the debates in global environmental governance. Even when various quantitative and qualitative projects have offered insights on how the conventions are being implemented[4], factors such as the

---

1 This chapter was developed as a result of the author's participation in the Academic Council of the United Nations System (ACUNS) Workshop in 2016. Its findings are part of the Environmental Conventions Initiative, a research project at the Center for Governance and Sustainability at the University of Massachusetts, Boston. Its development has been possible thanks to the support of the Center's Director Prof. Maria Ivanova, and the institutional engagement of Universidad EAFIT, the John W. McCormack Graduate School of Policy and Global Studies at UMass Boston, the Federal Office for the Environment of the Government of Switzerland and UN Environment
2 Haas, P.M., Keohane, R.O. and Levy, M.A. (1993), *Institutions for the Earth: Sources Of Effective International Environmental Protection*, Cambridge, MA: MIT Press; Mitchell, R.B. (2010), *International Politics and the Environment*, London / Thousand Oaks, CA / New Delhi / Singapore: Sage Publications Limited; Steiner, A., Kimball, L.A. and Scanlon, J. (2003), "Global Governance for the Environment and the Role of Multilateral Environmental Agreements in Conservation", *Oryx*, 37 (2), pp. 227–237
3 Mitchell, R.B. (2013), International Environmental Agreements (IEA) Database Project; http://iea.uoregon.edu/page.php?file=home.htm&query=static
4 Breitmeier, H., Young, O.R. and Zürn, M. (2006), *Analyzing International Environmental Regimes: From Case Study to Database*, Cambridge, MA: MIT University Press; Brown-Weiss, E. and Jacobson, H.K. (1998), *Engaging Countries: Strengthening Compliance with International*

lack of coherence on the obligations defined by the conventions, and difficulties with countries' reporting, make measuring implementation a difficult task, and existing studies are not by empirical assessment and verification. The literature has also identified multiple factors related to both countries and conventions as determinants of implementation, but there is no systematic assessment to evaluate their influence. Among those factors, facilitation implementation mechanisms deserve special attention as instruments to support state parties in the achievement of their international environmental obligations. Understanding how these mechanisms operate in global environmental conventions to influence implementation is the objective of this paper.

Using the case study of four conventions ⁻ the Ramsar Convention on Wetlands of International Importance (1971), the Convention on International Trade in Endangered Species CITES (1973), the Basel Convention on the Transboundary Movement of Hazardous Waste (1989), and the Stockholm Convention on Persistent Organic Pollutants POPs (2001) ⁻ this paper develops a taxonomy of their facilitation implementation mechanisms, explaining how they operate and how they are being used by state parties, as a way to understand the extent to which they are effectively influencing implementation. Beyond the traditional categorizations established for implementation mechanisms, comparing these elements across conventions allows for an empirical evaluation on how and why they work, and their actual contribution to implementation. It will also inform discussions about the policy areas that they should address in order to promote the behavioural change required at the national level to implement the conventions. Ultimately, understanding these mechanisms as determinants of implementation will contribute to the development of an analytical framework that allows countries, conventions, international organizations and donor countries to act together in co-ordinated, stronger, targeted, efficient and more collective programmes to promote and improve implementation and to strengthen the role of the conventions in the solution of global environmental problems.

Global environmental conventions, governance and international law

Global environmental conventions – also known as multilateral environmental agreements (MEAs) or international environmental regimes ⁻ go back to the 19th century. States created them as mechanisms for international co-operation and collective action to protect the environment[5]. Initially, these agreements

---

Environmental Accords, Cambridge, MA: MIT Press; Stokke, O.S. (2012), *Disaggregating International Regimes: A New Approach to Evaluation and Comparison*, MIT Press; Victor, D.G., Raustiala, K. and Skolnikoff, E.B. (1998), *The Implementation and Effectiveness of International Environmental Commitments: Theory and Practice*, Cambridge, MA: MIT Press; Young, O.R. (2003), "Determining regime effectiveness: a commentary on the Oslo-Potsdam solution", *Global Environmental Politics*, 3 (3), pp. 97–104

5 Mitchell, R.B. (2003), "International Environmental Agreements: a Survey of their Features, Formation, and Effects", *Annual Review of Environment and Resources*, 28, pp. 429–461

focused on the management of shared environmental resources[6]. However, during the 1970s they experienced two fundamental changes. First, the creation in 1972 of the United Nations Environment Programme (UNEP) constituted a new framework for global environmental governance. Second, agreements and conventions became useful instruments to share information, expertise and good practices, to mobilize resources, and to centralize commitments and innovations around environmental problems[7]. Since then, environmental conventions serve as institutional frameworks to address trans-boundary environmental problems[8]. They deliver various functions including agenda-setting, regulation of actions, socialization of environmental issues, reduction of uncertainty around regulation, and generation of policy responses[9] towards the 'control and prevention of environmental harm and the conservation and sustainable use of natural resources and ecosystems'[10]. They also contribute to policy specialization, opening spaces for the participation of civil society and for the use of innovative instruments to solve environmental challenges. In terms of functions, they focus on, as DeSombre explains, the agreements 'are rarely the end product, but instead create the framework and the process that guide responses to the environmental problem in question'[11].

The study of the conventions includes multiple issues ranging from the reasons why countries decide to work together to address specific environmental issues, to the way in which the agreements are structured as international law instruments[12]. Conventions are one of the most common sources of international law for the environment, creating multilateral rules and regimes that apply to all state parties[13]. They also create executive and subsidiary bodies in charge of

---

6 DeSombre, E.R. (2004), "The Evolution of International Environmental Cooperation", *Journal of International Law and International Relations*, 1 (1–2), pp. 75–88
7 Steiner, A., Kimball, L.A. and Scanlon, J. (2003), "Global Governance for the environment and the role of Multilateral Environmental Agreements in conservation", *Oryx*, 37 (2), pp. 227–237
8 Brown-Weiss and Jacobson (1998), see note 4; DeSombre, E.R. (2004), see note 6; Mitchell, R.B. (2003), see note 5
9 Brunée, J. (2006), "Enforcement mechanisms in International Law and International Environmental Law", in Beyerlin, U., Stoll, P.-T. and Wolfrum, R. (eds.), *Ensuring Compliance with Multilateral Environmental Agreements: A Dialogue Between Practitioners and Academia* (pp. 1–24), Leiden; Boston, MA: Martinus Nijhoff Publishers; Haas, P.M., Keohane, R.O. and Levy, M.A. (1993), *Institutions for the Earth: Sources of Effective International Environmental Protection*, Cambridge, MA: MIT Press; Mitchell, R.B. (2010), see note 2; Steiner *et al*. (2003), see note 7
10 Birnie, P., Boyle, A. and Redgwell, C. (2009), *International Law & the Environment*, New York, NY: Oxford University Press, p. 212
11 DeSombre, E.R. (2004), p. 84, see note 6
12 Bodansky, D. (2010), *The Art and Craft of International Environmental Law*, Harvard University Press; Dimitrov, R.S. (2003), "Knowledge, power, and interests in environmental regime formation", *International Studies Quarterly*, 47 (1), pp. 123–150; Susskind, L.E. (1994), *Environmental Diplomacy: Negotiating More Effective Global Agreements*, Oxford University Press, USA; Susskind, L.E., Dolin, E.J. and Breslin, J.W. (1992), *International Environmental Treaty Making*, Cambridge, MA: Program on Negotiation at Harvard Law School
13 Birnie *et al*. (2009), see note 10

the operation of each convention, working on issues such as decision-making, the provision of scientific recommendations, the review and verification of the regime, its application at the national level, and adjusting it to advance in the pursue of their objectives and purpose[14].

Agreements, however, are of little significance if they are not translated into national politics. Implementation and domestic compliance are fundamental to guarantee that environmental conventions effectively contribute to the solution of global environmental problems. In this process, multiple actors and factors intervene. Additionally, conventions design a series of mechanisms that both control and assist parties in the implementation of the dispositions established by each agreement. The next section summarizes the main concepts about the process of implementation, and the factors that influence it, as a framework to characterize the facilitation implementation mechanisms and their functioning in global environmental conventions.

Implementing environmental conventions: definition and determinants

Environmental conventions operate in a system with no hierarchical authority to co-ordinate or enforce them. That is why the question of its implementation, institutional performance and contribution to the solution of environmental problems is particularly relevant. This question revolves around three core concepts: compliance, implementation and effectiveness. *Compliance* refers to conformance to expectations, the adherence of state parties to the obligations that the agreement represents[15]. *Implementation* refers to the adoption of domestic regulations to facilitate compliance[16]. And *effectiveness* is related with the fulfilment of the goals of the agreement and the resolution of the environmental problem in question[17]. Departing from these definitions, it is possible to

14 Ibid.
15 Chayes, A. and Chayes, A.H. (1993). "On compliance", *International Organization*, 47 (02), pp.175–205; Hasenclever, A., Mayer, P. and Rittberger, V. (1997), *Theories of International Regimes*, Cambridge; New York: Cambridge University Press; Jacobson, H.K. and Brown-Weiss, E. (1997), "Compliance with international environmental accords: achievements and strategies", in Rolén, M., Sjöberg, H. and Svedin, U. (eds.), *International Governance on Environmental Issues* (pp. 78–110), Dordrecht; Boston, MA: Kluwer Academic Publishers; Kurukulasuriya, L. and Robinson, N.A. (2006), *Training Manual on International Environmental Law*, UNEP/Earthprint; Simmons, B.A. (2000), "International Law and State Behavior: Commitment and compliance in international monetary affairs", *American Political Science Review*,94 (4), pp. 819–835; Young, O.R. (1979), *Compliance and Public Authority: A Theory With International Applications*, Baltimore, MD: Johns Hopkins University Press.
16 Jacobson, H.K. and Brown-Weiss, E. (1995), "Strengthening Compliance with International Environmental Accords: Preliminary Observations from Collaborative Project", *Global Governance*, 1, p. 119; Mitchell, R.B. (2001), "Institutional aspects of implementation, compliance, and effectiveness", in Luterbacher, U. and Sprinz, D. F. (eds.), *International Relations and Global Climate Change* (pp. 221–244), Cambridge, MA: MIT Press; Simmons, B.A. (1998), "Compliance with International Agreements", *Annual Review of Political Science*, 1 (1), pp. 75–93; Young (1979), see note 15
17 Bernstein, S. and Cashore, B. (2012), "Complex global governance and domestic policies: four pathways of influence", *International Affairs*, 88 (3), pp. 585–604; Simmons (1998), see note 16

conceptualize that compliance – and by association implementation – refer to the consequences that international agreements have on states' behaviour either when they result in changes in foreign and domestic policies[18], or when states, contrary to expectations, do not change their behaviour, generating processes of non-compliance[19].

Overall, scholars agree that compliance – and to some extent implementation – of international law is high[20]. Environmental law, specifically, has been often used as an example of positive results in these matters[21]. However, measuring this is a difficult task. No specific standard to determine what constitutes 'good' compliance exists. Compliance is perceived as the result of a subjective evaluation in which measurement will depend on expectations and is rarely understood as a single variable[22]. This affects both its practical measurement and its empirical verification. The same applies for environmental policies. Existing results on implementation lack a common definition for measurement standards and do not offer a systematic empirical assessment demonstrating positive results. Even within the same environmental issues, treaties have different conceptions of what is acceptable behaviour from state parties[23]. Vague legal obligations, lack of common terminology, and difficulties with countries' reporting reduce available information to determine the extent to which countries are fulfilling their obligations and translating them into national policies, and the effectiveness of measures taken[24].

Multiple positive and negative drivers influence the process of implementing the conventions[25]. Initially, it is expected that states comply with international law based on the principle of *pacta sunt servanda* – agreements must be kept. Countries' motivation to engage in treaty formation processes is also relevant[26]. Under the framework of rational functionalism, states are more motivated to comply when they have a clear understanding of the reasons that support

---

18 Chayes and Chayes (1993), see note 15; Mitchell (2001), see note 16; Young (1979), see note 15; Young, O.R. (1994), *International Governance: Protecting The Environment in a Stateless Society*, Ithaca, NY: Cornell University Press
19 Downs, G.W., Rocke, D.M. and Barsoom, P.N. (1996), "Is the Good News about Compliance Good News about Cooperation?" *International Organization*, 50 (3), pp. 379–406; Simmons (1998), see note 16
20 Chayes and Chayes (1993), see note 15; Crossen, T.E. (2003), "Multilateral Environmental Agreements and the Compliance Continuum", *Bepress Legal Series* (Paper No. 36); Downs *et al.* (1996), see note 19
21 Chayes and Chayes (1993), see note 15; Jacobson and Brown-Weiss (1995), see note 16
22 Chayes and Chayes (1993), see note 15; Simmons (1998), see note 16
23 Beyerlin, U., Stoll, P.-T. and Wolfrum, R. (2006), *Ensuring Compliance with Multilateral Environmental Agreements: A Dialogue Between Practitioners and Academia*, Leiden; Boston, MA: Martinus Nijhoff Publishers; Jacobson and Brown-Weiss (1995), see note 16
24 Helm, C. and Sprinz, D. F. (2000), "Measuring the effectiveness of international environmental regimes", *Journal of Conflict Resolution*, 44 (5), pp. 630–652; Levy, M.A. (1995), "Is the environment a national security issue?" *International Security*, 20 (2), pp. 35–62
25 Fearon, J.D. (1998), "Bargaining, enforcement, and international cooperation", *International Organization*, 52 (02), pp. 269–305
26 Von Stein, J. (2005), "Do treaties constrain or screen? Selection bias and treaty compliance", *American Political Science Review*, 99 (04), pp. 611–622

their participation in the regime and the benefits they receive[27], including the improvement of their standards or reputation[28]. Foreign policy considerations also motivate countries to advance in developing the necessary conditions defined by environmental conventions[29]. Other factors – related to states' interaction and the practical role of international law – also bring other variables into consideration. Legitimacy, transnational interaction, and states' interests are considered as influences to the process of implementation[30].

But probably the most comprehensive categorization has been presented by Jacobson and Brown-Weiss[31] and by O'Neill[32] in their analyses about strengthening compliance and countries' commitment to environmental agreements. Their categorizations include factors associated to the nature of the environmental problems addressed by each agreement, the characteristics of the international system, countries' conditions, the agreements structural design and the role of the organizations and treaty bodies connected to each convention. These last two elements seem to be covered when scholars argue that implementation also depends on the mechanisms each agreement establishes to encourage or discourage it[33].

It is clear, then, than implementation is a multi-level, multi-actor process that goes beyond the legal nature of international agreements and incorporates domestic and international influences that should be considered in combination with governance capacities and attitudes. Specifically, the mechanisms established by each convention deserve special attention. The next section explores the literature on facilitation implementation mechanisms and how they are conceived as instruments to support state parties in the achievement of the objectives established by each agreement. Only by connecting these instruments to all the other factors that determine implementation, would it be possible to determine their interactions and contributions to the solution of common environmental problems, and how they can be used to address implementation gaps, generating

---

27 Simmons (1998), see note 16; Underdal, A. (1998), "Explaining compliance and defection: three models", *European Journal of International Relations*, 41 (1), pp. 5–30
28 Chayes and Chayes (1993), see note 15; Fearon (1998), see note 25
29 Chayes and Chayes (1993), see note 15; Guzman, A.T. (2002), "A Compliance-Based Theory of International Law", *California Law Review*, 90 (6), pp. 1823–1887; McLaughlin Mitchell, S. and Hensel, P.R. (2007), "International institutions and compliance with agreements", *American Journal of Political Science*, 51 (4), pp. 721–737; Simmons, B.A. and Hopkins, D.J. (2005). "The constraining power of international treaties: Theory and methods", *American Political Science Review*, 99 (04), pp. 623–631
30 Crossen (2003), see note 20; Guzman (2002), see note 29; Mitchell (2010), see note 2
31 Jacobson, H. K. and Brown-Weiss, E. (1995), "Strengthening Compliance with International Environmental Accords: Preliminary Observations from Collaborative Project", *Global Governance*, 1, 119; Brown-Weiss and Jacobson (1998), see note 4
32 O'Neill, K. (2009), *The Environment and International Relations*, Cambridge, UK / New York: Cambridge University Press
33 Chayes and Chayes (1993), see note 15; Chayes, A. and Chayes, A.H. (1995), *The New Sovereignty: Compliance With International Regulatory Agreements*, Cambridge, MA: Harvard University Press; Downs et al. (1996), see note 19

positive influences and improving the role and effectiveness of global environmental conventions.

Conceptual framework and categorization of facilitation implementation mechanisms

Ensuring compliance, implementation and effectiveness is one of the core challenges of the system of global environmental governance. Since its origins, the system of global environmental governance has acknowledged the importance of implementation and discussed strategies to promote the domestication of international environmental obligations. In 1992, the United Nations Conference on Environment and Development called for parties to international agreements to develop 'procedures and mechanisms to promote and review their effective, full and prompt implementation'[34], including capacity-building, information, science, technology, institutional arrangements and finances, among others. This approach was reinforced by the 2002 Plan of Implementation of the World Summit on Sustainable Development, referring not only to the implementation of environmental conventions, but also to the development agenda[35]. More recently, the 2012 United Nations Conference on Sustainable Development Rio+20 reaffirmed the previous discussion, and invited countries to make progress not only in implementation of their policy commitments but also in those associated to means of implementation, and recognized finance, technology, capacity-building, trade and information as decisive factors to achieve the sustainable development agenda[36].

Environmental conventions have also addressed implementation[37]. 'Once an agreement has come into force, compliance by the Parties with their commitments may be controlled by a variety of techniques developed under different environmental regimes'[38]. However, the process faces multiple challenges, including inadequate means, the existence of multiple environmental commitments at the country level, the collision of those commitments with countries' political and economic interests, and the multi-dimensional nature of some environmental threats. These factors raise concerns about non-compliance, implementation gaps, and ineffectiveness in the solution of global environmental problems[39].

---

34 UNCED (1992), *Agenda 21*, para. 39.8; retrieved from New York: www.un.org/esa/sustdev/documents/agenda21/english/Agenda21.pdf

35 WSSD (2002), A/CONF.199/L.1, Draft plan of implementation of the World Summit on Sustainable Development; retrieved from Johannesburg, South Africa: http://daccess-dds-ny.un.org/doc/UNDOC/LTD/N02/446/85/PDF/N0244685.pdf?OpenElement

36 United Nations (2012), A/RES/66/288, The Future We Want - Outcome Document from Rio+20, United Nations Conference on Sustainable Development, Rio de Janeiro

37 Beyerlin, U. and Marauhn, T. (2011), *International Environmental Law*, Bloomsbury Publishing; Sands, P. (2003), *Principles of International Environmental Law*, 2nd edition, Cambridge University Press

38 Sand, P.H. (1992), *The Effectiveness of International Environmental Agreements: A Survey of Existing Legal Instruments*, Grotius Publications, p. 13

39 Beyerlin and Marauhn (2011), see note 37; Kurukulasuriya, L. and Robinson, N.A. (2006), *Training Manual on International Environmental Law*, UNEP/Earthprint

That is why countries, as part of the development of international environmental law, put into place different mechanisms and techniques to guarantee compliance and consequently implementation, guaranteeing that countries adhere to the provisions of each convention through the definition of domestic policies and measures[40]. Literature on these mechanisms classifies them in two approaches: control and assistance or facilitation[41]. Even though both of them have the same ultimate goal of fulfilling the obligations of each agreement, conditions such as the existing relationship among the parties, and the level of pressure that wants to be exercised towards the observance of the treaty, are some of the factors to consider when deciding which approach to use.

Regarding facilitation implementation mechanisms, environmental conventions use them as instruments for promotion and prevention. Control is not enough, and countries face multiple challenges that require support not only in terms of fulfilling their environmental obligations, but also in connection to countries' general economic, development and geopolitical conditions[42]. This situation was acknowledged by the 1992 United Nations Conference on Environment and Development that, in its Rio Declaration, recognized that 'in view of the different contributions to global environmental degradation, states have common but differentiated responsibilities'[43] based on the specific role of developing countries in international environmental relations and the accountability of developed countries[44]. This points to the 'growing internationalization of the domestic implementation and legal process, and an awareness that international law will not achieve its objectives if it does not also take account of the need, and techniques available for improving domestic implementation of international environmental obligations'[45].

Different analyses have established typologies of implementation mechanisms. Furthermore, policy outcomes within the system of governance have specifically defined means of implementation under different categories. The following taxonomy attempts to group these approaches offering an overall picture of the available instruments to support countries in the implementation of international environmental law.

Reporting

Reporting constitutes the foundation of the mechanisms used by the conventions to support implementation, as national reports monitor and provide critical information on how countries are making progress to fulfil their global environmental

---

40 Beyerlin and Marauhn (2011), see note 37; Sands (2003), see note 37
41 Beyerlin and Marauhn (2011), see note 37
42 Beyerlin and Marauhn (2011), see note 37; Birnie *et al.* (2009), see note 10
43 United Nations (1992), Principle 7. A/CONF.151/26, Rio Declaration on Environment and Development, United Nations Conference on Environment and Development, Rio de Janeiro
44 Beyerlin and Marauhn (2011), see note 37
45 Sands (2003), p. 227, see note 37

commitments[46]. However, the process of national reporting faces multiple challenges. First, in some cases reporting systems are not comprehensive enough to address the multi-dimensional nature of the conventions. Second, reports are not analyzed or included in the scope of compliance and implementation systems[47]. State parties rarely obtain feedback on the information they provide. And third, questions persist about the extent to which countries are actually fulfilling their reporting requirements[48].

Institutional arrangements

As indicated in UNEP's *Training Manual on International Environmental Law*, 'for the purpose of facilitating implementation, most MEAs establish institutions such as Secretariats, COPs, and other technical bodies to oversee the implementation of the Convention, and to provide policy guidance'[49]. These institutional bodies are crucial to the process of implementation fulfilling both political and technical functions[50]. As they advance in the fulfilment of their mandates they both facilitate implementation and co-ordinate other mechanisms with the same purpose[51], strongly influencing institutional performance and policy outputs[52].

Executive and subsidiary bodies serve as agents to the decisions reached by the state parties, through the convening of meetings, monitoring, scientific assessment, assistance, capacity-building, connection to stakeholders and information and data collection[53]. Their capacity, autonomy, visibility, organizational structure, legitimacy, people and procedures are central to their capacity for action and the role they have in facilitating implementation[54].

46 Kiss, A. (2006), "Reporting obligations and assessment of reports", in Beyerlin *et al.*, pp. 229–246, see note 23
47 Ibid.
48 Beyerlin and Marauhn (2011), see note 37
49 Kurukulasuriya and Robinson (2006), p. 41, see note 39
50 Brown-Weiss and Jacobson (1998), see note 4; Sands, P. (2003), see note 37; Urquhart, B. (1995), "Selecting the World's CEO: Remembering the Secretaries-General", *Foreign Affairs*, 74 (3) (May–Jun), pp. 21–26
51 Bauer, S. (2006), "Does bureaucracy really matter? The authority of intergovernmental treaty secretariats in global environmental politics", *Global Environmental Politics*, 6 (1), pp. 23–49; Biermann, F. and Siebenhüner, B. (2013), "Problem Solving by International Bureaucracies: The Influence of International Secretariats on World Politics", in Reinalda, B. (ed.), *Handbook of International Organization* (pp. 149–161), New York, NY / Abingdon, UK: Routledge; Ege, J. and Bauer, M.W. (2013), "International Bureaucracies from a Public Administration and International Relations Perspective", in Reinalda, B. (ed.), *Handbook of International Organization* (pp. 135–148), New York, NY / Abingdon, UK: Routledge
52 Ivanova, M. (2010), "UNEP in global environmental governance: design, leadership, location", *Global Environmental Politics*, 10 (1), pp. 30–59; Trondal, J. (2013), "International Bureaucracy: Organizational Structure and Behavioural Implications", In Reinalda, B. (ed.), *Handbook of International Organization* (pp. 135–148), New York, NY / Abingdon, UK: Routledge
53 Biermann and Siebenhüner (2013), see note 51; Ege and Bauer (2013), see note 51
54 Andresen, S., & Skjærseth, J.B. (1999), "Can international environmental secretariats promote effective co-operation?" Paper presented at the United Nations University's International

Capacity-building and technology transfer

Discussions about means of implementation in the system of global environmental governance have always focused on capacity-building and access to technology for developing countries and economies in transition. Principle 9 of the Rio Declaration recognized the need to 'strengthen endogenous capacity-building' including 'the development, adaptation, diffusion and transfer of technologies'[55], and further outcomes from the key conferences reinforced this need, and established additional mechanisms to provide this type of support[56]. Furthermore, in 2002 the UNEP Governing Council recognized the urgent need to develop a strategic plan to provide these instruments, which led to the development of the Bali Strategic Plan for Technology Support and Capacity.

Capacity-building aims at enhancing the human, scientific, technological, organizational, institutional and resource capabilities of state parties to address the different obligations of the conventions, including the development of legal and institutional frameworks.[57] Technology transfer strategies are conceived to support the development and enhancement of technical capacities in state parties required for scientific assessment, monitoring, data processing and analysis[58].

Finance

Financial resources are central to multilateral diplomacy. They are not only required to support countries in the definition of national policies, but they also provide resources for the conventions to execute broader projects[59]. Conventions establish different financial mechanisms, funded by contributions – mandatory and voluntary – of state parties and other channels, designed to transfer the cost of implementation in developing countries to other developed state parties or international actors[60].

Conference on Synergies and Co-ordination between Multilateral Environmental Agreements, Tokyo, Japan; Bauer, S. (2006), pp. 23–49, see note 51; Biermann and Siebenhüner (2013), see note 51; Cox, R.W. (1969), "The executive head: an essay on leadership in international organization", *International Organization*, 23 (02), pp. 205–230; Grigorescu, A. (2013), "International Organizations and their Bureaucratic Oversight Mechanisms: The Democratic Deficit, Accountability and Transparency", in Reinalda, B. (ed), *Handbook of International Organization* (pp. 176–188), New York, NY / Abingdon, UK: Routledge

55 United Nations (1992), Principle 9. A/CONF.151/26, Rio Declaration on Environment and Development, United Nations Conference on Environment and Development, Rio de Janeiro
56 United Nations (2012), A/RES/66/288, The Future We Want – Outcome Document from Rio+20 United Nations Conference on Sustainable Development, Rio de Janeiro; WSSD (2002), A/CONF.199/L.1, Draft plan of implementation of the World Summit on Sustainable Development; retrieved from Johannesburg, South Africa: http://daccess-dds-ny.un.org/doc/UNDOC/LTD/N02/446/85/PDF/N0244685.pdf?OpenElement
57 Beyerlin and Marauhn (2011), see note 37
58 Ibid.
59 Kurukulasuriya and Robinson (2006), see note 39
60 Beyerlin and Marauhn (2011), see note 37

International environmental law 67

These mechanisms takes multiple forms including loans, credits, grants and funds, and operate not only as means to induce and restore compliance and implementation, but as instruments to face emergencies[61].

In some cases, the financial mechanisms are administered by third parties. In 1992 the Earth Summit established the Global Environment Facility to bring together the resources of various international organizations working on environmental issues and to serve as the funding mechanism for environmental conventions providing grants to developing countries, and countries with economies in transition, for projects that generate global environmental benefits within the context of sustainable development[62]. Some conventions also establish positive or negative economic incentives to promote implementation[63].

Countries are expected to make use of the facilitation implementation mechanisms offered by the conventions[64]. However, it is not clear how effective these mechanisms are. Despite some analyses[65], establishing causality between the use of these mechanisms and the level of implementation of a given convention in a given country is a complex task, especially when there is no standardized measure for progress in the fulfilment of international environmental obligations. Additionally, facilitation implementation mechanisms face fundamental challenges that prevent them from exercising positive influence on countries' compliance and domestic policies. Some of those challenges include:

- Lack of information to determine the best policy approaches and the kind of assistance that each country requires, and to establish priorities[66].
- Overlapping of mechanisms across different conventions. Interlinkages and synergies are required to improve efficiency in the process of facilitating implementation[67].
- Lack of participation from civil society, as the public is excluded from most compliance and implementation mechanisms which reduce possibilities to raise awareness and identify non-compliance situations and assistance requirements[68].

---

61 Boisson de Chazournes, L. (2006), "Technical and Financial Assistance and Compliance: The Interplay", (pp. 273–300) in Beyerlin *et al.*, see note 23
62 Kurukulasuriya and Robinson (2006), see note 39
63 Matz, N. (2006), "Technical and Financial Assistance and Compliance: The Interplay", (pp. 301–318) in Beyerlin *et al.*, see note 23
64 Kurukulasuriya and Robinson (2006), see note 39
65 Sand (1992), pp. 14–15, see note 38
66 Stahl, M.M. (2011), "Doing What's Important: Setting Priorities for Environmental Compliance and Enforcement Programs", in Paddock, L., Qun, D., Kotzé, L. J., Markell, D. L., Markowitz, K. J. and Zaelke, D. (eds.), *Compliance and Enforcement in Environmental Law: Toward More Effective Implementation*, Edward Elgar Publishing
67 Beyerlin and Marauhn (2011), see note 37
68 Paddock, L., Qun, D., Kotzé, L.J., Markell, D.L., Markowitz, K.J. and Zaelke, D. (2011), *Compliance and Enforcement in Environmental Law: Toward More Effective Implementation*, Edward Elgar Publishing

- Decisions about the application of specific facilitation implementation mechanisms are in some cases of interpretative nature[69] and are conditioned to other processes.

In general, environmental conventions facilitation implementation mechanisms 'have been innovative and have posed a variety of challenges'[70]. Countries, however, still have to determine the roles and responsibilities of the different actors involved in these instruments in order to guarantee effective co-ordination and action at the national level[71]. Furthermore, policy decisions need evidence that connects facilitation implementation mechanisms to the conventions, and that determines effectiveness through time. The next section offers a conceptual framework that categorizes the facilitation implementation mechanisms of the four conventions included in this study, connecting them to specific categories of obligations that countries are expected to adhere to, and reflecting on the main challenges that these mechanisms confront. Understanding these connections would support the adequate application of the available mechanisms to promote the conventions 'effective, full and prompt implementation'[72].

Facilitation implementation mechanisms in global environmental conventions

Environmental conventions define their facilitation implementation mechanisms through different paths. While most of the conventions' legal texts include the definition of some of these instruments, the Conferences of the Parties still have to put into place the necessary institutional frameworks to develop them. This is not a simple process and in some cases – such as the compliance mechanism of the Stockholm Convention – negotiations have taken more than 15 years and still remain open. Other implementation mechanisms are the result of consequent decisions of the Conferences of the Parties, or are included in the strategic plans and visions that the conventions develop. The 2015 reports of the latest COP meetings of the Basel (COP 12) and the Stockholm (COP 7) Conventions outline strategic frameworks to strengthen their means of implementation[73]. CITES Strategic Vision 2008–2020 was adopted in COP 16 and has as one of its main goals to 'ensure compliance with an implementation and enforcement of the Convention', and to 'secure the necessary financial resources and means

69 Ibid.
70 Beyerlin and Marauhn (2011), p. 357, see note 37
71 Kurukulasuriya and Robinson (2006), see note 39
72 Sand (1992), see note 38
73 Basel Convention (2015), UNEP/CHW.12/27, Report of the Conference of the Parties to the Basel Convention on the Control of Transboundary Movements of Hazardous Wastes and Their Disposal on the work of its twelfth meeting, Geneva (Switzerland), UNEP; Stockholm Convention (2015), UNEP/POPS/COP.7/36*, Report of the Conference of the Parties to the Stockholm Convention on Persistent Organic Pollutants on the work of its seventh meeting, Geneva (Switzerland), UNEP

for the operation and implementation of the Convention'[74]. In 2016, the Ramsar Convention published its Fourth Strategic Plan, outlining the convention's strategic and operational goals for the next decade[75]. Operational Goal 4 – Enhancing Implementation – has six different targets associated to the means of implementation of the convention, including scientific and technical guidance, regional initiatives, finances, communications and awareness, international co-operation and capacity-building.

Considering the paths presented above, this analysis is based on three sets of documents: the conventions' legal texts, their current strategic plans, and the reports of their Conferences of the Parties. Once the facilitation implementation mechanisms are identified, they are classified in terms of the type of obligations whose implementation they try to expedite, using the categories established by the Environmental Conventions Initiative, a research project of the Center for Governance and Sustainability at the University of Massachusetts Boston[76] (See Table 4.1). If some mechanisms point to multiple types of obligations, the analysis indicates so. The second part of the analysis presents evidence on how countries are using select mechanisms (See Table 4.2), based on data from the same project, to reflect on the extent of their relevance, the general functioning of each mechanism, and the availability of information about its operation. Overall, this taxonomy informs the extent to which facilitation implementation mechanisms are achieving their ultimate goal of supporting countries in the implementation of the conventions.

Analyzing the facilitation implementation mechanisms evidences both positive results and challenges in the support to state parties for their achievement of international environmental commitments. On the challenge side, reporting systems deserve special attention. Each of the four conventions included in this study has specific reporting requirements, and different turn-outs in terms of obtaining the information they request. In the case of the Ramsar Convention, for example, most countries have submitted every report they were obliged to since 2005, while in the cases of CITES and Stockholm they have only reported one third of the required times.

This raises questions about how reporting systems are designed and put into place. Further research shows that reporting systems that are comprehensive, include feedback and follow-up, are supported by institutional arrangements inside the conventions' secretariats, and explain how the information is being processed and used, have better return rates. However, this is not conclusive, if one considers the Stockholm Convention efforts to design electronic systems

---

74 CITES (2013), Res. Conf. 16.3 CITES Strategic Vision: 2008–2020, Bangkok (Thailand), UNEP
75 Ramsar Convention Secretariat (2016), The Fourth Ramsar Strategic Plan 2016–2024, *Ramsar Handbooks for the Wise Use of Wetlands*, 5th edition, Gland (Switzerland), Ramsar Convention
76 This project aims at assessing the extent to which countries are translating their commitments to environmental conventions into domestic politics and is currently under development. The results presented here are therefore preliminary. For more information, see http://environmentalgovernance.org/research/environmental-conventions-initiative/

Table 4.1 Facilitation implementation mechanisms in global environmental conventions

	General	Type of obligation				
		Financial	Technical	Information	Management / strategic	Regulation
RAMSAR	• Secretariat • Conference of the Parties • Standing Committee	• Ramsar Small Grants Fund • Subgroup for Finance in the Standing Committee • GEF collaboration through Convention on Biological Diversity	• Ramsar sites list • Ramsar advisory missions • Wetlands / ecosystems valuation strategies • Scientific & Technical Review Panel	• COP reporting system • Ramsar Communication, Education, and Awareness Programme • Information sharing and expertise	• Partnerships with IOPs • Partnerships with other conventions • Management Working Group on the Standing Committee • Regional initiatives • Strategic plan	
CITES	• Secretariat • Conference of the Parties	• Trust Fund • External Trust Fund	• Trade suspensions • Voluntary national export quotas • Species programmes • Issue-based programmes • Identification manuals and guidelines • CITES virtual college	• Annual and biannual reporting systems • Scientific assessment – Animals and Plants Committees • Trade databases	• Collaboration with UNEP and ITTO • Capacity-building programme and co-ordinator • Enforcement strategies • International Consortium on Combating Wildlife Crime • Strategic vision • Regional programmes	• National legislation project

BASEL	• Secretariat • Conference of the Parties • Open-ended Working Group • Implementation and Compliance Committee • Institutional partnerships (FAO, business and industry, NGOs, academia and research institutions)	• Trust fund • Technical co-operation trust fund	• Regional centres for assistance • Standards for notification and movement documents	• Annual electronic reporting system • Regional centres	• Co-operation with other agreements and collaboration with UNEP • Regional centres • Synergies process in the Chemicals & Waste cluster • SAICM
STOCKHOLM	• Secretariat • Conference of the Parties • Institutional mechanisms for compliance (still under negotiation) • Institutional partnerships (FAO, business and industry, NGOs, academia and research institutions)	• Trust fund • Special voluntary fund trust fund • Financial mechanism • The Global Environment Facility (GEF)	• POPs Review Committee • Toolkits for the identification of POPs	• Periodic reporting system • Global monitoring plan • Clearing house mechanism for information-exchange • Regional centres	• Co-operation with other agreements and collaboration with UNEP • National implementation plans • Regional centers • Synergies process in the Chemicals & Waste cluster • SAICM

72 *Natalia Escobar-Pemberthy*

Table 4.2 Implementation mechanisms included in the analysis by convention

Convention	Implementation mechanisms
Ramsar	- Reporting system – use of the reports to monitor implementation - Regional initiatives - Reception of development assistance - Reception of assistance from IOPs
CITES	- Reporting system - Submission of information for NLP - Activities for effectiveness enhancement - Participation in regional initiatives
Basel	- Reporting system - Documents templates
Stockholm	- Reporting system - Reception of financial assistance for NIP development and review - Reception of technical assistance

and to develop webinars and training sessions to support countries reporting activities. The influence of institutional arrangements is also relevant to other obligations besides data collection and analysis. As explained before, executive and subsidiary bodies not only are implementation mechanisms in themselves, but they also are in charge of putting in place other instruments to promote implementation and compliance. Executive Secretariats, Conferences of the Parties, Standing Committees and other specialized committees and panels are tasked with the design and execution of promotion implementation strategies. Among those, the case of international co-operation and co-ordination is very interesting, considering how conventions engage with external actors. Ramsar's partnerships with its International Organizations Partners (IOPs) and CITES role in the International Consortium on Combating Wildlife Crime (ICCWC) are examples of this[77].

Conventions have also made an effort to expand the outreach in terms of the support to the implementation efforts of state parties, bridging the gap between the location of the Executive Secretariats and the rest of the world. All the conventions included in this study have some form of regional strategy that brings implementation support closer to the needs and local circumstances of state parties. The Ramsar Convention, for example, has a regional initiatives programme orientated to support co-operation and capacity-building on wetland-related issues at the regional and sub-regional levels. The programme includes

77 CITES, ICPO-INTERPOL, UNODC, World Bank and WCO. (2010). Establishing the International Consortium on Combating Wildlife Crime, *Letter of Understanding*, St. Petersburg (Russia); Ramsar Convention (1999), Resolution VII.3: Partnerships with international organizations, 7th Conference of the Parties, San Jose (Costa Rica)

regional centres and regional networks[78]. A similar approach has been developed by the Basel and Stockholm Conventions. In accordance with Article 14, the Basel Convention created a network of Regional and Coordinating Centers (RC) created to provide training and technology transfer on the management of hazardous wastes[79]. Centres also work on awareness-raising and information dissemination to support the implementation of the convention. Currently there are 14 centres around the world. In its efforts to define its own technology transfer and capacity-building strategies, the Stockholm Convention developed similar terms of references to create its own centres and to collaborate with the ones existing for the Basel Convention. To date, there are 16 regional centres for the Stockholm Convention, six of them shared with the Basel Convention as part of the synergies process[80]. CITES also develops specific projects with different partners to address issues on wildlife trade, non-detriment findings, enforcement, monitoring and trade controls in regions such as Central and West Asia, South and Southeast Asia, Africa, and Central and South America[81].

CITES also exemplifies another interesting trend. It is the only convention that counts on implementation mechanisms to support legislative processes – to create new legislation or adjust existing ones. The convention's National Legislation Project, adopted by the Conference of the Parties in 1992, is CITES' main mechanism for 'encouraging and assisting Parties' legislative efforts', since adequate national legislation is required to establish the wildlife trade controls necessary to implement the convention[82]. However, the latest data from the project establishes that in more than 50% of the state parties, legislation regarding CITES still does not meet the requirements for the implementation of the convention either partially or totally[83]. None of the other conventions registers specific mechanisms to support countries in the implementation of their regulatory dimension, focusing instead on other technical and management aspects.

As with legislation for these conventions, in some of the other existing mechanisms there is still room for improvement. On one side, state parties need

---

78 Ramsar Convention (1999), Resolution VII.19, Guidelines for international cooperation under the Ramsar Convention, 7th Conference of the Parties, San Jose (Costa Rica); Ramsar Convention (2015), Resolution XII.8 Regional initiatives 2016–2018 in the framework of the Ramsar Convention, 12th Conference of the Parties, Punta del Este (Uruguay)
79 UNEP (1989), *Basel Convention on the Control of Transboundary Movements of Hazardous Wastes and their Disposal*; retrieved from http://basel.int/TheConvention/Overview/TextoftheConvention/tabid/1275/Default.aspx
80 Stockholm Convention (2015), UNEP/POPS/COP.7/17, Stockholm Convention regional and subregional centres for capacity-building and the transfer of technology, Geneva (Switzerland), UNEP.
81 CITES (2013), Res. Conf. 16.3 CITES Strategic Vision: 2008–2020. Bangkok (Thailand), UNEP; CITES Secretary General (2009), *Annual Report of the Secretariat 2008–2009*; retrieved from Geneva, Switzerland: http://cites.org/sites/default/files/document/2008-09.pdf.
82 CITES (2010), Res. Conf. 8.4 (Rev. CoP15) National laws for implementation of the Convention, Doha (Qatar), UNEP
83 CITES (2014), SC65 Doc. 22 Interpretation and implementation of the Convention Compliance and enforcement: National Laws for Implementation of the Convention, 65th Meeting of the Standing Committee, Geneva (Switzerland), UNEP

to overcome the obstacles to reach agreements about the actual structures of the mechanisms. On the other, existing mechanisms demand human resources and funding to maintain and improve their operations, and require updated information that reflects exactly the areas in which countries need support in the process of implementation in order to offer targeted solutions that are both efficient and effective. Existing mechanisms also need to define synergies among them, increasing their outreach and impact at the national level[84]. Efforts such as the one developed by the Basel and Stockholm Conventions in terms of their regional centres are an example of how this could work for other agreements.

But probably the most important challenge that these mechanisms currently meet is the lack of empiric evaluations on their actual effectiveness in terms of the improvement of implementation. The absence of such evaluations prevent the mechanisms to use countries' performance in previous support processes as criteria to decide on new assistance initiatives. An example of this refers to the debate about the evaluation of the national projects on the implementation of the conventions funded by the GEF[85]. As the conventions continue the definition of new strategies and instruments to advance in the process of implementation[86], they need input on the role of the existing ones and the extent to which they are being used, in order to guarantee that all their efforts point to the same result of implementing effective solutions at the national level. The next section offers data on the use of some facilitation implementation mechanisms in the four conventions included in this study as a way to inform this discussion.

Use of implementation mechanisms in global environmental conventions

Based on data from the Environmental Conventions Initiative, this section presents preliminary results on how state parties used select facilitation implementation mechanisms in each convention's latest national reporting cycle. The objective is to offer a snapshot on how the mechanisms are working, in order to reflect on their coverage, functioning and effectiveness. The number

---

84 Paddock *et al.* (2011), see note 68
85 Boisson de Chazournes, L. (2005), "The Global Environment Facility (GEF): A Unique and Crucial Institution", *Review of European Community & International Environmental Law*, 14 (3), pp. 193–201; Möhner, A. and Klein, R.J. (2007), "The Global Environment Facility: Funding for adaptation or adapting to funds", *Stockholm, Stockholm Environment Institute*
86 Basel Convention (2015), UNEP/CHW.12/27 Report of the Conference of the Parties to the Basel Convention on the Control of Transboundary Movements of Hazardous Wastes and Their Disposal on the work of its twelfth meeting, Geneva (Switzerland), UNEP; CITES (2013), Res. Conf. 16.3 CITES Strategic Vision: 2008–2020. Bangkok (Thailand), UNEP; Ramsar Convention Secretariat (2016), The Fourth Ramsar Strategic Plan 2016–2024 *Ramsar Handbooks for the Wise Use of Wetlands* (5th ed.), Gland (Switzerland): Ramsar Convention; Stockholm Convention (2015), UNEP/POPS/COP.7/36*, Report of the Conference of the Parties to the Stockholm Convention on Persistent Organic Pollutants on the work of its seventh meeting, Geneva (Switzerland), UNEP

International environmental law 75

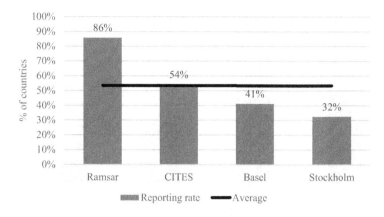

Figure 4.1 Reporting rate for global environmental conventions

of mechanisms analyzed for each convention differs since not all the existing mechanisms have information available on their use.

In terms of the reporting systems, it is clear that the conventions obtain different results (See Figure 4.1). While for the Ramsar Convention 86% of the countries obliged to submit the report to the last cycle (2015) did so, for the Stockholm Convention only 32% of the countries did. This raises questions about how the structure of the reporting system influences the response of countries to it. Factors such as reporting templates, the use given to the information collected, the training secretariats provide for the process, and the resources and data availability at the national level are relevant in this case.

Other implementation mechanisms register different levels of use across conventions. Unfortunately, data collected by the conventions on these instruments is not standardized, which prevents the development of across-conventions analyses for other mechanisms besides national reporting systems. However, within-conventions analyses offer important insights on how state parties are taking advantage of the options available to support them.

In the case of the Ramsar Convention, regional initiatives are the mechanisms state parties use the most, while most of them do not receive assistance from International Organizations Partners (IOPs). It is also relevant to notice that out of the countries eligible to receive development assistance from donor countries almost half of them do not get any of these resources, but it is not clear if this is because they are not requesting them or because their projects are not selected for funding initiatives.

CITES registers mixed results regarding its facilitation implementation mechanisms. While most countries (76%) use the National Legislation project as a mechanism to evaluate the fulfilment of their regulatory obligations (See Figure 4.3), for all the criteria evaluated on the legislation, about one third of the countries have inadequate or incomplete legislation, which suggests that the

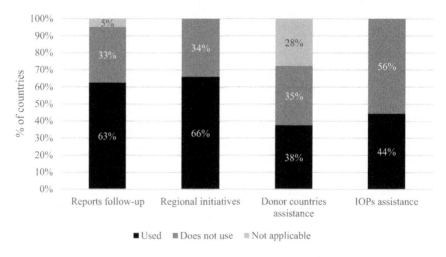

Figure 4.2 Facilitation implementation mechanisms for the Ramsar Convention

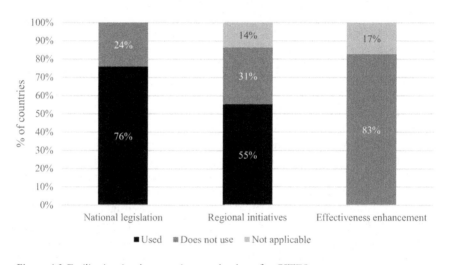

Figure 4.3 Facilitation implementation mechanisms for CITES

project should focus additional efforts on guaranteeing that state parties improve the legislation that they have already marked as insufficient. Fifty-five per cent of the countries take part in regional initiatives, but interestingly, none of them participates in effectiveness-enhancement activities. Even though further research would be required to explain the reasons for the lack of use of this mechanism, the complexity of the convention and the criminal dimension of the environmental problems it addresses can be considered as factors that prevent countries from using these mechanisms.

International environmental law 77

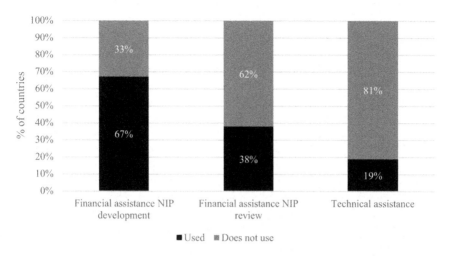

Figure 4.4 Facilitation implementation mechanisms for the Stockholm Convention

In the Chemicals & Waste cluster there are two important cases. On one side the Basel Convention's national reports do not evaluate the use of implementation mechanisms. The only data they collect is associated to the use of the standardized documents for the movement of hazardous waste, which is one of the mechanisms established by the convention to achieve information-related obligations. Those are used by 96% of the countries that submitted national reports in 2014. The rest of the data collected has to do with the actual implementation of the conventions. This poses a challenge in determining the functioning and effectiveness of the mechanisms available to the Basel Convention's state parties. On the other side, the Stockholm Convention takes into consideration some of its instruments to support countries in the questionnaire of its national reports. About two-thirds of the countries have received financial assistance for the development of their National Implementation Plans (NIPs), but only 38% (See Figure 4.4) use this support mechanism when it comes to the review and update of this obligation, which is in itself a mechanism to move forward with the prioritization and definition of strategies for the overall implementation of the convention. Furthermore, only 19% of the countries for which information is available use technical assistance as a mechanism to advance on the obligations established by the convention. This is particularly relevant when considering the highly technical nature of the convention, and the need for scientific assessment and data to advance in the identification and disposition of persistent organic pollutants (POPs).

Conclusions

International environmental law is in a constant state of development, not only in terms of the process of law formation but also regarding its implementation.

Through its evolution, it has been clear that, in order to achieve its objectives, the system of environmental governance has to 'take account of the need, and techniques available for improving domestic implementation'[87]. These techniques have traditionally been seen as mechanisms to control compliance. However, implementation mechanisms are a central component of the process of translating international obligations into domestic policies, especially when taking into account the specific role of developing countries in international environmental relations.

The study of these mechanisms is fundamental because of three reasons: first, because a better analytical framework about their operation certainly contributes to expand its scope and coverage. It is necessary to establish how mechanisms point to specific obligations in order to develop targeted approaches that guarantee effective and efficient solutions. Second, this understanding needs empirical data that supports it. National reports – as implementation mechanisms themselves – need to collect data on how the mechanisms work and their results in terms of both implementation and effectiveness. And third, mapping the existing facilitation implementation mechanisms opens the door to the identification of clusters and synergies that address the challenge presented by the proliferation of instruments[88].

Next steps in this project would include two research strands: one, a causality and correlation analysis that connects the use of implementation mechanisms and the actual improvement that countries register on the level of implementation of the conventions they are part of. Standardized measures of implementation from the Environmental Conventions Initiative would help in this process. And two, the inclusion of additional mechanisms and data beyond what is included in the national reports. As conventions move forward with the process of implementation, and countries expand their commitments on environmental and sustainable development issues, facilitation implementation mechanisms are even more critical. Understanding how they operate is then a requirement to expand their coverage, to identify best practices that can be exchanged across countries and conventions and the factors that determine their success, as a way to guarantee their effectiveness towards the solution of global environmental problems.

---

87 Sands (2003), p. 227, see note 37
88 Beyerlin and Marauhn (2011), see note 37

5 Strengthening compliance under the Convention on Biological Diversity

Comparing follow-up and review systems with the global climate regime

Ana María Ulloa and Sylvia I. Karlsson-Vinkhuyzen

Introduction

The Convention on Biological Diversity (CBD) and the United Nations Framework Convention on Climate Change (UNFCCC) were the two treaty outcomes of the United Nations Conference on Environment and Development (UNCED) – also referred to as the Earth Summit – held in Rio de Janeiro in 1992. Similar to the central role of the UNFCCC in the international community's response to climate change, the CBD has become a key instrument for addressing biodiversity loss in the international political arena. Twenty-five years later, the CBD remains the most comprehensive international treaty to address biodiversity issues, from conservation to sustainable use, including considerations on fairness and equity. Yet the CBD has been criticized as an ineffective instrument, both for failing to achieve its objectives and for having little impact on states' practice (i.e., showing a low degree of compliance)[1].

Assessing the effectiveness of the CBD in terms of solving the problem it was created to address (i.e., whether biodiversity is conserved at acceptable levels) is challenging. Biological systems are complex, and data needed to evaluate the (changing) status of biodiversity is often unavailable or out-dated[2]. Moreover, ecological processes occur over time, hence the impact of (stopping) detrimental activities is not necessarily observed as an immediate recovery, increase or further loss of biodiversity[3]. And finally, establishing cause-effect linkages between

---

1 See Morgera, E. and Tsiounami, E. (2011). "Yesterday, Today, and Tomorrow: Looking Afresh at the Convention on Biological Diversity", *Yearbook of International Environmental Law*, 21, pp. 3–40 for a detailed analysis on the trajectory of the CBD
2 See Collen, B. *et al.* (2013), "Biodiversity Monitoring and Conservation: Bridging the Gap between Global Commitment and Local Action", in Collen, B., Pettorelli, N., Baillie, J.E.M. and Durant, S.M. (eds.) *Biodiversity Monitoring and Conservation: Bridging the Gap between Global Commitment and Local Action,* John Wiley & Sons, for an overview of the challenges of monitoring and assessing the status of biodiversity
3 Jones, J.P.G. *et al.* (2011), "The why, what and how of global biodiversity indicators beyond the 2010 target", *Conservation Biology*, 25, pp. 450–457

a policy and impact is difficult and requires a combination of methodological approaches.[4] Despite these considerations, evidence consistently reveals that biodiversity is not only declining as a consequence of increasing human pressures[5], but also that biodiversity loss is occurring at unprecedented rates[6], and is not showing signs of relenting[7]. This clearly indicates the failure of the CBD in solving the biodiversity crisis. However, this poor record in overall outcome effectiveness does not necessarily mean that the CBD does not have any influence on the behaviour of states, such as in eliciting their compliance towards their commitments under the CBD.

Our objective in this chapter is to: analyze the strategies for reviewing compliance and implementation within the CBD, particularly under the Strategic Plan for Biodiversity 2011–2020; and, when appropriate, compare them to the approach followed in the UNFCCC, particularly the Paris Agreement adopted under the UNFCCC in 2015. Such comparison is timely considering that both the CBD and the PA are in a process of elaborating their modalities for follow-up and review.

Principles, obligations and institutional arrangements under the CBD

The CBD opened for signature at UNCED in 1992, and entered into force in 1993. The CBD has gained near worldwide participation, with 196 contracting parties[8]. The CBD is a treaty with three objectives: the conservation of biodiversity; the sustainable use of the components of biodiversity; and the equitable sharing of the benefits derived from the use of genetic resources (Article 1). These three core objectives are framed within the overarching principles of national sovereignty and international co-operation (Articles 3–5), and translated into binding operational commitments (Articles 6–22), as well as into arrangements for further institutional development and follow-up on implementation (Articles 23–42).

As reflected in the core objectives, the Convention has a very broad scope. This wide remit was the result of a troublesome North-South divide at the time

---

4 Young, O.R. (2011), "Effectiveness of international environmental regimes: Existing knowledge, cutting-edge themes, and research strategies", *Proceedings of the National Academy of Sciences,* 108 (50), pp. 19853–19860

5 See for example: Butchart, S.H.M. *et al.* (2010), "Global biodiversity: indicators of recent declines", *Science,* 328 (5982), pp. 1164–1168; and Mace, G.M. & Baillie, J.E.M. (2007) "The 2010 biodiversity indicators: challenges for science and policy", *Conservation Biology,* 21 (6), pp. 1406–1413

6 See for example: *Global Biodiversity Outlook 3* (2010), Secretariat of the Convention on Biological Diversity, Montréal, 94 pp.; and *Millennium Ecosystem Assessment (2005) Ecosystems and Human Wellbeing: Biodiversity Synthesis,* Washington DC: World Resources Institute

7 See for example: Pereira, H.M. *et al.* (2010), "Scenarios for global biodiversity in the 21st century", *Science,* 330 (6010), pp. 1496–1501

8 Status as at 12 November 2016

of framing of the CBD[9]. During the negotiation phase of the Convention, whilst developed countries were in favour of a strong conservation approach to halt biodiversity loss, developing countries advocated for their sovereign right to use biodiversity as a means to enhance social and economical development. To make the situation even more conflicting, whilst biodiversity tends to be abundant in developing countries, the technologies to exploit and profit from biodiversity are mostly owned by developed countries. Therefore, developing countries were also concerned about securing mechanisms for the transfer of financial resources and technology and the equitable sharing of benefits. Thus, if an agreement was to be reached for the CBD, a broad remit was needed in order to reunite opposed interests (i.e., exploitation vs. conservation approaches) of varied nature (i.e., environmental, social, economic) and sensitive character (i.e., fairness, transparency, sovereignty)[10,11]. A notable consequence of these conflicting interests was the reluctance of the United States to ratify the CBD after actively having participated in the negotiations[12]. Furthermore, although apparently reconciled in a vague and ambiguous text, the North-South divide has prevailed within the CBD, and has became even more evident during the strategic phase, when concrete action programmes for the implementation of the CBD have been developed[13]. These considerations will be revisited from several perspectives in the paragraphs below.

Obligations and responsibilities: Articles 6–22

The obligations contained in Articles 6 to 22 for the operationalization of the core objectives of the CBD, along with Article 26 on national reporting, define well-differentiated collective and individual-state responsibilities. In some cases collective responsibilities concern only parties from a certain kind of country on the basis of their capability to comply with those specific obligations (i.e., most often developed and developing countries). In doing so, the CBD implicitly recognises the principle of *common but differentiated responsibilities* for the conservation, sustainable use and equitable sharing of benefits arising from

---

9 For an account of the negotiations that led to adoption of the CBD, see: McConnell, F. (1996), *The Biodiversity Convention: A Negotiating History*, Kluwer Law International; and Koester, V. (1997), "The Biodiversity Convention Negotiation Process and Some Comments on the Outcome", *Environmental Policy and Law*, 27 (3), pp. 175-192
10 Eser, U. *et al.* (2014), "Prudence, Justice and the Good Life: A typology of ethical reasoning in selected European national biodiversity strategies", Bundesamt für Naturschutz (BfN) / Federal Agency for Nature Conservation, Bonn, Germany
11 Neßhöver, C. *et al.* (2015), "Biodiversity governance – A global perspective from the Convention on Biological Diversity", in Gasparatos, A. and Willis, K.J. (eds.) *Biodiversity in the Green Economy*, London: Taylor & Francis
12 See McConnell, F., see note 9
13 See Eser, U. *et al.*, see note 10, for an analysis of the implications of using the term 'biodiversity' as a boundary object; and Neßhöver, C. *et al.*, see note 11, for an account of the institutional trajectory of the CBD

biodiversity[14]. Accordingly, the various obligations in the treaty call upon: '*each Party*', '*the Parties*', or country parties with distinctive characteristics (e.g., environmentally vulnerable countries, countries rich in biodiversity, country owners of the means to make use and profit out of biodiversity, etc.) and/or capacities or needs (i.e., developed countries, developing countries, least-developed countries).

Obligations targeting parties on an individual-state basis refer to the commitments that each party shall fulfil at the national level in order to effectively comply with the Convention and thus contribute to the global achievement of the core objectives of the CBD[15]. Among such obligations for each party are those to: develop national strategies, plans or programmes including mainstreaming biodiversity conservation and sustainable use into relevant sectors (Article 6); monitor the state of biodiversity at the country level (Article 7); conserve biodiversity, including through the adoption of social and economic incentives (Article 8, 9 and 11); take measures for the sustainable use of biodiversity (Article 10); assess impacts of national programmes and policies on biodiversity and, where relevant, minimize those impacts with adverse consequences (Article 14); and report on measures taken towards the implementation of the Convention and their effectiveness (Article 26).

In comparison to individual-state obligations, collective responsibilities derive from the commitments of parties to co-operate with one another in ways that enable the achievement of the overarching objectives of the Convention. They include, but are not limited to, the responsibility that developed countries have towards developing countries to facilitate enabling means to comply with the commitments acquired under the Convention. Thus, obligations calling on 'the Parties' are hereby categorised within two different types: those entailing responsibilities that concern all parties alike, or *collective responsibilities*, and those encompassing distinctive responsibilities for a well-defined group of parties, or *common but differentiated responsibilities* (CBDRs). On the one hand, collective responsibilities among parties include obligations to: raise the profile of biodiversity through education and public awareness (Article 13); exchange information (i.e., technical, scientific and socio-economic research; on surveying programmes and techniques; specialised knowledge; indigenous and traditional knowledge) (Article 17); and to strengthen existing financial institutions to provide resources for the conservation and sustainable use of biodiversity (Article 21). On the other hand, examples of CBDRs in the CBD include obligations to: provide and facilitate research and training (Article 12), and provide financial resources (Article 20). Specifically, the CBDR principle is stressed in obligations concerning the third objective of the Convention on the equitable sharing of benefits arising from the use of biodiversity, such as: Article 15, on access to genetic resources;

14 See Rajamani, L. (2006), *Differential Treatment in International Environmental Law*, Oxford, Oxford University Press, for an overview of the use of this principle in different treaties

15 From a perspective of international law, binding obligations are only created from the verb 'shall' (ref). We, however, take a broader approach by looking at the variety of actions that

Article 16, on access to and transfer of technology; and Article 19, on handling of biotechnology and distribution of its benefits. In the CBD, CBDRs comprise additional responsibilities for developed countries; hence developed countries are expected to have more obligations than developing countries.

As mentioned above, a vague and ambiguous text was necessary to secure the adoption of the Convention[16]. However, besides vague and ambiguous, the text of the CBD has been characterised as being beleaguered by escape clauses[17]. According to Harrop and Pritchard[18], '*most articles of the CBD contain provisions which are expressed in imprecise language or over-qualified terms which enable member states to implement these provisions in virtually any manner they wish, whether challenging or not*'[19]. For instance, we have identified that most of the CBD's obligations that individually target states include statements such as '*as far as possible and as appropriate*', '*in accordance with its particular conditions and capabilities*', '*in accordance with its capabilities*', '*in accordance with national legislation and policies*' and '*subject to national legislation and international law*'. Similarly, Harrop and Pritchard also point out that the obligations specified in the text of the Convention – expressed in terms of 'shall' and 'will' – are often diluted by the concomitant use of 'subject to national legislation', 'subject to patent law', 'as far as possible' and 'as appropriate'. In all, these caveats reflect a low degree of obligation imposed on parties by the CBD.

Unlike the UNFCCC, the CBD is not explicitly called a framework convention; however, some scholars do refer to the CBD as such[20]. As with the UNFCCC, the framework character of the CBD concedes flexibility for country parties to decide, at their discretion, the specific measures they will adopt to translate the obligations of the Convention into the national context. This is reflected, for instance, in Article 6, which stipulates the development of Natural Biodiversity and Actions Plans (NBSAPs) – the key instrument to implement the CBD at the country level – subject to the 'particular conditions and capabilities' of each country party and addressing issues they consider 'relevant'.

Institutional arrangements

For the more than 20 years that the CBD has been in force, it has undergone continuous institutional development guided by the specifications contained in

---

are given as responsibilities to states in the treaty text and other CBD documents (e.g., COP decisions)
16 See Eser, U. *et al.* and Neßhöver, C. in note 13
17 Raustiala, K. (1997), "Domestic institutions and international regulatory cooperation: comparative responses to the convention on biological diversity", *World Politics*, 49 (4), pp. 482–509
18 Harrop, S.R. & Pritchard, D.J. (2011), "A hard instrument goes soft: The implications of the Convention on Biological Diversity's current trajectory", *Global Environmental Change*, 21, pp. 474–480
19 See Harrop, S.R. & Pritchard, D.J., p. 476, in note 18
20 See for example: Glowka, L. *et al.* (1994), "A guide to the Convention on Biological Diversity", Global Biodiversity Strategy Environmental Law and Policy paper No 30. IUCN Environmental

Articles 23 to 42. This includes, *inter alia*, the adoption of two supplementary legal agreements to the Convention in accordance with Article 28 on adoption of Protocols, namely: the *Cartagena Protocol on Biosafety* and the *Nagoya Protocol on Access and Benefit Sharing*. The former entered into force in 2003, and has been ratified by 170 parties[21]. Its overarching objective is the protection of biological diversity from the potential risks posed by the development and introduction of living modified organisms. The latter entered into force in 2014, and has been ratified by 89 parties[22]. It seeks fair and equitable sharing of the benefits arising from the utilisation of genetic resources, including the appropriate access to genetic resources and the appropriate transfer of relevant technologies.

However, it is not through these legal agreements that the CBD has primarily endeavoured to achieve its objectives. On the one hand, the Cartagena Protocol is highly narrow in its objective and, as a result, has been considered by some scholars as a disjointed process in relation to the overarching objectives of the Convention[23]. Although the Cartagena Protocol derives from Article 8(h) on the risks posed on biodiversity by alien species, it has been accused of being driven by commercial interests arising from biotechnology markets, rather than by the conservation concerns associated with the management of alien species (including genetically modified organisms). On the other hand, the Nagoya Protocol derives from the third objective of the CBD on the fair and equitable sharing of benefits. Specifically, it deals with the definition of intellectual property rights of biodiversity – a conflict deriving from abusive practices of 'bio-prospecting' or 'bio-piracy'[24]. Although it is early to assess the impact of the Nagoya Protocol – since it entered into force in 2014, and almost half of the parties are yet to ratify it – the Nagoya Protocol represents an important instrument for the operationalization of the third objective of the Convention[25]. However, as illustrated above, the Cartagena Protocol and the Nagoya Protocol both fail to comprehensively address the three core objectives of the CBD, particularly concerning the first objective on biodiversity conservation[26].

The international community expected that subsidiary protocols would come into being (as stipulated in Article 28) so as to back up the vague commitments scattered in the text of the Convention with more precise obligations for

---

Law Centre, IUCN Biodiversity Programme; Harrop, S.R. & Pritchard, D.J. in note 18; Sand, P.H. (1993), "International law after Rio", *European Journal of International Law*, 4, pp. 377–389

21 Status as at 12 November 2016
22 Status as at 12 November 2016
23 See Morgera, E. and Tsiounami, E. in note 1
24 See Harrop, S.R. and Pritchard, D.J. in note 18
25 For a detailed analysis of the implications of the Nagoya Protocol, see Morgera, E. *et al.* (2014), *Unraveling the Nagoya Protocol: A Commentary on the Nagoya Protocol on Access and Benefit-Sharing to the Convention on Biological Diversity*, Legal Studies on Access and Benefit-Sharing, Martinus Nijhoff; available at DOI: 10.1163/9789004217188
26 See Harrop, S.R. and Pritchard, D.J., in note 18

parties[27]; however, this has not materialised. As a matter of fact, the CBD has been characterised as relying on non-binding goals and targets for the implementation of its core objectives[28], even when targets or time-frames were not part of the text of the Convention nor were they envisaged as part of the institutional arrangements of the CBD[29].

In 2002, after almost ten years – in which the parties were primarily focused on negotiating and defining the operational rules of the Convention[30] – the Conference of the Parties finally moved towards the operationalization path with the adoption of the first Strategic Plan for Biodiversity[31]. Under the overarching (and very unspecific) goal of '[achieving] by 2010 a significant reduction of the current rate of biodiversity loss at the global, regional and national level as a contribution to poverty alleviation and to the benefit of all life on earth'[32], the Strategic Plan for Biodiversity 2002–2010 comprised 19 objectives grouped under four operational goals. The Strategic Plan for Biodiversity 2002–2010 was intended to guide an effective and coherent implementation of the three core objectives of the CBD at the national, regional and global levels; yet none of the objectives comprised numerical targets or deadlines. As such, the first strategic plan of the CBD was criticised on several fronts. From a biological standpoint, the scientific community argued that the achievement of the 2010 Biodiversity Target was compromised from the very beginning, owing to the short time span in which a *significant* reduction of biodiversity loss should be not only achieved, but also assessed[33]. Furthermore, no reference was made to the parameters against which the reduction of the *current*, yet unspecified, rate of biodiversity loss should be assessed[34]. From a policy perspective, the Strategic Plan was deemed as vague and unspecific (i.e., the goals and objectives were not measurable or verifiable), and not action-orientated[35]. The goals and targets largely repeated the already vague commitments of the Convention without contributing to add any strength[36]; and did not refer to the underlying drivers of biodiversity that needed to be addressed[37]. Ultimately, the prospect of

27 See for example: Bragdon, S. (1996), "The convention on biological diversity", *Global Environmental Change*, 6 (2), pp. 177–179; and Sand, P.H. in note 20
28 See Harrop, S.R. and Pritchard, D.J. in note 18; and Morgera, E. and Tsiounami, E. in note 1
29 See Glowka, L. *et al.* in note 20
30 See Neßhöver, C. *et al.* in note 11
31 Convention on Biological Diversity; Decision of the Conference of the Parties VI/26: *Strategic Plan for the Convention on Biological Diversity*. UNEP/CBD/COP/DEC/VI/26 (April 2002); available from: ww.cbd.int/doc/decisions/cop-10/cop-10-dec-02-en.pdf (accessed 12 November 2016)
32 See note 31: CBD/COP/DEC/VI/26, Annex, para. 11
33 See Collen, B. *et al.* in note 2
34 Mace, G.M. *et al.* (2010), "Biodiversity targets after 2010. Current Opinion", in *Environmental Sustainability*, 2, pp. 1–6
35 See Collen, B. *et al.* in note 2; and Harrop, S.R. and Pritchard, D. J. in note 18
36 See Harrop, S.R. and Pritchard, D.J. in note 18
37 See Collen, B. *et al.* in note 2

translating global commitments into national and local actions and measures was very remote[38].

Furthermore, with the adoption of the Strategic Plan for Biodiversity 2002–2010, the need to facilitate a mechanism for evaluation of progress in implementing the Convention (i.e., as in assessing and effectively communicating progress towards the 2010 target) was formally recognised for the first time. Accordingly, in decision VII/30[39], a framework to enhance the evaluation and effective communication of achievements and progress in implementation of the CBD, as well as trends in biodiversity. It involved the development of a limited number of trial indicators – for which data was available at the time – and the establishment of a process for identifying, developing, reviewing and/or testing indicators and for reporting progress (i.e., through the Global Biodiversity Outlook[40] – GBO). Parties were called to develop national targets that reflected national circumstances whilst at the same time being in line with the global targets of the framework. Parties were also called to integrate those country targets into their National Biodiversity Strategies and Action Plans (NBSAPs). However, this flexible framework did not achieve its purpose. By 2010 not only was the implementation of NBSAPs low, but also only a minority of countries had established national targets[41]. Moreover, many indicators were not developed in time[42]. Information arising from these processes was envisaged as a tool to identify obstacles encountered in the implementation of NBSAPs, and accordingly to provide supporting mechanisms to parties (i.e., capacity-building, and resource and technology transfer).

---

38 Mace, G.M. and Baillie, J.E.M. in note 5
39 Convention on Biological Diversity, Decision of the Conference of the Parties VII/30: *Strategic Plan: future evaluation of progress.* UNEP/CBD/COP/DEC/VII/30 (13 April 2004); available from: www.cbd.int/doc/decisions/cop-07/cop-07-dec-30-en.pdf (accessed: 12 November 2016). This framework for the monitoring of progress towards the achievement of the 2010 Biodiversity Target was refined in the eighth meeting of the Conference of the Parties held in 2006. For details see: Convention on Biological Diversity, Decision of the Conference of the Parties VIII/15: *Framework for monitoring implementation of the achievement of the 2010 target and integration of targets into the thematic programmes of work.* UNEP/CBD/COP/DEC/VIII/15 (15 June 2006); available from: www.cbd.int/doc/decisions/cop-08/cop-08-dec-15-en.pdf (accessed 12 November 2016)
40 The Global Biodiversity Outlook is the periodic flagship publication of the Secretariat of the Convention on Biological Diversity. It comprises a periodic report *inter alia* providing: a summary of the status and trends of biological diversity at the global and supranational regional level; an analysis of the global and regional trends in implementation of the objectives of the Convention; and a summary of the implementation of the Convention at the national level on the basis of the information contained in national reports and other up-to-date scientific data
41 Convention on Biological Diversity. Note by the Executive General: Implementation of the Convention and the Strategic Plan and Progress towards the 2010 Biodiversity Target UNEP/CBD/COP/10/8 (31 July 2010); available from: www.cbd.int/doc/meetings/cop/cop-10/official/cop-10-08-en.pdf (accessed 12 January 2016)
42 Walpole, M. *et al.* (2009), "Tracking progress toward the 2010", *Science*, 325 (5947), pp. 1503–1504

Articles 7 and 26 provide specifications for parties to monitor and assess the status of biodiversity and the impact of measures adopted, and subsequently report on progress and challenges, respectively. Despite this, follow-up mechanisms in the CBD are weak, and not systematic. The Conference of the Parties is the body responsible for reviewing the implementation of the Convention (Article 23). Unfortunately, its role has been rather passive: instead of reviewing national reports in plenary sessions, the COP has limited its actions to provide summaries of conclusions drawn from the syntheses of national reports prepared by the Secretariat and from the Global Biodiversity Outlook[43]. Although these summaries provide general feedback in the form of trends in implementation and indications to define the course of action, they do not entail the review by peers and/or other actors of the (lack of) actions taken by states to comply with the CBD. Thus, the COP has not provided an arena for asking parties about their (lack of) actions, nor have parties had the opportunity to openly explain and justify the reasons of such an outcome. In short, the CBD has not had an arena for enacting accountability among parties[44]. One of the consequences is that the COP is unable to identify countries in need of support, and peer-learning is not enabled.

In 2010, when the Strategic Plan for Biodiversity 2002–2010 was due, the third edition of the GBO[45] – based on national reports and the latest scientific data on status and trends of biodiversity – concluded that the 2010 Biodiversity Target had not been met at the global level. The publication also assessed the causes for the failure, analyzed future scenarios for biodiversity, and reviewed possible actions that might be undertaken to reduce future loss. It specifically stated that the scale of actions taken until that moment was not sufficient to reduce the underlying drivers of biodiversity loss as a consequence of inappropriate integration of biodiversity issues into broader policies, strategies, programmes and actions. Furthermore, most parties identified a lack of financial, human and technical resources as a factor limiting the appropriate implementation of the Convention. For example, technology transfer was considered to be very limited and scientific information for policy- and decision-making insufficient[46]. The conclusions drawn from the GBO 3 pointed out the underpinning role of biodiversity in ecosystem functioning and the provision of ecosystem services essential for human well-being, which in turn was related to the

---

43 See Morgera, R. and Tsounami, E. in note 1
44 For an overview on the relational character of accountability dynamics, see: Bovens, M. (2007), "Analysing and Assessing Accountability: A Conceptual Framework", *European Law Journal*, 13, pp. 447–468; Mashaw, J.L. (2006), "Accountability and Institutional Design: Some Thoughts on the Grammar of Governance", in: Michael, D. (ed.) *Public Accountability: Designs, Dilemmas and Experiences,* Cambridge University Press; and Steffek, J. (2010), "Public Accountability and the Public Sphere of International Governance", *Ethics & International Affairs,* 24, pp. 45–68
45 See GBO 3, in note 6
46 See CBD/COP/10/8 in note 41

achievement of the MDGs (millennium developments goals), including poverty reduction. These considerations were crucial elements in shaping the subsequent Strategic Plan for Biodiversity 2011–2020.

After the failure in achieving the Biodiversity Goal 2010, a successive strategic plan was adopted in 2010 for the period 2011–2020[47]. The CBD considers the Strategic Plan for Biodiversity 2011–2020 as a milestone in achieving its long-term vision of 'Living in harmony with nature', where 'By 2050, biodiversity is valued, conserved, restored and wisely used, maintaining ecosystem services, sustaining a healthy planet and delivering benefits essential for all people'[48]. For that purpose, the Strategic Plan for Biodiversity 2011–2020, similar to its predecessor, was envisaged as an instrument to promote effective implementation of the Convention through a strategic approach. In this case, the strategy relied on five strategic goals, 20 targets (the 'Aichi Biodiversity Targets') and a set of 168 indicators (those established under decisions VII/30 and VIII/15[49], until revised indicators were available).

Given the poor outcome of the Strategic Plan 2010, some scholars questioned why the CBD, for the second time, directed its institutional efforts towards adopting an approach based on non-binding goals and targets[50]. Some other scholars focused on the lessons learned from the process and how future approaches could be built upon[51]. In this regard, it becomes relevant to review whether the shortcomings identified in the Strategic Plan 2010 were, at least to some extent, addressed in the new Strategic Plan 2020. Referring to the Strategic Plan 2010, Harrop and Pritchard had pointed out, 'the product of a pre-negotiation agreement rather than the outcome of an established convention'[52]; however, this does not seem to be the case with the new plan, which appears more robust on paper. It contains a separate section on implementation, monitoring, review and evaluation, as well as a section on supporting mechanisms. Moreover, all the Aichi targets explicitly refer to a desired 'end point' by 2020 – although only three of them set numerical standards or refer to measurable rates and comparable baselines to define 'success' by 2020. Thus, although most of the Aichi targets struggle to classify as 'specific' or 'measurable', they at least refer to the achievement of points where ecosystems are functional, constantly emphasizing the link between biodiversity, ecosystem services and human well-being – a welcome approach[53]. Similarly, the Strategic Plan does not only aim to reduce

---

47 Convention on Biological Diversity, Decision of the Conference of the Parties X/2: *The Strategic Plan for Biodiversity 2011–2020 and the Aichi Biodiversity Targets*. UNEP/CBD/COP/DEC/X/2 (29 October 2010); available from: www.cbd.int/doc/decisions/cop-10/cop-10-dec-02-en.pdf (accessed 12 November 2016)
48 See note 47: CBD/COP/DEC/X/2, Annex, p. 7, para. 11
49 See CBD/COP/DEC/VII/30 and CBD/COP/DEC/VIII/15 in note 39
50 See Harrop, S.R. and Pritchard, D.J. in note 18
51 See Jones, J.P.G. *et al.* in note 3; and Mace, G.M. *et al.* in note 34
52 See Harrop, S.R. and Pritchard, D.J., p. 477, in note 18
53 Mace, G.M. *et al.* (2013), "Science to Policy Linkages for the Post-2010 Biodiversity Targets", in Collen, B., Pettorelli, N., Baillie, J.E.M. & Durant, S.M. (eds.), *Biodiversity Monitoring and*

pressures on biodiversity but also refers – although in a broad manner – to the drivers of biodiversity loss[54]. This means that the new targets openly addressed sensitive, but critical issues avoided by the CBD until that moment (i.e., sustainable management of commercial fisheries, and the regulation of incentives with impacts on biodiversity)[55]. Moreover, Target 20 refers to increasing the mobilisation of resources for effectively implementing the Strategic Plan for Biodiversity 2011–2020, and specifically refers to the Strategy on Resource Mobilisation[56].

It must also be noted that while some aspects of the Strategic Plan, other than those mentioned above, were 'updated', they still remained very similar to the features of its predecessor. For instance, recognising differential national circumstances and capabilities across parties, the Strategic Plan provided a flexible framework consisting of goals, targets and indicators. Parties were invited to set targets at the national or regional levels, based on national needs and priorities, which were to contribute to the achievement of the global targets. As in the Strategic Plan 2010, the need to continue strengthening the ability to monitor biodiversity at all levels was reinforced[57]. Accordingly, while updating NBSAPs and in national reporting thereafter, parties were encouraged to use indicators that were ready for application at the global level[58]; consequently an updated list of indicators was provided in 2012[59]. Concerning the latter, availability of relevant, credible and solid data-grounded indicators – with specific links to individual targets and clear links to biodiversity status – increased when compared to the Strategic Plan 2010[60]. The flexible framework of targets and indicators reflecting/adapted to national circumstances was proposed to be used by parties not only for monitoring and assessing the status of biodiversity, but also in national reporting. Had 'tailored' targets and indicators been explicitly used in national reports, as envisaged by the CBD, the COP would have had a better opportunity to follow up on national progress and identify challenges encountered by parties.

Conservation: Bridging the Gap between Global Commitment and Local Action, John Wiley & Sons, for an overview of the challenges of monitoring and assessing the status of biodiversity

54 See Mace, G.M. *et al.* in note 53
55 See Harrop, S.R. and Pritchard, D.J. in note 18
56 Convention on Biological Diversity, Decision of the Conference of the Parties IX/31: Financial Mechanism. UNEP/CBD/COP/DEC/IX/31 (9 October 2008); available from: www.cbd.int/doc/decisions/cop-10/cop-10-dec-02-en.pdf (accessed 12 November 2016)
57 In accordance with decision X/7. See Convention on Biological Diversity, Decision of the Conference of the Parties X/7: *Examination of the outcome-oriented goals and targets (and associated indicators) and consideration of their possible adjustment for the period beyond 2010*. UNEP/CBD/COP/DEC/X/7 (29 October 2010); available from: www.cbd.int/doc/decisions/cop-10/cop-10-dec-07-en.pdf (accessed 12 November 2016)
58 See note 39: headline indicators as defined in CBD/COP/DEC/VII/30 and CBD/COP/DEC/VIII/15
59 Convention on Biological Diversity, Decision of the Conference of the Parties XI/3: *Monitoring progress in implementation of the Strategic Plan for Biodiversity 2011–2020 and the Aichi Biodiversity Targets*. UNEP/CBD/COP/DEC/XI/3 (5 December 2012); available from: www.cbd.int/doc/decisions/cop-11/cop-11-dec-03-en.pdf (accessed 12 November 2016)
60 Tittensor, D.P. *et al.* (2014), "A mid-term analysis of progress toward international biodiversity targets", *Science*, 346, pp. 241–244

As national reports are to be written in accordance with agreed guidelines[61], information provided in national reports, although 'tailored' to reflect national circumstances, would still provide a common ground for the review of progress at the regional and global levels (i.e., through the GBO, which in turn heavily relies on information provided in national reports to identify global trends). On the basis of the principle of *'adaptive management through active learning'*[62], findings arising from these processes should allow: sharing experiences on implementation, making recommendations on means to address obstacles encountered, and strengthening the mechanisms to support implementation, monitoring and review. However, apart from the resolution to use those indicators in the fifth national reports – due for submission in 2014 for consideration at the twelfth meeting of the Conference of the Parties – as a mid-term review of progress towards the achievement of the Aichi Biodiversity Targets[63], the COP did little to strengthen the already weak follow-up process of the Convention. Collective evaluations of progress, through the consideration of the GBO and the synthesis reports of the Secretariat, were to be performed, as usual, on a quadrennial basis at the corresponding meetings of the COP. As such, national reports were to continue being sources of information for the aggregation of data that allow tracking progress at regional and local levels, rather than material for the active peer review of country parties' performance. In addition to the deficit, in relation to the fact that this condition presupposes that states can be held accountable under the CBD[64], there is another major challenge: despite the fact that national reporting is mandatory (under Article 26), and constitutes the building block of the follow-up architecture of the CBD, national reporting rates have been consistently low. The percentage of parties that have submitted national reports by the due date for consideration at the corresponding meeting of the COP has been as low as 1.6% and has never been above 15.5%. By November 2016, a percentage of parties, varying between 2.6% and 23.9% across the five national reports that have been agreed by the COP, has never submitted theirs for at least one of the specific deadlines[65].

Translating global goals and targets into concrete national actions and measures: the challenge ahead

The road up to 2014

NBSAPs are the principal instruments for implementing the Convention at the national level. In accordance with Article 6, the Convention requires parties to

---

61 Convention on Biological Diversity, Decision of the Conference of the Parties X/10: *National reporting: review of experience and proposals for the fifth national report*. UNEP/CBD/COP/DEC/X/10 (29 October 2010); available from: www.cbd.int/doc/decisions/cop-10/cop-10-dec-10-en.pdf (accessed 12 November 2016)
62 See note 47: CBD/COP/DEC/X/2, p. 11, para. 19
63 See CBD/COP/DEC/XI/3 in note 59
64 See Bovens, M., Mashaw, J.L. and Steffek, J. in note 44 for a review on accountability as a relational concept, based on the giving and demanding of reason of conduct between social actors
65 Data collected from the official website of the CBD (www.cbd.int/reports/search/)

prepare a national biodiversity strategy (or equivalent instrument), and ensure that this strategy is mainstreamed into the planning and activities of all those sectors whose activities can have an impact (positive and negative) on biodiversity. NBSAPs should reflect the measures that need to be taken in light of specific national circumstances. States are generally reluctant to subject themselves to detailed prescriptions in global instruments for what national management and policies of domestic resources, such as biodiversity, should be[66]. Moreover, because the management of biodiversity encompasses the participation of multiple stakeholders on the ground, national planning requires public support and engagement[67]. However, specifically in the case of the CBD, scholars have a critical view on the lack of obligation Article 6 imposes on parties, which leaves NBSAPs devoid of strong commitment to action and transforms them into merely declarations of intention[68]. Indeed, the commitment to develop NBSAPs is subject to national 'particular conditions and capabilities' (as specified in Article 6). Similarly, the Strategic Plan 2020 '*urges*' rather than '*requires*' parties, for instance, to 'review, and as appropriate update and revise, their national biodiversity strategies and action plans'.

The development of NBSAPs has been inconsistent, particularly concerning the call to revise and update NBSAPs after adoption of the Strategic Plan 2020. According to a report released by the Executive General in preparation for the thirteenth meeting of the Conference of the Parties held in December 2016: out of the 196 country parties, seven (4%) have not submitted their first NBSAP in 23 years; and out of the 189 NBSAPs submitted, only 121 (62%) have been revised at least once. Concerning Aichi Target 17, which called on parties to revise, update and implement their NBSAPs by December 2015, only 69 parties (35%) had revised/updated their NBSAPs after the adoption of the Strategic Plan 2020. A year later, when the report was issued, a total of 131 (67%) parties had done so, and another six were awaiting final domestic approval. Thus, there are still 48 parties (24%) in the process of revising and/or updating their NBSAPs, and 11 (6%) parties that have not yet started or do not plan to do so in the near future or have provided no information in this regard[69].

In summary: are there any effective mechanisms that can enhance implementation and compliance of parties with their obligations and responsibilities under the CBD and the Strategic Plan 2020? For its first two decades, the answer is largely 'no' for the CBD: national reporting rates are low[70]; revision, update and even development of NBSAPs is inconsistent[71]; the 2010 Global Biodiversity

---

66 See Harrop, S.R. and Pritchard, D.J. in note 18
67 See Glowka, L. *et al.* in note 20
68 See Glowka, L. *et al.* in note 20; and Harrop, S.R. and Pritchard, D.J. in note 18
69 Convention on Biological Diversity, Note by the Executive General: Update on progress in revising/updating and implementing National Biodiversity and Action Plans, including national targets. UNEP/CBD/COP/13/8/Add.1/Rev.1 (24 November 2016); available from: www.cbd.int/doc/meetings/cop/cop-13/official/cop-13-08-add1-rev1-en.pdf (accessed 12 January 2016)
70 See rate of submission of national reports referred to in note 65
71 See the official report on submission, revision, updating and implementation of NBSAPs referred to in note 69

Target was not achieved[72]; and in the mid-term evaluation of progress towards the Aichi Targets performed in 2014, it was already acknowledged that the achievement of all the targets will not be met in 2020 without urgent action to scale up implementation[73].

Inconsistency of states to comply with multilateral environmental agreements, which involve significant political and economic investments – whether legally binding or not – has been well documented[74]. Because of the absence of global-level enforcement bodies, and taking into account that states commit to international agreements on a voluntary basis, compliance with international (environmental) norms is claimed to be also the result of reciprocity processes, reputational sanctions, learnt-lessons dynamics over time and capacity-building[75]. Therefore, despite the non-legally binding character of the Strategic Plan and the allied Aichi targets, they have the potential to generate compliance if backed up by adequate mechanisms[76].

Follow-up mechanisms that seek to track progress on implementation of international agreements comprise valuable tools for scrutinising states' behaviour; the object is to hold states to their word, for the commitments that they voluntarily made[77]. In the next sub-section, the most recent developments concerning the follow-up arrangements of the CBD are presented and their implications discussed.

Developments post-2014 at the twelfth and thirteenth meetings of the COP

The CBD has used the formal evaluations of collective progress (i.e., towards the implementation of the SPB-2010 and SPB-2020, through GBO 3 and 4, respectively) to, *inter alia*, identify the challenges faced by parties, feed back on the results, and accordingly adopt measures to meet the shortcomings. For instance, lack of

---

72 See for example the conclusions of: Butchart, S.H.M. *et al.* in note 5; GBO 3, in note 6; Mace, G.M. *et al.* in note 34
73 See for example: *Global Biodiversity Outlook 4* (2014), Secretariat of the Convention on Biological Diversity, Montréal, 155pp.; and Tittensor D.P. *et al.* in note 60
74 See for example: Mitchell, R.B. (2003), "International Environmental Agreements. A Survey of Their Features, Formation, and Effects", *Annual Review of Environmental Resources*, 28, pp. 429–61; and Oberthur, S. and Lefeber, R. (2010), "Holding countries to account: The Kyoto Protocol's compliance system revisited after four years of experience", *Climate Law*, 1, pp. 133–158
75 See Abbott, K.W. and Snidal, D. (2000), "Hard and Soft Law in International Governance", *International Organization*, 54, pp. 421–456; and Raustiala, K. (2000), "Compliance and Effectiveness in International Regulatory Cooperation", *Case Western Reserve Journal of International Law*, 32, p. 387
76 For an analysis of the mechanisms through which hard and soft can influence the behaviour of states, see: Guzman, A.T. and Meyer, T.L. (2010), "International Soft Law", *Journal of Legal Analysis*, 2; Karlsson-Vinkhuyzen, S.I. and Vihma, A. (2009), "Comparing the legitimacy and effectiveness of global hard and soft law: An analytical framework", *Regulation & Governance*, 3, pp. 400–420; Raustiala, K. in note 75; Tallberg, J. (2002), "Paths to Compliance: Enforcement, Management, and the European Union", *International Organization*, 56, pp. 609–643
77 See Raustiala, K. in note 75

financial, human and technical capacity were identified as limiting factors for the implementation of the objectives of the Convention and the Global Biodiversity Target 2010[78]. In response, the COP stressed that the fulfilment of biodiversity targets and obligations by developing countries partly depends on the implementation of the provisions of the Convention by developed countries, to facilitate access to and transfer technology, financial resources, and financial mechanisms (in accordance with Articles 16, 20 and 21, respectively). Therefore, previous decisions on capacity-building were recalled, in order to overcome the financial, human and technical limitations that ultimately undermine the efforts of states to fully implement the Convention. An existing strategy on resource mobilisation originally called on developed countries to provide new and additional financial resources to enable developing countries to meet the incremental implementation costs of complying with the SPB-2020[79]. Building upon this resolution, the COP remarkably – in the view of some scholars[80] – resolved to strengthen the strategy on resource mobilisation by adopting a follow-up mechanism (i.e., global monitoring reports), so as to track the status and trends in the provision of financial resources[81]. Similarly, aiming to promote effective implementation of the Convention, the Strategic Plan for Biodiversity 2011–2020 specifically involved the enhancement of support mechanisms to parties[82], such as: capacity-building (i.e., for the revision and updating of NBSAPs and for the development of indicators at the national level); the Clearing-House Mechanism[83] (CHM) and technology transfer; financial resources; and partnerships and initiatives to enhance co-operation at all levels. Furthermore, acknowledging the discouraging conclusions of the formal mid-term review of progress towards the Aichi targets[84], the Subsidiary Body on Implementation (SBI) was established[85], and the COP decided that progress on implementation of the Strategic Plan would

---

78 See GBO 3, in note 6
79 See note 56: CBD/COP/DEC/IX/31
80 See Morgera, E. and Tsiounami, E. in note 1
81 Convention on Biological Diversity, Decision of the Conference of the Parties X/3: *Strategy for resource mobilization in support of the achievement of the Convention's three objectives*. UNEP/CBD/COP/DEC/X/3 (29 October 2010); available from: www.cbd.int/doc/decisions/cop-10/cop-10-dec-05-en.pdf (accessed 12 November 2016)
82 In accordance with decision X/5. See Convention on Biological Diversity, Decision of the Conference of the Parties X/5: *Implementation of the Convention and the Strategic Plan*. UNEP/CBD/COP/DEC/X/5 (29 October 2010); available from: www.cbd.int/doc/decisions/cop-10/cop-10-dec-05-en.pdf (accessed 12 November 2016)
83 The Clearing-House Mechanism was established in response to Article 18.3 on technical and scientific co-operation of the Convention. It has been further developed and refined in several decisions. Currently, its mission is to contribute to the implementation of the Convention (and its Strategic Plan for Biodiversity 2011–2020) at the national and global level, through effective information services in order to promote and facilitate scientific and technical co-operation, knowledge sharing and information exchange, and to establish a fully operational network of parties and partners.
84 See GBO 4 in note 73
85 The Subsidiary Body on Implementation (SBI) was established in order to replace the Ad Hoc Open-ended Working Group on Review of Implementation of the Convention. The SBI has the mandate to support the Conference of the Parties in keeping under review the implementation of

be reviewed at every MCOP; beginning at MCOP 13 in 2016, and continuing until 2020[86]. These reviews, along with the information provided in the national reports, and including information from scientific assessments, were envisaged as a mechanism to guide the COP in defining the actions to be taken in order to support implementation (i.e., enhancement of capacity-building, technical and scientific co-operation), and to provide general advice to all states for policy development (i.e., for reviewing, updating and revising NBSAPs and for adopting indicators at the national level).

However, the institutional approach followed by the CBD presents several shortcomings, namely: first, follow-up mechanisms have not been used to scrutinise states' behaviour, as in assessing individual party compliance; and second, institutional efforts within the CBD have been directed towards strengthening capacity-building, but not towards encouraging unwilling actors to act. Concerning the former consideration, although formal assessments of progress made towards the implementation of the SPB-2010 and SPB-2020 identified lack of capacity as the main reason for failure[87], informal actors have also pointed out lack of political will as a critical factor. For instance, renowned environmental NGOs consider lack of political will as one of the main challenges to overcome for the successful implementation of the CBD (and also of the UN 2030 Agenda for Sustainable Development adopted in September 2015, which specifically addresses biodiversity in Goal 15)[88]. In this context, mechanisms to enable capacity-building – such as those on which the CBD has focused its institutional efforts – are not by themselves enough to overcome the limitations so far faced to achieve the objectives of the CBD. On the other hand, with reference to the lack of robust and systematic follow-up systems, the CBD has established responsibilities for parties on monitoring and reporting (Articles 7 and 26, respectively). However, the development of indicators has been acknowledged as a slow process[89] and a challenging task[90] (particularly for least-developed country parties, those which are economies in transition and those which are particularly environmentally vulnerable[91]), whilst national reporting has been inconsistent[92]. Moreover, the review

---

the Convention. See: Convention on Biological Diversity, Decision of the Conference of the Parties XII/26: *Improving the efficiency of structures and processes of the Convention: Subsidiary Body on Implementation*. UNEP/CBD/COP/DEC/XII/26. (17 October 2014); available from: www.cbd.int/doc/decisions/cop-12/cop-12-dec-26-en.pdf (accessed 12 November 2016)

86 Convention on Biological Diversity, Decision of the Conference of the Parties XII/31: *Multi-year programme of work of the Conference of the Parties up to 2020*. UNEP/CBD/COP/DEC/XII/31 (17 October 2014); available from: www.cbd.int/doc/decisions/cop-12/cop-12-dec-31-en.pdf (accessed 12 November 2016)

87 See GBO 3, in note 6, for an analysis of the causes that prevented the achievement of the Global Biodiversity Target 2010, and the GBO 4, in note 73, for analysis of the factors limiting adequate progress towards the Aichi Biodiversity Targets

88 Interviews by Ulloa, A.M. (2016), "The Role of NGOs in Holding States Accountable: Considerations on Global Biodiversity Governance", a Master thesis at the Technical University of Munich

89 See Walpole, M. *et al.* in note 42

90 See Collen, B. *et al.* in note 2

91 See Harrop, S.R. and Pritchard, D.J. in note 18

92 See data referred to in note 65

process (which is the responsibility of the COP as stipulated under Article 26) has been limited to the collective evaluations of progress through consideration of the GBO and synthesis reports of the Secretariat during plenary sessions[93]. As such, the CBD is devoid of a mechanism that allows for a true review of the progress, achievements and/or challenges faced by individual parties during the implementation of the objectives of the Convention.

On the bright side, in spite of 'political reservations', the need to formally strengthen the review system within the CBD has been increasingly acknowledged. Although not explicitly addressed in the Strategic Plan 2020, the CBD seems to have recognised the shortcoming to effectively follow up progress in achieving the objectives of the Convention. It has taken more comprehensive measures to address both lack of capacity and, more discreetly, the unwillingness of states to implement the Convention. For instance, the strengthening of the strategy on resource mobilisation not only addresses an increase in the provision of financial resources – an enabling precondition for developing countries to comply with the CBD, that is at the core of the principle of common but differentiated responsibilities[94] – but it also involves the adoption of a follow-up system[95]. This fact shows the disposition of parties to, if not fully engage in stronger accountability dynamics, at least discuss the need for stricter follow-up systems. Furthermore, the *modus operandi* of the SBI was adopted in the thirteenth Conference of the Parties held in December 2016[96]. It involves: reviewing progress in implementation and achievement of targets; contributing towards the definition of strategic actions to enhance implementation; identifying and developing recommendations to overcome obstacles encountered in the implementation process, as well as developing recommendations on how to strengthen the means of implementation; and reviewing the impacts and effectiveness of existing processes under the Convention in order to increase efficiencies (i.e., in areas such as resource mobilisation, guidance to the financial mechanism, capacity-building, national reporting, technical and scientific co-operation and the clearing-house mechanism, and communication, education and public awareness). In comparison to the climate regime, where an analogous body has a well-established role under the UNFCCC, some environmental NGOs consider its follow-up processes and structures stricter and more robust than the ones of the CBD. Therefore, the establishment of the SBI under the CBD has been welcome by the international community, as a favourable step towards strengthening compliance CBD[97].

Furthermore, since 2008 the CBD has been discussing the establishment of a peer-review process for the development and implementation of NBSAPs.

---

93 See Morgera, E. and Tsiounami, E. in note 1
94 See Morgera, E. and Tsiounami, E. in note 1
95 In accordance with decision X/3 of the COP referred to note 81
96 Convention on Biological Diversity, Decision of the Conference of the Parties XIII/25: *Modus operandi of the Subsidiary Body on Implementation and mechanisms to support review of implementation*. UNEP/CBD/COP/DEC/XIII/25 (9 December 2016); available from: www.cbd.int/doc/decisions/cop-13/cop-13-dec-25-en.pdf (accessed 12 January 2016)
97 See note 88

The methodology for voluntary peer-review for the exchange of best practices and lessons learned from the preparation, updating and implementation of NBSAPs was put under consideration of the SBI in 2014[98]. In accordance with the methodology under consideration, the main goal of the peer-review system is to help parties to improve their individual and collective capacity so as to more effectively implement the CBD[99]. The peer-review system is intended as a mechanism: to assess the development and implementation of NBSAPs in the context of the Strategic Plan for Biodiversity 2011–2020, and produce specific recommendations for parties under review; provide opportunities for peer-learning for parties directly involved and for other parties; and create greater transparency and accountability for NBSAP development and implementation to the public and other parties[100] (i.e., by aiming for broad participation of relevant governmental institutions and stakeholders in the review process[101]). Peer reviews are envisaged as mechanisms to stimulate mutual experience-sharing, learning and capacity-building by sharing information (within the CBD but also across other biodiversity-related multilateral environmental agreements and to the broader public) about what measures lead to progress, which ones do not, and/or which ones present a continuous challenge in the management of biodiversity[102]. Additionally, the methodology specifies that countries under review are to be allowed to consider how to respond to recommendations, and how to use the review report[103].

Scholars have argued that review processes are mechanisms through which compliance can be strengthened and promoted because they allow the identification of non-compliance (and non-compliant actors) and its roots (i.e., incapability or unwillingness). Accordingly, transparency is enhanced, and causes of non-compliance can be addressed[104]. More importantly, if review processes are open, active and dynamic, they have the potential to put pressure on states to justify their (lack of) actions, also in the absence of legal sanctions[105]. If enough criticism is mobilised, active and dynamic review processes have the potential to encourage non-compliant actors to justify their choices or to clarify or defend their positions. By comprising an arena were feedback can be given, open and dynamic reviews offer the opportunity for states to self-reflect on conduct,

---

98 In accordance with decision XII/26 of the COP, referred to in note 85
99 United Nations Environmental Programme, Convention on Biological Diversity. Note by the Executive General: *Voluntary Peer-Review Process for the National Biodiversity Strategies and Action Plans: Progress Report and Updated Methodology*. UNEP/CBD/COP/13/19 (27 September 2016); available from: www.cbd.int/doc/meetings/cop/cop-13/official/cop-13-19-en.pdf (accessed 12 November 2016)
100 See note 99: CBD/COP/13/19, para. 2(a–c)
101 See note 99: CBD/COP/13/19, para. 6(f)
102 See note 99: CBD/COP/13/19, para. 5
103 See note 99: CBD/COP/13/19, para. 6(e)
104 van Asselt, H. *et al.* (2015), "Assessment and Review under a 2015 Climate Change Agreement", Denmark, Nordic Council of Ministers
105 See Steffek, J. in note 44

promote catharsis, and subsequently encourage the search for strategies (i.e., triggering a switch in governance arrangements from 'routine mode' to 'crisis mode'[106]). Moreover, some authors have further argued that if active and dynamic review processes are aimed towards improving individual and collective performance of country parties, rather than at pointing out wrongful individual behaviour, as is in the case of the CBD, they may have a prophylactic role in deterring con-compliance[107]. Specifically, by comprising arenas where mutual learning, trust, co-operation and stewardship are promoted, open and dynamic reviews have the potential to influence the behaviour of states before failure occurs – *ex ante*.

Comparison with the UNFCCC: discussion and conclusion

The UNFCCC has the same starting date as the CBD, which in itself makes for an interesting comparison on how obligations and institutional arrangements have evolved over time. That, in turn, may provide for learning across the regimes. We can here only make a brief journey through the key aspects of the UNFCCC and the agreements that have followed under its 'shadow', highlight features linked to the legal nature of the obligations and arrangements for follow-up and review, and put them in perspective with the institutional arrangements of the CBD portrayed in this chapter.

In order to facilitate the analysis of the 23-years-long institutional development process of the CBD, some authors divide it in three blocks: phase I ($\approx$1992–2000), characterised by the definition of operational rules; phase II ($\approx$2000–2005), characterised by the formulation of the first strategic plan (2002–2010); and phase III (2005–present), characterised by the formulation of the strategic plan 2011–2020 and its allied Aichi targets and the mainstreaming of the concept of ecosystem services (which directly links human well-being to biodiversity) into the CBD[108]. Despite the recent emphasis on the importance of conserving biodiversity in order to achieve sustainable development, eradicating poverty and improving the well-being of people around the globe, the profile of biodiversity is still low in the global political agenda, as well as in many national agendas. It has been argued that governments are more likely to take action on urgent affairs with implications in the short-term, hence the lack of interest in responding to biodiversity loss[109].

The agenda to tackle climate change and regulate greenhouse gas emissions developed under the UNFCCC faces similar challenges; however, it is interesting to note that despite this, it has received more attention by far[110].

106 See Bovens, M.and Steffek, J. in note 44
107 See Mashaw, J. and Steffek, J. in note 44
108 See Neßhöver, C. *et al.* in note 11
109 Balmford, A. *et al.* (2005), "The Convention on Biological Diversity's 2010 targets", *Science*, 5707, pp. 212-213
110 Gilbert, N. (2010), "Biodiversity hope faces extinction", *Nature*, 467, p. 764

Biodiversity continues declining[111], and as illustrated in this chapter the CBD has so far failed to deliver on its objectives. Despite the multiple institutional arrangements adopted since the Convention entered into force, states have consistently failed with the responsibility to report on national progress; the degree of national implementation remains low as reflected in the inconsistent development, updating and/or implementation of NBSAPs (and above all, the declining status of biodiversity worldwide). Some authors argue that this is because of the softer character that the CBD has acquired over time[112]. However, we argue the opposite: that the CBD has never been hard, and that over time has put increasing institutional effort into developing more precise commitments, and more effective mechanisms to enhance compliance, making the obligations harder (even if within the soft spectrum of the legalization continuum[113]). For instance, having a very vague and ambiguous text as a starting-point, the CBD moved on to developing a set of very imprecise goals and targets contained in the Strategic Plan 2010, which were ultimately revised and refined (i.e., made more specific) in the Strategic Plan 2020. So it is true that the CBD has not given priority to the development of subsidiary legal protocols but rather focused on non-binding goals and targets for the operationalization of its objectives. Whilst goals and targets are considered not very useful tools for resource management (i.e., the development of concrete and implementable policies and legislations[114]), they are effective in mobilising political efforts and raising the profile of political agendas[115]. If precise and measurable, targets may assist in keeping implementation in focus[116], yet too much emphasis on formal compliance may be counterproductive as it can detract attention for overarching objectives[117]. In this regard, the Strategic Plan 2020 is noteworthily more ambitious than its predecessor – addressing biodiversity conservation as a cross-cutting issue for human well-being – and its goals and targets more precise – referring to time-bounded objectives, in some cases to measurable rates and comparable baselines, and specifically referring to sensitive (i.e., economical aspects) but urgent issues (i.e., drivers of change) with detrimental impacts on biodiversity.

Both the CBD and the UNFCCC regimes have had similarly highly ambitious, but very vague objectives from the very beginning. In both cases, the objectives became somewhat specified only 18 years after the regime was adopted. For

---

111 See Butchart, S.H.M. *et al.* and Mace, G.M. & Baillie, J.E.M. in note 5; and Pereira, H.M. in note 7
112 See Harrop, S.R. and Pritchard, D.J. in note 18
113 See Abbott, K.W. *et al.* (2000), "The Concept of Legalization", *International Organization*, 54, pp. 401–419, for a categorization of hard and soft law
114 See Harrop, S.R. and Pritchard, D.J. in note 18
115 See for example: Mace, G.M. and Baillie, J.E.M. in note 5; and Sachs, J.D. (2012), "From Millennium Development Goals to Sustainable Development Goals", *The Lancet*, 379, pp. 2206–2211
116 See for example: Maxwell, S. (1999), "International targets for poverty reduction and food security: a mildly skeptical but resolutely pragmatic view with a call for greater subsidiarity", *IDS Bulletin*, 30 (2), pp. 92–105.; and Sachs, J.D. in note 115
117 See Mace, G.M. *et al.* in note 53

the first time, in 2010, the UNFCCC agreed on a more specific objective – in this case a temperature target (2°C[118]). This target was adopted in a non-legal COP decision (as the Aichi targets); but in 2015 was even further sharpened through the Paris Agreement by referring to '[h]olding the increase in the global average temperature to well below 2°C above pre-industrial levels and to pursue efforts to limit the temperature increase to 1.5°C above pre-industrial levels' (Article 2.1a)[119], as well as by formulations on the 'aim to reach global peaking of greenhouse gas emissions as soon as possible' and to 'achieve a balance between anthropogenic emissions by sources and removals by sinks of greenhouse gases in the second half of this century' (Article 4.1). The 'apportioning' or 'allocation' of responsibilities for achieving this objective among countries have also followed a similar pattern to the CBD – indeed that it has been left entirely to countries to decide on what their respective responsibilities are.

The implementation of the objectives of the CBD formally relies on the translation of goals and targets into NBSAPs. However, as a framework convention, the CBD allows flexibility to country parties to decide on the means to do so, and on how ambitious the goals and targets are within NBSAPs. As for the SBP-2020, this specifically involves the review and updates of NBSAPs to integrate the values of biodiversity and ecosystem services into government decision-making. In turn, this depends on the engagement of heads of state, local governments and parliamentarians to gain the political support necessary to translate vague goals and targets into concrete country policy instruments. Since governments are more likely to take action on affairs they deem relevant for their own interests[120], and conservation measures have proved to have a greater impact when relevant stakeholders are involved on the ground[121], allowing parties to define their own national priorities and accordingly plan relevant measurements for the management of natural resources may encourage action.

As with the NSPABs, the UNFCCC obliges countries to '[f]ormulate, implement, publish and regularly update national and, where appropriate, regional programmes containing measures to mitigate climate change…' (Article 4b) and to report on their implementation efforts through their national communications which are obligatory for all countries albeit with different frequency. In addition all countries are obliged to periodically send in greenhouse gas inventories (Article 4a). This has in the Paris Agreement been upgraded into the obligation to send in every five years a country's Nationally Determined Contributions

---

118 United Nations (2011), Report of the Conference of the Parties on its sixteenth session, held in Cancun from 29 November to 10 December 2010, Decision 1/CP.16 *The Cancun Agreements: Outcome of the work of the Ad Hoc Working Group on Long-term Cooperative Action under the Convention*, Cancun, Mexico, United Nations Framework Convention on Climate Change.
119 UNFCCC (2015), Conference of the Parties,. Twenty-first session, Paris, 30 November to 11 December 2015, *Agenda item 4(b) Adoption of the Paris Agreement. Paris, United Nations Framework Convention on Climate Change*
120 See Harrop, S.R. and Pritchard, D.J. in note 18
121 See Glowka, L. *et al.* in note 20

(NDCs). However, the content of the UNFCCC mitigation programmes or the Paris Agreement's NDCs are entirely up to countries to determine. Some observers argue that the Kyoto Protocol adopted in 1997 (entered into force 2004) was significantly different in this regard, as it included specific emission reduction obligations for all developed country parties to it, and have referred to these obligations as being adopted 'top-down'[122]. It is easy to provide strong counter-arguments to this top-down notion. First, the obligations that countries had under the Kyoto Protocol were largely identical to what they themselves had put on the negotiation table. Second, the total emission reductions of countries under the Protocol was far away from the required measures to reach the objective of the UNFCCC. Third, international law is, per definition, not 'top-down', as it is voluntary for countries to sign on to them (and countries can also withdraw, as was done by the United States of America, Canada and Australia vis-à-vis the Kyoto Protocol).

The parties to the Paris Agreement are expected to formulate NDCs that 'reflect its highest possible ambition, reflecting its common but differentiated responsibilities and respective capabilities, in the light of different national circumstances' (article 4.3), thus leaving it to countries to determine what such highest possible ambition means. Interestingly, the Agreement includes an obligation for all countries to consider the outcome of the global stocktake every five years when they revise their NDCs: 'The outcome of the global stocktake shall inform Parties in updating and enhancing, in a nationally determined manner, their actions and support in accordance with the relevant provisions of this Agreement, as well as in enhancing international cooperation for climate action' (Article 14.3). It is indeed only a procedural obligation but the transparency framework for the Paris Agreement does prescribe that each party shall regularly provide information that is '…necessary to track progress made in implementing and achieving its nationally determined contribution' (Article 13.7b). The flexibility for countries to adopt their own targets remains in the climate regime – but the procedural obligations to do so on a regular basis – and the explicit obligation that successive NDCs have to be more ambitious than previous ones (Article 4.3) within the context of a legally binding agreement (in comparison to the CBD COP decisions for its Strategic Plan) bodes for at least higher political accountability, if not legal accountability for the climate regime.

In this regard, the CBD has also provided a flexible framework for implementation (i.e., global goals, targets and indicators for monitoring and reporting), so that parties have a stance to define how they will contribute to the achievement of global goals according to their own national priorities and circumstances. Most importantly, the CBD has also put effort into strengthening and establishing mechanisms to enable and promote compliance. Concerning mechanisms to encourage action of states beyond their capabilities, the CBD has: backed up the Strategy on Resource Mobilisation with a follow-up mechanism; established the SBI and given the mandate

---

122 Bodansky, D. (2016), "The Paris Climate Change Agreement: A New Hope?", *American Journal of International Law,* 110 (2), pp. 288–319

to support the COP in the review of the Convention; increased the frequency of interim evaluations on progress towards meeting the Aichi targets; and emphasised the need to assess compliance at the national level with the consideration of a peer-review process on the development, update and implementation of NBSAPs. From the lens of accountability, the latter is a much-needed process – even if proposed on a voluntary basis – and a very welcome one when it begins to be adopted.

The UNFCCC regime, in contrast with the CBD, set up its Subsidiary Body on Implementation (SBI) from the beginning; this body meets twice every year, since 1997. Its mandate is to assist the COP in the 'assessment and review of the effective implementation of the Convention' (UNFCCC article 10.1) and under the guidance of the COP, it shall 'assess the overall aggregated effect of the steps taken by the Parties in the light of the latest scientific assessments concerning climate change' (UNFCCC Article 10.2a). The follow-up and review of individual countries' actions under the regime was limited even if there were reporting requirements for all countries (with differentiated frequencies) to send in national communications on actions taken.

In 2010, the COP of the UNFCCC adopted a more detailed approach, the Monitoring, Reporting and Verification (MRV) system; it included all countries, albeit in a bifurcated manner. This was the result of negotiations launched through the mandate agreed upon in 2007 to set up an MRV system meant also for developing countries. As per the system, all countries are asked to submit reports biannually (in addition to their national communications), which will be subject to technical review. However, the reports have different remits for developed and developing country parties[123]. Subsequently, developed countries go through an International Assessment and Review (IAR) process, and developing countries a considerably lighter International Consultation and Analysis (ICA)[124]. The IAR and ICA take the form of each country making a public presentation of their reports at the SBI meetings, and a process of submission of written questions by other parties has preceded this[125]. For developing countries important elements of the report, in addition to greenhouse gas inventories and mitigation actions, are constraints and gaps, including support needed and received.

The provisions for follow-up and review of individual country's actions in the Paris Agreement are described primarily in Articles 13 and 15. On the one hand, Article 13 on the transparency framework outlines that 'each Party shall participate in a facilitative, multilateral consideration of progress with respect to efforts under Article 9 [for developed country parties, this concerns their financial contributions], and its respective implementation and achievement of its nationally determined contribution' (Article 13.11). Such a multilateral consideration is based on information provided by parties on mitigation and

---

123 For details see http://unfccc.int/national_reports/items/1408.php
124 UNFCCC (2011), Report of the Conference of the Parties on its sixteenth session, held in Cancun from 29 November to 10 December 2010, Cancun, United Nations Framework Convention on Climate Change, paras. 42–63
125 We do not describe the procedures for reporting and accounting of developed country parties under the Kyoto Protocol here

finance (the latter only for developed countries), information that will undergo technical review. The IAR and ICA have only been in place for a few years yet these processes, together with the 'older' elements of the follow-up under the UNFCCC 'shall form part of the experience drawn upon for the development of the modalities, procedures and guidelines' of the transparency framework for the Paris Agreement (Article 13.4).

In addition to the process under the transparency framework (Article 13), Article 15 outlines the mandate of an expert-based committee as being to 'enhance implementation and promote compliance' (Article 15). This committee will operate in a way that is 'facilitative, non-intrusive, non-punitive and respectful of national sovereignty' (Article 15). While there is considerable work remaining to provide the operational details of how these processes will be institutionalized, it seems that follow-up and review of individual countries will take place both in a political and an expert-based arena, both of which will be underpinned by the ethos of facilitation (rather than sanctions). In addition, there is a review framework of progress towards the global goals based on stocktaking every five years, assessing 'collective progress towards achieving the purpose... and its long-term goal' (Article 14.1).

The existence of an 'implementation and compliance committee' under the Paris Agreement was not an obvious outcome of the negotiations, particularly not one where there is no differentiation included in its mandate between different categories of countries. When the last two weeks of negotiations started in Paris in December 2015 there were still a wide range of options on how to deal with compliance on the table[126]. The very name of the committee also reveals the intention that it will deal with not only the explicitly legally binding elements of the Paris Agreement – but also the other elements for which the term implementation is used[127]. Rajamani considers that the Paris Agreement, 'establishes a rigorous system of oversight to ensure effective implementation of the many requirements it places on Parties'[128]. It is difficult to judge if this system will indeed be so rigorous, and whether it will be able to facilitate implementation and compliance.

In all, despite similarities in the framework character of both Conventions – which as described in this chapter concedes flexibility to parties to decide both on the 'size' of their obligations/commitments and on the means to comply with their obligations and/or implement their commitments – the UNFCCC is several steps ahead of the CBD. We ground this conclusion not only on the basis of the 'harder' legal approach followed by the UNFCCC – for we have argued that compliance with international norms is also the result of reciprocity processes, reputational sanctions, learnt-lessons dynamics over time and capacity-building[129] – but because the UNFCCC has evolved more robust mechanisms

---

126 Voigt, C. (2016), "The Compliance and Implementation Mechanism of the Paris Agreement", *RECEIL*, 25 (2)

127 Ibid.

128 See Rajamani, L. (2016), "Ambition and Differentiation in the 2015 Paris Agreement: Interpretive Possibilities and Underlying Politics", *International and Comparative Law Quarterly*, pp. 1–25

129 See note 75

Strengthening compliance 103

to enhance compliance, and at a faster pace than the CBD. For instance, whilst the CBD is still developing its first review system (i.e., the voluntary peer-review system for the development, update and implementation of NBSAPs under consideration of the recently established SBI), the UNFCCC has through its SBI had a longer emphasis on review of implementation, as well as a review system for both developing and developed countries (until the Paris Agreement with well-differentiated responsibilities and pathways). Yet we emphasise that both the CBD and the UNFCCC have undertaken important institutional measures to strengthen compliance and/or implementation of parties in each regime. As for the CBD, although moving slowly and following more of a soft-law track, it seems to be directing efforts towards: the recognition of its own institutional limitations and needs; learning from its own experience and that of other regimes, such as the UNFCCC[130]; and accordingly shaping further mechanisms to enhance compliance. Despite these significant advances, the new measures are probably still far from sufficient to catalyse a real shift towards states' compliance and/or implementation. Therefore, whether these efforts are adequate to address the increasing pressures on biodiversity – and the threats these pose for the well-being of people worldwide – remains to be seen.

---

130 For instance, the methodology of the peer-review mechanism was based on, *inter alia*, the United Nations Framework Convention on Climate Change In-Depth Reviews of National Communications and specific national review processes, and the United Nations Human Rights Council universal periodic review as noted in COP/12/25/Add.3 and COP/12/INF/24. For detailed information see: United Nations Environmental Programme, Convention on Biological Diversity, Note by the Executive General: *Voluntary Peer-Review Mechanism for National Biodiversity Strategies and Action Plans.* UNEP/CBD/SBI/1/10/Add.1 (18 March 2016); available from: www.cbd.int/doc/meetings/sbi/sbi-01/official/sbi-01-10-add1-en.pdf (accessed 12 November 2016)

6 Five short words and a moral reckoning

The Paris regime's CMA-APA equity stocktake process

Hugh Breakey

The Paris Agreement developed at the 21st session of the Conference of the Parties (COP 21) in December 2015 included the prospect of a 'Global Stocktake' in 2023. The stocktake will assess the collective progress towards achieving the Agreement's purpose *in the light of equity*[1]. These five short words constitute the strongest-ever indication that some type of formal, equity-based consideration of states' Nationally Determined Contributions (NDCs) could occur. In simple terms, the stocktake will officially consider how well countries' climate commitments accord with the Convention's principles. Are countries doing what they promised? *And are they promising enough?*

I will argue that the stocktake may make up a significant part of the Paris Agreement's 'pledge and review' system, allowing moral suasion to get laggards to lift their game, while showcasing the best practices of the top performers. Achieving these benefits will not be easy: such a process threatens serious pitfalls as much as it promises vital advantages.

This chapter investigates how the stocktake might harness the power of moral language and argument – and avoid the risks of over-moralization, divisiveness and acrimony. I begin in Section 1 by explaining the official mandate the Paris Agreement sets down for the global stocktake, focusing especially on its equity dimensions. In Section 2, I explore a potential tension in the mandate between assessing 'collective progress' and 'progress in the light of equity'. To resolve the tension, Section 3 considers three desiderata for the stocktake, and argues that avoiding any serious engagement with the equity-dimensions of any individual party's NDCs will condemn the stocktake to irrelevance. However, serious engagement with such equity issues poses significant challenges. Section 4 considers the key challenges before Section 5 considers some promising ways forward.

Section 1: the stocktake mandated by the Paris Agreement

The 2015 Paris Agreement aims, by enhancing implementation of the existing 1992 Framework Convention[2], 'to strengthen the global response to the threat

---

1 Conference of the Parties (COP), "COP 21 Paris Agreement", Paris: UNFCCC, 2015, Art. 24(1)
2 United Nations (UN), "United Nations Framework Convention on Climate Change (UNFCCC)", Rio de Janeiro, 1992

of climate change, in the context of sustainable development and efforts to eradicate poverty'[3]. As well as increasing capacities for adaptation and resilience, the Agreement sets a 'temperature goal'[4] aiming to hold 'the increase in the global average temperature to well below 2°C above pre-industrial levels and to pursue efforts to limit the temperature increase to 1.5°C above pre-industrial levels'. Echoing the Framework Convention's Objective and Principles[5], the Paris Agreement declares that it 'will be implemented to reflect equity and the principle of common but differentiated responsibilities and respective capabilities, in the light of different national circumstances'.

While endorsing the pre-existing climate regime's goals and principles, the Paris Agreement nevertheless ushered in a new model for climate action. The previous regime, based on the Kyoto Protocol, was a 'top-down' model where decisions about mitigation targets were (to a significant extent) made by the collective as a whole. The regime established by the Paris Agreement instead works on a 'bottom-up' basis, where each country (as a party to the Agreement) is entitled to decide its own level of contribution to the global response to climate change. Under the regime, each party publicly declares and submits its own 'Nationally Determined Contribution' (NDC), which it is then obliged to implement domestically. The over-arching model is thus a 'pledge and review' system, whereby each party makes initial pledges, which over time are ratcheted up – and never down[6].

The global stocktake plays a potentially significant role within this pledge-and-review system. The Paris Agreement declares that the relevant body (the CMA[7]):

> Shall periodically take stock of the implementation of this Agreement to assess the collective progress towards achieving the purpose of this Agreement and its long-term goals (referred to as the "global stocktake"). It shall do so in a comprehensive and facilitative manner, considering mitigation, adaptation and the means of implementation and support, and in the light of equity and the best available science.[8]

The following sections will explore Article 14's notions of 'collective progress', 'equity' and 'facilitative manner'. For now, the point is that the stocktake is intended to play a role in parties updating and enhancing their national

---

3 (COP), "COP 21 Paris Agreement", Art. 2
4 This is the language used in, e.g., "COP 21 Paris Agreement", Art. 4(7)
5 (UN), "United Nations Framework Convention on Climate Change (UNFCCC)", Arts 2 & 3. The Convention's Objectives include the goals of food production and sustainable development. Its Principles refer to many of the equity principles discussed in Section 2 below
6 (COP), "COP 21 Paris Agreement", Art. 4(3)
7 The CMA is the 'Conference of the Parties serving as the meeting of the Parties to the Paris Agreement'. The Paris Agreement was concluded by the COP (the Conference of the Parties to the Framework Convention), which is the supreme decision-making body of the 1992 Convention
8 (COP), "COP 21 Paris Agreement", Art. 14(1)

actions and support of the Paris Agreement[9]. It thus assists the pledge-and-review regime's 'ratchet'.

The stocktake's enumerated tasks emphasize issues of accounting, science and methodological consistency[10]. However, the global stocktake also implicates considerations about equity. There are several ways in which the assessment is infused with equity-considerations:

1) The stocktake's mandate explicitly requires that its assessment will be done 'in the light of equity'[11].
2) The stocktake's mandate requires it assesses the collective progress towards achieving the Paris Agreement's purpose:
   a) The Agreement's purpose includes equity-based considerations such as sustainable development and efforts to eradicate poverty[12].
   b) The Paris Agreement aims to enhance the implementation of the Framework Convention, whose objective and principles include explicit reference to equity principles[13].

These three considerations provide a *prima facie* case for thinking that the global stocktake will include significant equity-based features and assessments. (We will shortly turn to a complicating consideration in the mandate.)

In terms of implementation, the Paris outcome documents enrol several sub-institutions in the global stocktake. Centrally, the Ad Hoc Working Group on the Paris Agreement (APA)[14] is directed to identify 'sources of input' for the global stocktake – including (but not limited to) IPCC reports[15]. Figure 6.1 summarizes the parts of the regime and their interactions in the stocktake[16].

Given these institutions and their roles in the stocktake – and the above observations about the role of equity within the stocktake – I will refer to the

---

9 "COP 21 Paris Agreement", Art. 14(3)
10 This is clear in the COP 21 Decision on the adoption of the Paris Agreement (hereafter 'Paris COP Decision'), where none of the global stocktake paragraphs explicitly reference equity. "Paris COP Decision", Paris: UNFCCC, 2015, paras. 99–101
11 "COP 21 Paris Agreement", Art. 14
12 "COP 21 Paris Agreement", Art. 2(1). Article 2 goes on to state in its second clause that 'the Agreement will be implemented to reflect equity and the principle of common but differentiated responsibilities and respective capabilities, in the light of different national circumstances'
13 "COP21 Paris Agreement", Art.2 (1). The Framework Convention's Objectives and Principles are given at: UNFCCC, "United Nations Framework Convention on Climate Change", Rio: United Nations, 1992, Arts. 2, 3
14 (COP), "Paris COP Decision", para. 7
15 "Paris COP Decision", para. 99. The APA is also tasked with developing 'features of the nationally determined contributions for consideration and adoption...' by the CMA "Paris COP Decision", para. 26. This task could help inform the stocktake.
16 Figure 6.1 does not include the 2018 Facilitative Dialogue mandated in the Paris COP 21 Decision (para. 20). The Dialogue's mandate is not explicitly linked to the stocktake. However, its function is similar ('to take stock of the collective efforts of Parties in relation to the long-term goal...'). In what follows, I hope to show that a lengthy, structured process is called for, and therefore that the 2018 Facilitative Dialogue could play an important role. (See especially text to note 65)

The CMA-APA equity stocktake process 107

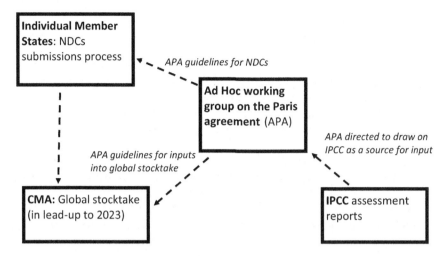

Figure 6.1 Official actors in the stocktake process

equity-based part of the stocktake as the 'CMA-APA equity stocktake process' – or 'equity stocktake process' for short[17].

Section 2: equity vs. 'collective progress'

The stocktake is mandated to assess the collective progress towards achieving the Agreement's goals[18]. This reference to 'collective progress' suggests that the stocktake will focus less on the contributions and situations of individual parties, and more on an overall question of how the collective is going in progressing towards the Agreement's aims.

This directive, however, appears to clash with the stocktake's mandate for considering equity issues. To see why this is so, we need to better understand the meaning(s) of 'equity'.

The meaning of 'equity'

Equity is a contested term. While significant to the climate regime, the term is nowhere given an authoritative definition[19]. Despite this contested quality, a set of staple meanings does arise in the literature[20]. Distinct normative principles falling

---

17 I term it a 'process' because, as we will see, the task may require a staged progression of action, dialogue and deliberation – rather than a 'snapshot' assessment. See below note 74
18 (COP), "COP 21 Paris Agreement"
19 See, e.g., its use at: UNFCCC, "United Nations Framework Convention on Climate Change", Art. 3(1)
20 See, e.g., Mattoo, A. and Subramanian, A. (2012), "Equity in Climate Change: An Analytical Review", *World Development,* 40, (6); Pickering, J., Vanderheiden, S. and Miller, S. (2012),

under the banner of equity, each bolstered by various parts of the UNFCCC documents and outcomes, include:[21]

I) *Intergenerational equity*[22]: Intergenerational equity involves respect for future generations' needs and entitlements. It links closely with the 'sustainability' part of 'sustainable development', and with mitigation and long-term adaptation ambitions – for these are what will equip future generations with a world fit for human flourishing.

II) *Intra-generational equity*: Intra-generational equity governs fair, distributive allocations of burdens and/or entitlements to existing people or nations. One popular understanding of intra-generational equity involves granting each individual person an equal-per-capita emissions entitlement (which is usually then granted to their nation on their behalf).

III) *Need-based equity*: Need-based equity, also termed 'development rights', highlights that some individuals and countries have a more desperate need for economic development (and its corollary carbon-emissions). States with crushing poverty need urgent development just to ensure basic human rights.

IV) *Capacity-based equity*: The capacity principle holds that those who currently enjoy the resources and wealth allowing them to more easily respond to the problem should shoulder the greatest load in terms of climate burdens. This principle reflects the widespread use of 'progressive' taxation, where higher incomes in principle attract higher proportional rates of tax contribution.

V) *Historical responsibility*: Historical responsibility requires that the aggregate of previous emissions be factored into current emissions allocations, in the sense that those actors who have already used up large parts of the

---

'"If Equity's in, We're Out': Scope for Fairness in the Next Global Climate Agreement", *Ethics & International Affairs*, 26 (04); Lange, A., Vogt, C. and Ziegler, A. (2007), "On the Importance of Equity in International Climate Policy: An Empirical Analysis", *Energy Economics*, 29; Intergovernmental Panel on Climate Change (IPCC) (2014), "Fifth Assessment Report – Synthesis Report", Chs. 3 and 4. Generally these principles are understood as moral rather than legal, though see also: Maguire, R. (2012), "Incorporating International Environmental Legal Principles into Future Climate Change Instruments", *Carbon and Climate Law Review*, 4 (2); International Law Association (ILA) (2014), "Legal Principles Relating to Climate Change", in *Report of the International Law Association's Committee on Legal Principles Relating to Climate Change*, Washington: ILA

21 Some commentators include an additional climate equity principle: 'equitable adjustment costs'. (E.g., Mattoo and Subramanian, "Equity in Climate Change" p. 1086, see note 20) This principle aligns with proposals and NDCs that consider mitigation cuts from an earlier (e.g., 1990) baseline. The implied principle may be presented thus: *Equitable adjustment costs*: historical path-dependency makes a shift to low-carbon economies harder for many countries. While all states must eventually converge to a fair distribution, current high-emitters need the space to avoid substantial socio-economic disruption triggered by an over-hasty transition to low-carbon. I resist, however, including this principle in the main list above, as there is little support in the FCCC instruments themselves linking this approach with invocations of equity.

22 Inter-generational equity is the only type of equity explicitly mentioned in Paris Agreement (Preamble)

global atmospheric sink (or those subjects who inherited benefits from their antecedents doing so) must now be restricted in their use[23].

Some of these principles align well with one another. For example, capacity-based equity functions as a normative silhouette of needs-based equity; the former picks out rich people to carry the main burdens, and the latter picks out poor people to hold the main entitlements. However, other principles provide at least somewhat conflicting allocations of burdens and entitlements[24].

Collective progress versus equity's differentiations

The above equity principles share one thing in common: they are all norms of *distributive justice*. That is, they all speak about how burdens, benefits, entitlements and risks should be shared (distributed) across a continuing population. A given state of affairs will accord with equity when entitlements and burdens are shared across the population's individuals in accordance with equity's prescriptions[25]. It follows that an assessment *in the light of equity* must take a granular view, looking at how burdens and entitlements fall upon each member of the group.

It is here that a *prima facie* tension arises with the stocktake's mandated assessment of collective progress. This phrase suggests an aggregated – and not granular – assessment of how the overall group is performing. For example, an assessment of collective progress towards the temperature goal could simply tally up all nations' contributions to mitigation, and assess how far these contributions cumulatively progressed the world towards achieving the temperature goal.

Yet the meaning of 'collective progress' in the *light of equity*, and towards *objectives that are themselves infused with equity*, is much murkier. Because of this ambiguity, I will presume in what follows that the notion of *collective* progress creates enough 'wiggle-room' for the CMA – if it so decides – to effectively neuter any serious engagement with equity's distributive demands. In the following section, I will argue that there are good reasons to pursue the reverse course, and to seriously (but intelligently) engage with equity.

Before doing so, however, it needs to be stressed that there is only a potential tension – and not a flat contradiction – in calling for an assessment of collective progress towards an equity-based distribution. For example, it is quite coherent to ask how a nation is currently progressing – as a collective – towards the morally worthy goal of eradicating extreme poverty. The ensuing analysis would perforce

---

23 Debate surrounds whether previous emissions should be counted from 1990 (when the problem become widely acknowledged) or earlier (e.g., 1850)
24 Mattoo and Subramanian, "Equity in Climate Change", pp. 1090–92, see note 20
25 Instead of 'individuals', it may sometimes be appropriate to refer to 'groups' if (and *only if*) that group shares similar characteristics as viewed through the lens of the particular equity principle in question. Existing blocs of nation-states, however, should not be assumed to share the relevant similarities. E.g., there are striking variations within developing countries on their historical contributions to carbon emissions, see Weisbach, D.A. (2012), "Negligence, Strict Liability, and Responsibility for Climate Change", *Iowa Law Review*, 97 (2)

need to look at the situation of individual people – it would not do to simply note that the nation's gross domestic product (GDP) is improving, which may just mean that the rich are getting richer. Instead, the analysis would have to look at individual wealth: the more individual people in the extreme-poverty financial bracket, the worse the situation. The assessment of collective progress will thus hinge on the nation's overall achievements in lowering the number of individuals in that bracket.

Similarly, assessing collective progress in the light of equity, and towards equity-infused objectives, requires looking at individual nations, and assessing whether their existing burdens, entitlements, benefits and risks accord with equity's prescriptions. These myriad granular assessments would then be used to develop, explain and justify the stocktake's assessment of how the overall collective is progressing towards its stated objectives. Because its ultimate assessment is on how the collective is performing, the stocktake would not explicitly 'name and shame' individual nations performing poorly in equity terms. Instead, the assessment would diffuse responsibility for poor performance across the group; in assessing collective progress, it would speak about how *we* (the international community as a whole) are performing and about what *we* can do to improve things – such as supporting those countries struggling to play their role. Within the development, explanation and justification of the stocktake's overall assessment, however, would lie information about individual countries and the way their equity-related activities contributed to – or impacted deleteriously on – the final assessment. In other words, the assessment would say that the *collective* is performing poorly overall because (individual parties) X, Y and Z have failed to help (other individual parties) A, B and C.

If this elucidation of what it means to assess 'collective progress towards an equity-infused (distributive) goal' is correct, then an attempt to strictly follow the Paris Agreement's mandate will largely accord with the results of my prescriptive arguments below. In the following sections, I argue that the best way forward for the stocktake requires engagement with individual countries' equity-related efforts, but that explicit, individualized allocations of blame – such as 'naming and shaming' – should be avoided.

Section 3: ways forward and three desiderata

While the stocktake's nature may be clear enough in general terms, there is much yet to be resolved in terms of the logistics and processes by which it will function. As well, as we saw in the previous section, the stocktake's engagement with equity is unclear.

In assessing how the stocktake's equity stocktake process should be implemented, consider the following desiderata for judging its success[26]:

26 The three desiderata might be achieved directly, stemming from activities within the CMA-APA equity stocktake process itself, or indirectly – for example if the process stimulated increased domestic dialogue and awareness of the moral issues resonant in climate change policies

A) *Improved ambition:* The process leads to an overall improved effort in mitigation and adaptation efforts and for increased finance for loss-and-damage.
B) *Improved ethics:* The process provides an improved moral dialogue; it is more transparent, inclusive and deliberative. Furthermore, it leads to fairer outcomes, such as that: vulnerable countries are supported in transitioning their economies, developing countries improve governance to ensure benefits filter down to their citizenry, and developed countries shoulder the largest burdens of mitigation and financing.
C) *Increased subjective legitimacy:* The process leads to a widespread belief in the overall fairness of the climate regime. National delegates and domestic constituents feel that other countries are engaging in good faith and on the basis of reasonable (if differing) moral perspectives. They think that the agreement, even if not perfectly fair, nevertheless is legitimate.

These three desiderata all accord with, and promote, the objectives of both the Paris Agreement and the over arching Framework Convention. If the stocktake could fulfil these desiderata, then it would play an important role within the larger pledge-and-review system. After all, any pledge-and-review model requires a process of review as much as a mechanism for pledges. Within the new climate regime, the stocktake stands as the major formal process where such review could take place in a systematic and principled way. For this reason, it may turn out that the stocktake fulfilling these desiderata constitutes a *sine qua non* of an effective climate regime. As is well-known, the present situation is not satisfactory: the cumulative result of all the NDCs submitted before Paris – *even if they were all implemented and achieved* – would still leave the world headed for warming well above 2°C. Meanwhile, scientists argue that warming even well below 2°C may not prove safe[27]. As such, the improved ambition, fairness and subjective legitimacy of the regime, driven by an effective and constructive review process, may prove vital to global efforts to quell dangerous climate warming.

Suppose, then, that we agree on the reasonableness of these three desiderata. How may the stocktake – and in particular its engagement with equity – achieve these good results? Simplifying, there are three sorts of potential outcomes from the CMA-APA equity stocktake process: namely, that the process is destructive, beneficial or irrelevant[28].

Destructive: The CMA-APA equity stocktake process could set back each of the above desiderata if it foments divisions and encourages unproductive blame and accusation. This result would not be surprising. The shift from the top-down Kyoto model (where allocations were largely determined by the collective) to the current bottom-up Paris regime (where each party selects its own level of contribution) was

---

27 Hansen, J. *et al.* (2013), "Assessing 'Dangerous Climate Change': Required Reduction of Carbon Emissions to Protect Young People, Future Generations and Nature", *PLoS ONE*, 8 (12)
28 Of course, the stocktake process will likely have mixed results; good in some respects, and bad in others

motivated precisely to avoid divisive decisions on a fair allocation. The equity dialogue might resurrect such fruitless wrangling.

This divisive result would be especially disastrous if the Paris regime was actually working. In the lead-up to 2023, most countries may have achieved their initial 2015 NDC commitments, ratcheted up from these in 2020, and be preparing for a further increase (in 2025). While this may not put the world on track for containing global warming at 1.5°C (or even 2.0°C), the collective achievements may have generated momentum and a social context supportive of future efforts[29]. After all, nothing succeeds like success. In such a scenario, all states may judge that it is in their interests to contribute[30]. A process of dialogue that reignites moral accusation and calls attention to how far efforts fall short of objective moral demands might undermine this momentum[31].

Beneficial: The CMA-APA equity stocktake process has the potential to improve all three of the desiderata (ambition, ethics and subjective legitimacy). These benefits might begin well before 2023. With an eye towards preparing for the eventual stocktake, states may ramp up their efforts, and direct their attention to areas that an equity-based process will foreground. For developing countries, these areas will include undertaking improvements in governance, transparency and respect for human rights in undertaking mitigation and adaptation activities. For developed countries, these areas may centre on finance and technology-transfer as much as domestic mitigation. Later, in the lead-up to 2023, the dialogue and deliberation surrounding the stocktake process might encourage states to see that others are engaging in good faith on the basis of reasonable positions.

Irrelevant: The CMA-APA equity stocktake process could fail to impact upon the Paris/NDC regime. Irrelevance could occur for different reasons. If it threatened divisiveness, influential actors might strategically ignore the entire process. Alternatively, the equity stocktake process might prove irrelevant because

---

29 A belief that changing an outcome is possible impacts on the willingness to consider that outcome from a moral perspective. So, too, language of collective effort and risk can be less fraught than ethical language. (See Pickering, J. (2016), "Moral Language in Climate Politics", in *Climate Justice in a Non-Ideal World*, Roser, D. and Heyward, C. (eds.), Oxford: Oxford University Press; and Täuber, S., van Zomeren, M. and Kutlaca, M. (2015), "Should the Moral Core of Climate Issues Be Emphasized or Downplayed in Public Discourse? Three Ways to Successfully Manage the Double-Edged Sword of Moral Communication", *Climatic Change*, 130.) Both these reasons suggest that in the best-case scenario where the Paris regime is working, there may be something to be said for letting sleeping ethical dogs lie

30 I.e. fulfilling the Paretian condition motivating many international treaties. See Idil Boran, I. (2016 In Press), "Principles of Public Reason in the UNFCCC: Rethinking the Equity Framework", *Science and Engineering Ethics*. That is, even if parties consider the agreement is less than what justice requires, they may still benefit from it as compared with the status quo

31 In terms I have previously employed, the worry would be that in an effort to secure normative motivation from objective and deliberative grounds for climate norms and policies, we lose the drive for compliance (that by 2023 might be) supplied by communitarian and pragmatic grounds. See Breakey, H. (2015), "Heating up Climate Change Norms – Lessons from Human Rights", in *Ethical Values and the Integrity of the Climate Change Regime*, Breakey, H., Popovski, V. and Maguire, R. (eds.), Surrey: Ashgate, pp. 234–36

parties perceive it as naïve and idealistic (unable to confront the problems of an imperfect world) or as technical and abstruse (an exercise in esoteric and/or technical moral philosophy).

Another way of condemning the process to irrelevance would be to avoid any substantive engagement with equity principles and their applicability to individual parties. I turn now to consider this possibility.

Collective progress and equity-irrelevance

As observed earlier, the reference to assessing 'collective progress' may create enough 'wiggle-room' for the CMA to avoid any serious engagement with equity issues. Equity's inherent controversies may motivate many parties to press for such avoidance. On the other hand, many *other* parties may judge that they would benefit from the process focusing more on equity issues. For example, since all the equity principles require early and significant action by developed countries, developing countries may take the view that (on both moral and practical grounds) the stocktake must highlight equity-based demands.

It is not easy to foresee where different blocs will stand on this issue, or the importance to which they will grant it. What we can say with some certainty is that a refusal to engage seriously with issues of equity will render the stocktake largely ineffectual on all the above desiderata. If it confines itself to purely aggregative considerations, the stocktake will presumably work rather like the ADP's 2015 *Synthesis Report* on Intended Nationally Determined Contributions (INDCs) provided in the lead-up to COP 21[32]. The *Synthesis Report* provided an aggregate estimate of projected emission levels given the implementation of existing INDCs. Consistent with its mandate, the *Report* only considered equity and fairness in the sense of synthesizing and relating information on fairness that countries provided with their INDCs[33]. Given the needs of the international community at that time, the *Report* helpfully highlighted how far the collective was from setting a path towards its temperature goals. Future *Synthesis Reports*, as well as future *IPCC Assessment Reports*, will doubtless perform a similar function. If the stocktake process is to improve on this business-as-usual situation, especially in terms of improving the Paris regime's fairness and perceived legitimacy, then following its mandate and engaging seriously with equity issues stands as a key mechanism to this end. The obvious weakness of a voluntary 'bottom-up' system is that laggards can free-ride on others' efforts while confecting specious claims about their own righteousness. As well as their own lack of contribution, their behaviour can strip momentum from global efforts and stunt the overall level of ambition. Engaging seriously with equity offers the promise of a review mechanism

---

32 Ad Hoc Working Group on the Durban Platform for Enhanced Action (ADP), "Synthesis Report on the Aggregate Effect of the Intended Nationally Determined Contributions", UNFCCC (2015)
33 "Synthesis Report on the Aggregate Effect of the Intended Nationally Determined Contributions", paras. 163–70

114 Hugh Breakey

capable of drawing attention – in a principled and reasonable way, and based on agreed criteria – to the best practices of top performers, and to those who are not pulling their weight.

But is there genuine promise in focusing on equity? In the best-case scenario, the process could help the regime capture the desirable parts of both bottom-up and top-down models[34]. Top-down models tend to do well on substantive (for example, distributive justice) legitimacy grounds, as they can approach an overall collective determination of fair burden-sharing. The same models, however, can perform poorly on other grounds, as they may say little about procedural, consensual and deliberative legitimacy[35]. The converse holds for bottom-up models, which can showcase inclusive, consensual and procedural virtues – but at the cost of eschewing substantive equity outcomes.

Adding the CMA-APA equity stocktake process to the existing superstructure, the resulting regime could hope to boast virtues from both models:

- *Bottom-up virtues*: The existing NDC system of individual pledges allows improved inclusiveness, participation, flexibility, dynamism and sovereign consent.
- *Top-down virtues*: The process will create pressure for parties to augment the influence of equity principles by focusing on improved distributive outcomes.
- *Combined virtues*: Situated at the juncture of bottom-up independence and top-down moral suasion, the stocktake process could improve accountability and transparency. By standardizing requirements about explanations of fairness within NDCs, and then subjecting them to interrogation (for example, how well does the party's NDC commitment actually accord with its invoked equity principle?), the stocktake process can make states' commitments more transparent, and hold them accountable for equity-failures. So, too, the stocktake process could boast improved deliberative legitimacy[36]. Instead of parties merely invoking equity principles as self-serving rhetoric, they could be required to justify the interpretation of their equity principles, leading to actual deliberation about the proper scope and demands of fairness[37].

---

34 See Pickering, J. (2015), "Top-Down Proposals for Sharing the Global Climate Policy Effort Fairly: Lost in Translation in a Bottom-up World?", in *Ethical Values and the Integrity of the Climate Change Regime*, Breakey, H., Popovski, V. and Maguire, R. (eds.), Surrey: Ashgate; Boran, "Principles of Public Reason.", see note 30; Breakey, "Heating up Climate Change Norms", see note 31

35 As Boran notes, even though the bottom-up model was chosen on pragmatic grounds of political feasibility, it can, nevertheless, capture key procedural virtues that eluded the top-down model. Boran, "Principles of Public Reason", see note 30. On the various modes of legitimacy available to international norms, see Breakey, "Heating up Climate Change Norms", see note 31; Bodansky, D. (1999), "The Legitimacy of International Governance: A Coming Challenge for International Environmental Law?", *American Journal of International Law*, 93 (3)

36 Parties might even agree on reasonable procedural-deliberative processes. See Boran, "Principles of Public Reason", see note 30

37 Thomas Risse would call this a 'logic of arguing' rather than a 'logic of appropriateness'. Risse, T. (2000), '"Let's Argue!': Communicative Action in World Politics", *International Organization*,

This reasoning might all seem too good to be true. Rather than seizing the virtues of both approaches, perhaps the hybrid model will suffer from both models' inherent *problems*. In particular, why wouldn't the introduction of equity principles inherent in the top-down approach necessarily reignite all the problems bedevilling the previous (Kyoto) top-down approach?

While I list some finer-grained challenges and recommendations in the following two sections, here I highlight three features of the equity stocktake process that mitigate the main problems plaguing top-down models.

1) Procedural virtues tend to be easier to secure agreement on, as compared to distributive principles[38]. This offers hope for agreements on how a deliberative process about equity should work, even while parties retained differences over substantive equity principles[39].

2) The Kyoto model required committing to a specific outcome, which necessarily privileged some (combination of) equity principles[40]. The equity stocktake process could still improve outcomes even if it allowed a plurality of equity principles, from which each country could select those they view as most reasonable.

3) States are notoriously sensitive about being bound by externally enforced, black-letter, hard law obligations[41]. Conversely, they prove more open to mere moral censure, even on the basis of agreed standards and in formal institutional settings.[42] Because the stocktake process can – even in its most thoroughgoing application – only lay down moral prescriptions rather than legal duties, states may prove more willing to engage with it.

---

54 (1). See more generally, Crawford, N.C. (2009), "Homo Politicus and Argument (Nearly) All the Way Down: Persuasion in Politics", *Perspectives on Politics*, 7 (1), and on this process in the context of climate change, Brown, D. (2015), "How to Assure That Nations Consider Ethics and Justice in Climate Change Policy Formulation", in *Ethical Values and the Integrity of the Climate Change Regime*, Breakey, H., Popovski, V. and Maguire, R. (eds.), Surrey: Ashgate

38 This may be because claims of objective distributive justice seem to carry a stronger metaphysical baggage (more philosophical depth and specificity) than procedural/discursive principles. This claim is sometimes suggested in human rights contexts: see Cohen, J. (2004), "Minimalism About Human Rights: The Most We Can Hope For?", *The Journal of Political Philosophy*, 12 (2), p. 193. For our purposes here, I merely appeal to the empirical fact that it often turns out to be easier to secure agreement on procedural rather than substantive justice. E.g., mainstream political parties in liberal democracies differ across the political spectrum on questions of distributive justice – but they usually acknowledge that democratic decision-making constitutes the appropriate way to manage those disagreements. (See below text to note 44)

39 See Boran, "Principles of Public Reason", note 30; Pickering, "Top-Down Proposals", note 34

40 However, it did not explicitly set down the particular principles driving the result, nor require parties to justify their targets in equity terms. See the below discussion of *Completely Theorized Agreement* as to why this may have been a helpful feature of the Kyoto regime

41 See Breakey, H. (2014), "Parsing UN Security Council Resolutions: A Five-Dimensional Taxonomy of Normative Properties", in *The Security Council as Global Legislator*, eds. Vesselin Popovski and Trudy Fraser, Abingdon: Routledge

42 See, in the context of human rights, Breakey, H. (2015), "What Human Rights Aren't For: Human Rights Function as Moral, Political and Legal Standards – but Not as Intervention-Conditions", *Research in Ethical Issues in Organizations*, 13

116 Hugh Breakey

For these three reasons, the CMA-APA equity stocktake process might prove tolerable, even for parties who rejected the Kyoto model.

The next two sections proceed on this assumption. Given that there are genuine benefits to be garnered by an equity-infused stocktake process, how can these benefits be realized, and the risks managed?

Section 4: challenges to a facilitative dialogue and a beneficial outcome

An array of challenges faces those trying to achieve constructive dialogue and beneficial outcomes. The main challenges confronting the implementation of a facilitative equity stocktake process include:

- *Moral Disagreement*: Countries disagree on what equity requires. Inasmuch as the stocktake process needs to weigh into such debates, it may be perceived as favouring one 'side' against another.
- *The Blame Game*: Disagreements on ethics prove particularly divisive when one side places blame on another (for example, delegations faulting developed countries for their early historical emissions). Such accusations can encourage out-group outrage rather than in-group change[43].
- *Everybody fails:* Perhaps on many reasonable conceptions of equity, *no* party (or, at least, no developed country party) even approaches what equity demands. This deflating finding might stifle momentum that had been built by proactive countries that had successfully implemented their NDC targets.
- *Completely theorized disagreement:* Moralizing the issue might undermine the possibility of what Cass Sunstein calls 'incompletely theorized agreements'[44]. These constitute agreements on particular outcomes, without agreement on the underlying principles justifying those outcomes[45]. Climate change responses may be an area where it is possible to get practical agreement, but only if each party resists trying to gain acceptance for its preferred *reason* for the justifiability of that agreement. In exposing moral reasons for each position, the dialogue might damage rather than encourage commitment and compliance.
- *One voice for all:* The equity stocktake process, as part of an official, UNFCCC-based regime, may need to speak in 'one voice'. While it has the capacity to acknowledge different equity perspectives (see below), it might

---

43 Täuber, van Zomeren and Kutlaca, "Double-Edged Sword", pp. 456–57, see note 29
44 Sunstein, C.R. (1996), *Legal Reasoning and Political Conflict,* New York: Oxford University Press, Ch. 2
45 Sunstein's idea somewhat parallels Rawls' 'overlapping consensus'. Rawls, J. (1993/2005), *Political Liberalism*, expanded edition, New York: Columbia University Press. However, Rawls uses overlapping consensus to arrive at large-scale political-constitutional principles. The current problem bedevilling the climate regime is precisely the absence of such principles. The hope based on Sunstein's insufficiently theorized agreements would rest on agreement on particular policies despite disagreement on the deeper reasons legitimizing those policies.

still be limited in its capacity to speak in different voices to different constituencies. Social science literature suggests the benefits of tailoring different types of ethical framing towards different parties. Sometimes the language used to persuade a non-believer differs from 'preaching to the choir'. So, too, sometimes it is better invoke actors' identities rather than their values[46]. The stocktake process may have to speak similarly to all audiences, stymieing strategic, targeted uses of moral language.

- *Untargeted discussion*: It may prove difficult to target the discussion to the areas where it would be most constructive. For example, different languages (collective risk, economic prudence, etc.) might be used to inform the question of who-should-do-what – while another language might better motivate compliance[47]. The process might struggle to apply different languages to areas that possess the greatest promise.

One final concern warrants special emphasis: the make-up of both the APA and the CMA may not be logistically conducive to the sorts of activities that would lead to a constructive dialogue. The APA is modelled on the now-retired Ad Hoc Working Group on the Durban Platform (ADP). The ADP worked like a smaller-scale COP, with hundreds of delegates governed by a chair and co-chairs, meeting a handful of times between the COPs. With its large numbers, the ADP increasingly employed smaller groups with specific tasks, who would then report back to the larger plenaries. For its part, the CMA will parallel the make-up of the COP. In short, both bodies are very large, with hundreds of delegates.

Previous models of constructive and productive dialogues about ethical norms (such as the drafting process for the *Universal Declaration of Human Rights*[48]), operated with far fewer delegates, and employed liberal use of tiny sub-committees – even to the point of setting analytical, quasi-legal tasks for specific individuals[49]. This allowed space for prolonged question-and-answer dynamics and genuine reflection and persuasion. To be successful, the CMA-APA may need to similarly make use of small but representative sub-committees and expert working groups.

Section 5: ways forward

Given the challenges just described, this section outlines the most promising ways forward for the stocktake process in terms of achieving a constructive dialogue and seizing the desired benefits.

---

46 Täuber, van Zomeren and Kutlaca, "Double-Edged Sword", pp. 459–60, 62, see note 29
47 For example, Pickering notes that concerns with *risk* are highly motivating – yet unhelpful in assigning burdens. Contrariwise, moral thinking can inform us about proper burdens – yet fail to motivate action. Pickering, "Moral Language in Climate Politics", p. 270, see note 29
48 See Breakey, H. (2015), "COP 20's Ethical Fallout: The Perils of Principles without Dialogue", *Ethics, Policy & Environment*, 18 (2)
49 Glendon, M.A. (2001), *A World Made New: Eleanor Roosevelt and the Universal Declaration of Human Rights*, New York: Random House

Logistical recommendations

Timing and roll-out: Any moral stocktaking does its best work *before* it comes out, by motivating actors to work out how to shine by the time the assessments become public knowledge, and by giving them a sense of ownership and inclusiveness in the process. For this reason, it is crucial to get parties thinking well ahead of 2023 about how they would like to input into the process, and how they can best craft and communicate their NDCs.

Moral entrepreneurs: Sometimes moral suasion occurs less through the force of argument than through the force of the speaker's personality. Actors who are seen to possess integrity, fair-mindedness, and an appreciation of others' perspectives, are likely to prove better communicators of potentially tough moral lessons (a little charisma might help too). The stocktake process might even be able to use different entrepreneurs in speaking to different blocs.

Allocating and executing specific tasks: Success may depend on having smaller groups capable of dealing with difficult technical and conceptual tasks, creating resources for larger debate.

For example, given its mandated role in the global stocktake, the APA could task select sub-committees with developing guidance on claims of equity and fairness – with the goal of ensuring standardization, clear categorization, comparability, methodological consistency and quantification of NDCs' invocations of equity principles. These guidelines could help parties' NDCs explain how their invoked equity-principles can work to justify their mitigation, adaptation and finance pledges.

Similarly, the Paris COP Decision sets down the IPCC as one source of input for the APA. Since its very beginnings the IPCC has consistently – and in its *Fifth Assessment Report* explicitly and comprehensively – considered matters of ethics and equity. This work, and any similarly themed future IPCC work, could inform the APA in developing inputs for the equity stocktake process.

Practical encouragement

Staying positive: Ethics is about aspiration as much as obligation, and moral ideals can prove effective in changing behaviour[50]. As such, the equity stocktake process might elect to work out ways of showcasing and lauding those who the process assesses as doing well, rather than focusing on poor performers. It might be best left to domestic civil society organizations to seize upon a country's poor rankings to agitate locally for improved efforts[51].

---

50 Täuber, van Zomeren and Kutlaca, "Double-Edged Sword", p. 458, see note 29
51 Another aspect best left to domestic civil society might be the operation of the principles of equity *as they apply domestically.* This issue might be too contentious for the CMA-APA equity stocktake process to explicitly confront, but the logic of each equity principle is bound to carry consequences for intra-state distribution of burdens and entitlements. Local actors may be well-placed to highlight these consequences, and states' failure to respect them

Different foci might also prove less contentious. Moral assessment of NDCs can apply to, a) the *norms* the Party selects and puts forward; b) the *interpretation and application* of those norms that lead to the Party's stated NDC commitments; and, c) the *failure* of the Party to implement those commitments. Each of these areas may allow for more or less constructive engagement. For example, (c) may occur for different reasons. It may be because despite the Party's best intentions, it encountered greater-than-expected challenges in achieving its goals, in which case it might require assistance and expertize. Alternatively, the Party may have failed because it did not invest the sustained political effort required – in which cases its failure is one of integrity. Rather than interrogating its principles or its interpretation of its principles (both issues of substantive justice), a focus on integrity would simply ask whether a Party was living up to *its own publicly stated standards*.[52]

Practical support: The belief that it is possible to respond effectively to a problem impacts on whether a party will alter their behaviour[53]. Often, a poor-performing state party will face genuine challenges in implementing climate initiatives (including eliciting public support, ensuring institutional representation for climate issues, overcoming corporate resistance, and so on). The more ethical concerns highlight practical problem-solving practices, the better their chances of success[54].

Comparability and quantification

Crunching the numbers: Many of the existing norms and protocols around the NDCs aim to clarify what the country is doing, such that its efforts can be quantified and compared to others. At present, the claims of fairness and ambition that accompany most NDCs remain comparatively opaque. It can be unclear what exact moral principle the party is invoking and how that principle determined (or helped determine) their specific climate policy and mitigation target[55]. The stocktake process could create an analytic framework that clarified equity-principles and linked them to specific (for example, mitigation) outcomes. Much of this work has been begun by civil society actors, including NGOs and scholars[56]. Extending this existing work, the analytic framework could encompass

---

52 On integrity in this context see: Breakey, H. and Cadman, T. (2015), "A Comprehensive Framework for Evaluating the Integrity of the Climate Regime Complex", in *Ethical Values and the Integrity of the Climate Change Regime*, Breakey, H., Popovski, V. and Maguire, R. (eds.), Surrey: Ashgate. Breakey, H., Cadman, T. and Sampford, C. (2015), "Conceptualizing Personal and Institutional Integrity: The Comprehensive Integrity Framework", *Research in Ethical Issues in Organizations*, 14
53 Täuber, van Zomeren and Kutlaca, "Double-Edged Sword", pp. 457–61, see note 29
54 Equally, phrasing the problem in terms of overcoming difficult challenges removes some of the moral blame for failing one's obligations
55 Brown, D.A. and Taylor, P. (eds.) (2014), *Ethics and Climate Change. A Study of National Commitments*, Gland: IUCN
56 See, e.g., Civil Society Organizations (CSO) (2015), "Fair Shares: A Civil Society Equity Review of INDCs", Stockholm: EcoEquity; Mattoo and Subramanian, "Equity in Climate Change", see note 20

all the equity principles the CMA-APA feels significant, and incorporate flexibility in allowing parties to interpret different parts of the principles in different ways[57]. This framework would allow the stocktake process to gesture towards ball-park figures that a particular equity-principle would require of a particular country, given that country's features (for example, population, GDP-per-capita, emissions-per-capita, aggregate historical emissions, etcetera).

Such an analytic framework could be developed without the CMA-APA making any individual-country moral assessments or demanding substantive commitments. Rather than forcing any party to employ this or that equity-principle, the framework would simply assist parties in ensuring their NDC efforts accorded with their own asserted moral principles[58]. Similarly, it would assist other parties and civil society in making their own determinations of a party's efforts.

Comparisons: Many countries' NDCs gestured towards other similarly positioned parties, and compared their mitigation commitments[59]. The stocktake process may help shed some light on these comparisons, including in evaluating which parties should be 'grouped' together, and highlighting who is ahead of (or behind) the packs categorized in this way[60].

Plurality and inclusiveness

Pluralism: The analytic framework mentioned earlier on quantifying equity-principles shows how much can be accomplished even while allowing member-states considerable plurality and flexibility across the equity principles they invoke. The Civil Society Review of NDCs launched on the eve of the Paris COP employed two principles (capacity/need and historical responsibility), and allowed a range of interpretation for each principle[61]. Earlier, the work of Mattoo and Subramanian provided emissions outputs for given countries on the basis of

---

57 The result would extend the existing work (see note 56 above) by being: a) more diverse in terms of equity principles; b) allowing for different weightings to each principle in positions based on a mix of principles; and c) capable of responding to different but reasonable choices on sub-issues within the principles. For an in-depth discussion of the decisions on sub-issues of equity principles that should inform countries' development of their NDCs (and others' evaluations of those NDCs), see Brown, D., Breakey, H., Burdon, P., Mackey, B. and Taylor, P. (2018), "A four-step process for formulating and evaluating legal commitments under the Paris Agreement" *Carbon and Climate Law Review* 2, pp. 1–12

58 This would allow each country to live up to its own standards – 'consistency-integrity'. See Breakey and Cadman, "Integrity of the Climate Regime Complex", note 52; Breakey, Cadman and Sampford, "Comprehensive Integrity Framework", note 52

59 E.g. Australian Government (2015), "Australia's Intended Nationally Determined Contribution to a New Climate Change Agreement", UNFCCC

60 While this judgment might be misleading in terms of objective justice, a process whereby those at the back of the pack were encouraged to move to its centre could drive the ratcheting ambition that sits at the core of the bottom-up NDC process

61 (CSO), "Fair Shares: A Civil Society Equity Review of INDCs", see note 56. The Climate Equity Reference Project, which provided analytic support for the CSO Review, was capable of spanning over an even broader range of variables

an even wider array of equity principles – and allowed weighted combinations of the principles. In their technical representation of these combinations, the proportions overall summed to 1, meaning (for example) a state could crunch the numbers on its strong weighting for capacity (0.4) and equal shares (0.4), alongside its respect for needs (0.1) and historical responsibility (0.1).

An important outcome of both projects is that even when allowing parties considerable flexibility in their choice of principle, many countries *could still be shown to be not doing enough*. As such, the stocktake process could remain flexible and pluralist about the equity principles without sacrificing the capacity to make potentially tough assessments about parties' NDCs.

Narrowed pluralism: Even within its requirement to remain facilitative, the CMA-APA could decide to *somewhat* narrow the weights that can be given to each equity principle. After all, each country has signed up to all the equity principles, inasmuch as each of the agreements since 1992 have helped fill out these principles. As such, the process could consider the possibility of setting bounds on the weightings, such as requiring that each party must include at least a 0.1 weighting (on Mattoo and Subramanian's scale) to each of the equity principles, or to require that no single principle could be given more than a 0.5 weighting. Perhaps it could even be argued that the consistent concern for development, poverty and vulnerability running through almost every agreement's objectives and principles means that the *needs-based* equity principle could never be ranked below 0.2. These suggestions are highly speculative, of course, but they hint at some of the ways the stocktake process may, over time, seek more convergence across the available scope of reasonable ethical positions.

Construction and deliberation

Q&A to explore parties' reasons and rationales: The Paris Decision encourages all parties submitting their NDCs to explain how they consider their commitments are fair and ambitious. An overwhelming majority of parties in fact followed this recommendation. While the information provided was no doubt self-serving rhetoric aiming to ward off critique[62], exploring these explanations, and teasing out the substance of the principles (and the way interpretations and applications of those principles could create a specific quantified contribution), may prove a valuable endeavour.

One way this could happen is through a formal 'question and answer' process, whereby members of the (CMA-APA/sub-committee) group pose questions for the party about the moral reasoning inherent in their NDC[63]. The party is required to answer, and follow-up questions can delve deeper into their thinking.

---

62 Empirical research suggests most parties focused on what level of carbon mitigation was economically manageable in developing their NDCs, meaning their subsequent invocations of equity and fairness may be mere window-dressing Brown and Taylor, *Ethics and Climate Change*, see note 55
63 See Brown, "Climate Change Policy Formulation", note 37; Breakey, "Principles without Dialogue", note 48

122 *Hugh Breakey*

Clear guidelines and effective chairing could aim to keep questions constructive and based on rational enquiry, rather than emotive accusations[64]. For example, understanding and reflection can be encouraged by common-sense ethical questions such as, 'What would happen if everyone in the world followed your chosen policy?' and, 'Does your country domestically respect in its own laws the equity-principle you invoke in your NDC?'

A suitable venue for this process would be the facilitative dialogue mandated to take place in 2018. This dialogue aims to take stock, 'of the collective efforts of Parties in relation to progress' towards the Agreement's long-term goal[65]. The facilitative dialogue's mandate neither definitively rules in, nor rules out, equity-based considerations of NDCs. Including a constructive Q&A process in the dialogue could be a helpful learning experience, teasing out the equity-principles parties have in mind (that can then be employed in fashioning the 2023 global stocktake), while encouraging parties to start thinking about the relation of equity to their NDCs long before 2023 arrives.

Deliberative and functional norms: At present, the equity principles listed above (in §2) suffer from tensions with one another, and are only portrayed in general terms. To proceed, the stocktake process may need to do considerable work in deciding the appropriate formulation(s) of each principle, and the variables within it (for example, the date historical responsibility begins). So, too, it may be helpful for the stocktake process to try to iron out some of the more extreme interpretations of these principles. This would effectively construct a selection of determinate functional mid-level principles from which parties will be later free to choose. On this footing, the stocktake process would draw on established resources to pin down and operationalize the various principles of equity, and then invite (or guide) parties to employ these principles as they move forward with their NDCs[66].

The stocktake process might work to develop, informally or formally, a genuine, inclusive, structured deliberation to tease out these important questions[67]. Such a process – perhaps based around structures employed in the

---

64 An unstructured dialogue might well collapse into no genuine dialogue at all. See "Principles without Dialogue", note 48. The aim of the Q&A is not to simply rehearse the unhelpful moral rhetoric from the COPs. Such rhetoric includes developed countries prating of fairness when no reasonable moral positions would excuse their inaction (see Brown, D.A. (2013), *Climate Change Ethics: Navigating the Perfect Moral Storm*, London and New York: Routledge, p. 176), and developing countries seizing on historical responsibility without applying the principle consistently to members of their own group (see Weisbach, "Responsibility for Climate Change", note 25)

65 (COP), "Paris COP Decision", para. 20

66 Pickering considers such evolution and its prospects: Pickering, "Moral Language in Climate Politics", see note 29. We could consider this approach the development of *functional* norms (justified by appropriateness and efficacy in solving the problem) rather than *deliberative* norms (justified by construction through inclusive and reasonable dialogue). See Breakey, "Heating up Climate Norms", p. 228, see note 31

67 The process would use the logic of argument (working out what the norms are) rather than appropriateness (working out whether an action falls under the norm): see Risse, '"Let's Argue!"', note 37. More generally see: Breakey, "Principles without Dialogue", note 48; Boran, "Principles of

The CMA-APA equity stocktake process 123

drafting of the *Universal Declaration of Human Rights* – would require a significant period of development and careful attention to the rules of drafting and interventions. Again, the goal would be not to choose a single principle on a single interpretation, but to aim to narrow the legitimate diversity across invoked positions.

To harness deliberative legitimacy from their decisions, the CMA-APA would need to debate, discuss and democratically decide these issues. However, (hearkening back to the logistical recommendations noted above) it may be that more preliminary and technical work could be done by various groups of experts brought on board to perform specific, discrete tasks[68]. These could include:

- Helping to establish a settled collective carbon budget, or at least determine its upper limit in the light of the Paris temperature goal. Establishing the remaining carbon budget that would (with an appropriate probability of success) meet this target will ensure that each party is at least applying the equity-based based parts of their NDCs to the *same* overall figure[69].
- Guidelines for setting the procedural rules for CMA-APA debate on equity and NDCs. These could include canvassing common-sense questions to help tease out countries' positions on equity principles and how they relate to their NDCs[70].
- Analytic clarifying of the relevant equity principles (for example, to see how they relate to each other[71]) and how legal and ethical theorists have delineated those principles using resources from existing law, moral philosophy and the UNFCCC texts[72].
- Helping the CMA-APA construct the analytic framework described above – a framework that moves from specific equity-principles, applied to given

---

Public Reason", note 30. After all, declaring what justice demands of someone else is likely to elicit a different response compared to trying to draw them into a process of deliberation about what justice demands

68 While some of these tasks are quite technical, requiring considerable expertise, too much power should not be ceded to expert groups (or small sub-committees), lest parties feel alienated from the equity stocktake process, losing their sense of consent, ownership and deliberative legitimacy

69 In other words, while there would no doubt be dispute as to how the overall pie should be cut up and distributed, it may at least be possible to get agreement on the *size* of the existing pie. As Donald Brown puts it, each NDC is 'implicitly a position on two important ethical issues. These issues are: (a) an atmospheric GHG concentration goal and (b) the nation's fair share of safe global GHG emissions.' Brown, "Climate Change Policy Formulation", p. 63, see note 37. Given there will be inevitable disagreement over what equity requires for each nation's fair share (Brown's second issue), it makes sense to see if the first issue, of the overall target and corollary carbon budget, can be settled by the collective

70 See Brown and Taylor, *Ethics and Climate Change*, note 55. See also: http://ethicsandclimate.org/category/questions-about-climate-change-ethics/

71 See, e.g., Mattoo and Subramanian, "Equity in Climate Change", note 20

72 See, e.g., (ILA), "Legal Principles Relating to Climate Change", note 20; Maguire, "Incorporating International Environmental Legal Principles", note 20; Pickering, "Top-Down Proposals", note 34

124 Hugh Breakey

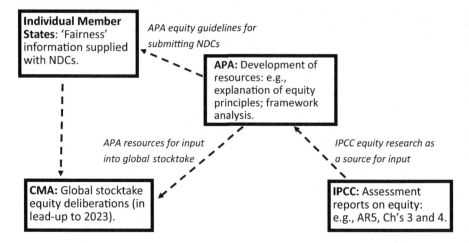

Figure 6.2 Potential actors in the CMA-APA equity stocktake process

national circumstances, to arrive at quantitative ball-park emissions-reduction estimates[73].

The harnessing of deliberative legitimacy underscores why I termed the CMA-APA equity stocktake process as a *process* rather than merely a task. Rather than concentrating on substantive questions of distributive justice, ethical attention can instead focus on deliberative *procedural* values (and the time, work and clear-eyed strategy required to realise them), governing how the collective can constructively reflect upon the ethical issues at stake[74].

The foregoing proposals allow Figure 6.1 to be refigured to draw attention to the specifically equity-based ways each institution can play a role in the CMA-APA equity stocktake process – as Figure 6.2 illustrates.

Conclusion

When so much is at stake, and differences in views so pronounced, there can be no idle hope that a process of equity-based consideration of NDCs will inevitably improve matters. To the contrary, there is every reason to suspect that, without thoughtful and far-sighted planning, things could go wrong.

73 See note 56
74 This emphasis stands in tension with many prevailing accounts of climate justice, including those foregrounded in the IPCC Reports, which imply a 'snapshot' methodology, where moral appraisals are read off from the straightforward application of objective moral theories to existing circumstances. On some of the issues arising here between substantive and procedural justice, see Pickering, "Top-Down Proposals", note 34; Boran, "Principles of Public Reason", note 20

Still, the mere fact that the COP 21 delegates decided to include this process – and those five words: 'in the light of equity' – in the Paris Agreement offers some cause for hope. Perhaps the collective realized that a certain amount of common-sense moral suasion was a necessary element in a system where bottom-up commitments must be increasingly ratcheted up to achieve common objectives. In line with that hope, this chapter has argued that there are many ways the CMA-APA equity stocktake process could be implemented. Naturally, not all the portfolio of recommendations noted above need to be acted upon: depending upon the overall situation, various avenues will prove more promising than others. However, if the CMA-APA equity stocktake process is crafted with a clear-eyed awareness of potential pitfalls, the process could help improve the justice and the efficacy of climate efforts over the coming decades[75].

---

75 This chapter has benefited from comments from workshop participants at: *From Commitment to Implementation: Carbon Integrity post Paris?* New Delhi, India, 14 March 2016. I would also like to especially thank Donald Brown, Idil Boran and Jonathan Pickering for providing helpfully penetrating critiques of previous drafts

7 Equity in the global stocktake

Swapna Pathak and Siddharth Pathak[1]

Introduction

On the day when 196 member parties signed the Paris Agreement, Jean-Claude Junker, the President of the European Commission, commemorating the event, said: 'today the world gets a lifeline… This robust agreement will steer the world towards a global clean energy transition.'[2] The Agreement lays out a firm global commitment towards mitigation, and even adaptation, to the effects of climate change; however, the key to the ultimate robustness of the agreement lies buried in Article 14, on the global stocktake. The first section of Article 14 states: 'The Conference of the Parties serving as the meeting of the Parties to this Agreement shall periodically take stock of the implementation of this Agreement to assess the collective progress towards achieving the purpose of this Agreement and its long-term goals (referred to as the "global stocktake"). It shall do so in a comprehensive and facilitative manner, considering mitigation, adaptation and the means of implementation and support, and in the light of equity and the best available science.'[3] If the Paris Agreement creates a roadmap to limit the global temperature increase to 1.5°C by the end of the century, the process of the global stocktake is the critical torch to navigate that path. A substantial number of academic and policy analyses of equity in the climate change regime have debated the fairness of mitigation targets, and more recently, adaptation-related challenges. These two issues can be categorized as the end-goal of the treaty. Now that the Paris Agreement has been universally agreed upon, we feel that it is prudent to start now focusing on its implementation. This article will consider issues of equity that go beyond the end-goal of the Agreement, towards its execution that will be facilitated by the process of global stocktake.

Global stocktake is an iterative exercise scheduled to occur every five years during the commitment period of the Agreement. The stocktake will assess the mitigation efforts made by all the countries within the context of the scientific

---

[1] Please note the views presented by the authors in this chapter are personal
[2] www.reuters.com/article/uk-climatechange-summit-reaction-factbox-idUKKBN0TV0 QK20151212
[3] Paris Agreement, 2015, Art. 14, para. 1

reality of climate change. The stocktake will also evaluate the financial and technical support offered toward mitigation and adaptation efforts, and make recommendations for scaling up national ambitions to reach the desired goal of the Agreement. Therefore, the global stocktake is the key to maintaining the longevity and robustness of the Paris Agreement in the milieu of any change in climate science, technological innovation, and economic and political fluctuations. The first comprehensive stocktake outlined in the Paris Agreement will be conducted in 2023. However, to ensure that the process of stocktaking gets a head-start, the United Nations Framework Convention on Climate Change (UNFCCC) will conduct a facilitative dialogue that will provide a framework for countries to assess their progress and identify opportunities for enhancing ambition in their Intended Nationally Determined Contributions (INDCs) before the commitment period for the agreement begins in 2020. During the run-up to COP 22 in Morocco, some parties envisioned the global stocktake process as a tool to assess the collective action toward achieving the long-term goals of the Paris Agreement. Most parties also agreed that the global stocktake should be facilitative and non-prescriptive. While it is clear that the global stocktake will be a key tool toward increasing ambition across all elements at regular intervals, governments still need to explore how this process will unfold equitably. The EU was one of the few parties that alluded to this concern and mentioned in its submission that 'the global stocktake (GST) will be undertaken in light of best available science and of equity'[4]. While the GST is not aimed to settle science- and equity-related questions directly, different aspects of both science (for example, through the IPCC) and equity can be relevant to the GST's deliberations.

Equity in the climate change regime

Since the time when the UNFCCC was signed, the primary roadblock towards reaching a global agreement to curb climate change had been the issue of 'equity', where the word embodied a sense of fairness and justice given the vast amount of difference in historical contribution toward the problem of climate change between the developed and the late industrializing economies. It was from this sense of equity that countries including China, India, Brazil and Indonesia were absolved of any legal obligations toward CO_2 reductions under the Kyoto Protocol adopted in 1996.

However, this approach toward equity was challenged by the United States in the light of projected overall emissions from China (and India) that were slated to overtake those of the US. In 2009, COP 19 in Copenhagen introduced a new form of equity by proposing that each party could voluntarily pledge the amount of CO_2 they would be able to reduce during the post-Kyoto commitment period. Unfortunately, the aggregate of voluntary commitments made by all the countries fell far short of what would be required to stay within the 2°C temperature

---

[4] FCCC/APA/2016/INF.4/Add.1

increase. It was then realized that, while the climate change regime needed to maintain the aspect of voluntary pledges to garner a broad support base, including that of the United States and several developing countries, at the same time there needed to be some form of centralized process to systematically increase the ambition toward aggregate mitigation.

Based on some of these realizations, the Paris Agreement creates a different form of equity where all the nations have an obligation to reduce their CO_2 emissions based on 'common but differentiated responsibilities and respective capabilities, in the light of different national circumstances'[5]. Articles 2 and 4 of the Paris Agreement create *ex ante* equity toward the fulfilment of the agreement by outlining four key conditions: first, the language of the Agreement makes it legally binding for all the member parties to submit their mitigation ambition in the form of INDCs every five years. Second, all the parties have to maintain transparency and accuracy with respect to their emissions-related data. Third, the language of the Agreement creates a strong obligation for developed countries to lead the mitigation efforts. And four, the developed countries are required to facilitate emissions reduction and adaptation efforts in developing countries by providing necessary finance and technology.

On the other hand, Article 14 of the Agreement, on the global stocktake, provides an opportunity to ensure *post ante* equity. This means that if we assume the objective of the Agreement as equitable (as outlined in Articles 2 and 4), then Article 14 on the global stocktake can be utilized to create a framework to ensure equity in the process of reaching that goal.

Operationalizing equity so far

Negotiators and political leaders from developed countries have often resisted references to equity at collective platforms of the UNFCCC. One of the most notable controversies around this issues occurred at the COP 17 meeting in Durban, when the lead negotiator from the US, Todd Stern, announced that 'if equity is in, we're out'[6]. Historically, the debate on equity in the climate change regime has centred around 'emissions allocation', which means how much CO_2 each country should be allowed to emit as part of an international agreement. The primary point of contention in such debates is that, even though the developed countries have historically contributed about 76% of CO_2 emissions, and have substantially higher levels of per capita CO_2 emissions, countries like China and India will continue to increase their CO_2 emissions as their economies expand. Given this reality, scholars have been labouring to reconcile the past with the future to create an equitable climate change regime. In 2010, the World Bank conducted a comprehensive meta-study of various analyses on ensuring equity for the next climate

---

5 Art. 2, Paris Agreement
6 Pickering, J., Vanderheiden, S. and Miller, S. (2012), '"If Equity's In, We're Out": Scope for Fairness in the Next Global Climate Agreement", *Ethics and International Affairs*, 26 (4), p. 423

change agreement. The study summed up four suggestions for establishing equity (primarily focused on emissions allocation) based on the existing literature: 1) equal allocation on a per capita basis; 2) historical contributions; 3) ability to bear the cost; and 4) future fossil fuel-dependent growth. Cost of adaptation, and loss and damage already being experienced by many countries, were still politically peripheral back then. Now, of course, any discussion of equity has to take into account these two variables that will extract a heavy cost in the future[7].

Since the World Bank analysis, a couple of prominent studies have emerged to operationalize equity in emissions reduction. During COP 19 in Lima, Brazil introduced an innovative approach toward equity in the next agreement of 2015. Brazil proposed a move away from the binary categorization of countries under the Kyoto Protocol into Annex I – the group of developed countries – and Non-Annex countries – developing countries – to a more dynamic structure of differentiation. The 'concentric model of differentiation' put forward by Brazil suggested that every country needed to commit to lowering its CO_2 emissions, but the form of commitment could vary based on national capabilities and circumstances. The model proposed a hierarchy of CO_2 reduction commitments represented in the form of three concentric circles. The Annex I countries were placed at the centre and were required to propose nationwide emission reduction targets. The middle circle included Non-Annex I countries that could propose reduction targets in terms of 1) economy-wide intensity targets; 2) economy-wide per capita targets; and 3) economy-wide departure from business-as-usual targets. The outermost circle included countries that were required to commit to some kind of mitigation projects but not necessarily to nationwide hard targets. The proposal also mentioned that as time goes on, and countries' circumstances change, they should aspire to move inward through the circles. It was also expected that no outward movement would occur. Which means that once a country is set on a trajectory toward better mitigation, it will establish a certain amount of 'path dependency' that will prevent countries from sliding back. The Brazilian proposal also called for 'South-South' co-operation in terms of monetary flow so that the developed countries were not solely responsible for financial aid to the developing countries[8].

The structure of INDCs submitted by countries before COP 21 somewhat followed these recommendations, whereby countries like India, Brazil, Indonesia and so on submitted economy-wide intensity targets, and the United States and the EU, for instance, submitted nationwide emission reduction targets.

Another significant proposal toward operationalizing equity was proposed in 2015 by a study conducted by civil society groups, including several non-governmental organizations from developing countries. The Civil Society Equity Review allocates a 'fair share' of mitigation targets for all the countries based on historical responsibility and capability (contingent on the percentage of population

---

7 Mattoo, A. and Subramanian, A. (2012), "Equity in Climate Change: An Analytical Review", *World Development*, 40 (6), pp. 1083–1097
8 http://paperroom.ipsa.org/papers/paper52285.pdf

above a certain income threshold). The Equity Review has created three different benchmarks – high progressivity, medium progressivity and low progressivity – contingent on different levels of historical responsibility, and evolving capabilities of different countries. The report presents a glaring gap in terms of mitigation required, and the aggregate national mitigation pledges from different countries.

The report puts the onus on developed countries to switch to 100% renewable energy, and to provide climate finance to developing countries. However, it does not absolve developing countries from making mitigation efforts. In fact, the report recommends developing countries set ambitious mitigation targets, and to possibly 'leapfrog to zero-carbon societies'[9].

While the negotiators from different countries have raised the issue of equity to safeguard their national interests, academics have been debating the trade-offs between issues of equity and effectiveness. Even as recently as 2016, the political scientist Robert Keohane, who has written extensively on international regimes, argued that focusing on equity in negotiations can generate serious trade-offs with the effectiveness of the climate regime. For academics, Keohane suggested, focusing on equity in the Paris Agreement will distract them from studying the actual politics of the process, and that normative concerns for equity could compromise the rigour of the research that is required. His comments at the Berlin conference were met with sharp criticism from several academics working on the climate change regime. In response to Keohane's views, Sonja Klinsky *et al.* noted that one of the fundamental reasons for focusing on equity is to understand the very parameters of the trade-offs that Keohane is talking about[10]. Which means that unless we outline what an equitable agreement and its implementation looks like, we will not be able to understand how the agreement departs from that normative concern of equity.

Ensuring equity in the global stocktake

To ensure that the global stocktake furthers the idea of equity as per Article 14 of the Paris Agreement, two issues need to be taken into account. First, the scope as well as the treatment of elements (outlined in the Agreement) within the stocktake and, second, how the observations and results from the collective stocktaking exercise are operationalized to further equity amongst countries.

Scope and issues

In 2007, the Bali Action Plan fleshed out a few core issues that would be crucial in addressing climate change[11]. These ranged from the long-term global goal for emissions reduction, to enhancing national and international action on mitigation

9 http://civilsocietyreview.org/wp-content/uploads/2015/11/CSOFullReport.pdf
10 Klinsky, S. *et al.* (2017), "Why equity is fundamental in climate change policy research", *Global Environmental Change*, 44, pp. 170–173
11 1/CP.13, para. 1; http://unfccc.int/resource/docs/2007/cop13/eng/06a01, p. 3

Table 7.1 Elements of implementation of the Paris Agreement and their institutional homes

	Element	Institutional home
I	Mitigation	None
II	Adaptation	Adaptation Committee
III	Technology	Technology Executive Committee + Climate Technology Centre and Network
IV	Finance	Standing Committee on Finance + Green Climate Fund + Adaptation Fund + Least Developed Country Fund + Special Climate Change Fund
V	Capacity-building	Paris Committee on Capacity-Building
VI	Loss and damage	Executive Committee on Loss and Damage + Warsaw International Mechanism on Loss and Damage

that included action on emissions from land use, adaptation, economic diversification, loss and damage, technology transfer, finance and capacity-building. Following that plan, the Paris Agreement focuses on six elements to construct a comprehensive climate change regime. These six elements are: mitigation, adaptation, technology, finance, capacity-building, and loss and damage. The global stocktake, in order to be a vehicle of ambition and equity, will need to address all of these elements of the Agreement[12]. A repeated assessment of mitigation targets as part of the stocktake will strengthen collective mitigation goals. However, a sustained evaluation of adaptation and loss and damage conditions will highlight the individual needs and vulnerabilities of less-developed countries.

Over the course of years, most of these elements now have institutional homes within the UNFCCC. Table 7.1 identifies the various institutional homes for these elements. Note that there is no specific institutional home for mitigation. This is because the Paris Agreement has left the mitigation ambition to be decided by individual countries through their nationally determined contribution (NDC).

Although that might be the case, these NDCs will undergo review and verification through the new enhanced transparency framework agreed upon in the Paris Agreement. The design of the global stocktake exercise will need to ensure that the relevant institutions of the UNFCCC dealing with these elements provide the necessary input for the assessment. The findings and recommendations from the stocktake then need to be relayed back to these institutions for operationalization and implementation.

The Paris decision does give a non-exhaustive list of elements[13]. However, as international circumstances evolve, the list of elements will need to expand further to possibly include issues like climate-related migration, which will affect the most vulnerable countries. All of these institutions can facilitate and enable

---

12 Paris Agreement, Article 14, para. 1
13 1/CP. 21, para. 99; https://unfccc.int/resource/docs/2015/cop21/eng/10a01.pdf

the implementation of commitments made by countries within the UNFCCC. However, these institutions are not necessarily under the authority of the COP, which is the supreme body for the UNFCCC. The relationships shared by these institutions differ on a case-by-case basis with financial instruments like the Green Climate Fund being the weakest link[14].

This will pose a serious challenge for co-ordinating input and output for the global stocktaking process. Therefore, the framework of stocktaking needs to specifically identify the institutions that will be responsible for providing input for assessment for an element and then implement the recommendations from the stocktake.

This will help in developing a common understanding of the progress made with respect to the relevant element along with ensuring that the outputs from the global stocktake are directed towards the respective institution rather than being generic observations that are just taken note of in future COP decisions.

An additional challenge to the framework of the global stocktake process is that it is only meant to look at the collective progress and assess this against the benchmarks set in the Paris Agreement. This can undermine equity concerns where data on individual country level input is inadequate. The global stocktake, therefore, should take into account disaggregated data on these elements to heed to the underlying principle of the climate change regime: 'common but differentiated responsibility and respective capabilities in light of different national circumstances.'

Science and mitigation

As identified in Article 2 of the Paris Agreement – 'holding the increase in the global average temperature to well below 2°C above pre-industrial levels and pursuing efforts to limit the temperature increase to 1.5°C above pre-industrial levels, recognizing that this would significantly reduce the risks and impacts of climate change' – the temperature threshold is an important benchmark against which the stocktake will assess this collective progress. The IPCC, through its assessment reports, will provide the global collective trajectory that is required to ensure holding the global average temperature below 1.5°C. However, there are many other inputs that will be critical to supplement the IPCC assessment.

The UNFCCC synthesis report on NDCs will be one of those supplementary inputs to the IPCC[15]. The previous UNFCCC synthesis reports have focused on the aggregated effects in intended nationally determined contributions (INDCs). However, a more effective stocktake will require more disaggregated and precise data from country INDCs. The INDCs submitted by each country usually contain two components – voluntary mitigation targets and conditional mitigation targets. Voluntary mitigation targets are CO_2 reductions that a country pledges independently. Conditional mitigation targets are CO_2 reduction pledges made by

---

14 www.odi.org/sites/odi.org.uk/_les/odi-assets/publications-opinion-_les/7918.pdf
15 http://unfccc.int/resource/docs/2016/cop22/eng/02.pdf

a country if it received the required financial and technological support from the international community. The previous UNFCCC synthesis reports have focused on the aggregate effect of voluntary pledges in the INDCs. These reports need to include the conditional elements of the INDCs submitted by countries and disaggregate them with respect to their conditionality. This information is critical to identify the needs of developing countries to reduce their CO_2 emissions if they are expected to share the burden of mitigation. The synthesis reports will also need to consider the increase in mitigation targets outlined in the INDC from the previous NDC submitted by a specific country, to assess progression in ambition from their previous contributions as stipulated in the Paris Agreement[16].

In addition to the above recommendations, the framework for the global stocktake should also include assessment from external sectors like aviation and shipping. These sectors currently lie outside the ambit of the IPCC. These sectors are also not directly addressed within the Paris Agreement, nor in other international agreements like the Montreal Protocol that are addressing short-lived climate pollutants. Inputs like the UNEP Gap report could also provide insight on where the barriers to increasing mitigation ambition are, and what steps can be taken to address them. Additionally, inputs from businesses as well as other non-state actors will also help in including different dimensions of increasing ambition and addressing concerns of equity while assessing overall progress against the global temperature benchmark.

These multitude of outside inputs will give a more comprehensive picture of the global mitigation effort while providing the specific details that will help in leveraging further ambition over and above what countries have committed to at the national level.

Adaptation

After a hard-fought battle at COP 20 in Lima to include adaptation within the scope of NDCs, it was finally agreed in decision 1/CP.20 that governments could also consider putting their actions on adaptation-planning within their contributions.

The UNFCCC synthesis report observed that 137 out of 181 countries that have submitted their NDCs included an adaptation component[17]. Previously, countries had been reporting their adaptation efforts through many different channels of communication – for example, COP 7 established the National Adaptation Programs for Action for the least-developed countries.

Similarly, the Cancun Agreements at COP 16 established the Cancun Adaptation Framework that mandated the development of National Adaptation Plans (NAPs)[18]. Article 7, paras. 10 and 11 of the Paris Agreement create another channel for adaptation-related communication, where it states that 'adaptation

---

16 Paris Agreement, Art. 4.3
17 http://unfccc.int/resource/docs/2016/cop22/eng/02.pdf – Section E, para. 59
18 http://unfccc.int/resource/docs/2010/cop16/eng/07a01. p. 4

communication referred to in paragraph 10 of this Article shall be, as appropriate, submitted and updated periodically, as a component of or in conjunction with other communications or documents, including a national adaptation plan, a nationally determined contribution as referred to in Article 4, paragraph 2, and/or a national communication'[19].

Such dispersed information on adaptation efforts will prove a significant challenge to global stocktaking. Additionally, there is no global benchmark against which individual adaptation plans and actions will be collectively measured. Therefore, it is critical to streamline adaptation communication as well as to set collective global benchmarks for adaptation, if adaptation is to be dealt with effectively in the global stocktake.

The equity dimension within the element of adaptation is not only limited to the provision of finance for adaptation actions at the national level, but by the establishment of global benchmarks for adaptation. Moreover, since adaptation is a very critical issue for developing and most vulnerable countries, it needs to be prioritized within the various processes of the UNFCCC, and should be accorded the same weight as mitigation in the global stocktake.

Finance

Finance is a core enabler for action globally, and is a crucial element within the larger discourse on equity in the climate change regime. In the run-up to COP 21, the World Resources Institute (WRI) made several recommendations, and two of the top three recommendations for creating an effective Agreement concerned finance[20]. The two primary finance-related recommendations were as follows: first, provide upfront investment for low-carbon pathways and adaptation efforts that are designed to enhance equity and build capabilities, including for equitably designed energy policies. Second, ensure that finance is accessible to those who need it, including non-traditional banking populations, to undertake innovative and locally appropriate climate action.

Although everyone recognizes the need for the aggregate global flows for climate finance, their quantum and quality are heavily debated. The OECD CPI report on Climate Finance in 2013–14, and the $100 billion goal published in the run-up to COP 21, stated that 'climate finance reached USD 62 billion in 2014, and USD 52 billion in 2013, equivalent to an annual average of USD 57 billion over the two years'[21]. However, there was heavy criticism from various quarters, including some counter-reports stating that the methods and reports of counting and tagging were seriously questionable on both theory and on facts[22]. One year after the Paris Agreement, Australia and the UK came together

---

19 Decision 5/CP.7; http://unfccc.int/resource/docs/cop7/13a01, p. 32
20 www.wri.org/sites/default/_les/building-climate-equity-ES.pdf
21 www.oecd.org/env/cc/Climate-Finance-in-2013-14-and-the-USD-billion-goal.pdf
22 http://pibphoto.nic.in/documents/rlink/2015/nov/p2015112901.pdf

to lead the donor countries in presenting a $100-billion 'roadmap' that aimed to provide increased predictability and transparency about how the annual goal of $100 billion by 2020 will be reached, and what it will take for countries to meet it[23].

These steps are headed in the right direction, but the rules for accounting for climate finance need to be made clearer. In order to ensure equity and effectiveness in terms of climate financing, the regime needs to do the following: (1) Currently the financing framework stipulates transparency in financing from donor countries. A more effective framework needs to seek regular reports from the recipient countries in order to ensure that the money allocated had actually been received; (2) In the present system, only a minuscule amount of climate finance that is owing is being routed through the existing financial mechanism of the UNFCCC. For the process of global stocktake, the accounting of climate finance needs to expand beyond the standing committee on finance within the UNFCCC. The input for the finance element needs to invite information from other global financial institutions like the multilateral development banks and investor networks that are channelling climate finance to developing countries; (3) It is critical to develop clear and comparable guidelines to track finance. This will make it easier to disaggregate data from different countries and assess their contribution to climate finance based on their 'respective capabilities and different national circumstances'; (4) The climate regime also needs to finalize a post-2020 goal for finance.

The need to have a post-2020 goal has already been recognized in the Paris Agreement under 1/CP.21, para. 53. However, a specific number will provide the necessary benchmark to assess the overall climate financial flows.

Outcome from the global stocktake

The core responsibility for the global stocktake is the need to take a bird's-eye view of the collective progress as well as to identify the implementation barriers towards the collective goals agreed within the Paris Agreement. Article 14 of the Paris Agreement states that 'The outcome of the global stocktake shall inform Parties in updating and enhancing, in a nationally determined manner, their actions and support in accordance with the relevant provisions of this Agreement, as well as in enhancing international cooperation for climate action'.

The legal language adopted in this stipulation indicates that the global stocktake cannot be prescriptive toward individual countries and undermine their national sovereignty. The recommendations of the stocktake need to be directed toward the collective, which in turn can inform individual countries to enhance their ambition.

---

23 http://dfat.gov.au/international-relations/themes/climate-change/Documents/climate-_nance-roadmap-to-us100-billion.pdf

This implies that the mandate of the global stocktake is only limited to making recommendations and not necessarily ensuring that those recommendations are actually followed through. The UNFCCC had conducted similar review exercises in the past such as the 2013 and 2015 reviews of the technical process for raising mitigation ambition before 2020. However, those exercises suffered from a similar design flaw which is that they did not have the mandate to actually carry out their prescription and most of the recommendations from these processes were left unaddressed. Therefore, an important challenge in making the global stocktake more effective is to ensure that its recommendations are actually executed by the parties.

There are some steps that could be taken to make the global stocktake more effective and further equity without overstepping the limited mandate for implementation.

First, the potential framework of the global stocktake can be designed to build synergy with Article 15.1 of the Paris Agreement. This provision of the Agreement establishes 'a mechanism to facilitate implementation of and promote compliance with the provisions of this Agreement. The mechanism is meant to be facilitative in nature and function in a manner that is transparent, non-adversarial and non-punitive. The committee shall pay particular attention to the respective national capabilities and circumstances of Parties'. The clear mandate given to this mechanism to facilitate implementation could be used as vehicle through which the follow-up of the outcome from the global stocktake could be conducted. Second, in most multilateral environmental agreements, the facilitative and compliance mechanisms focus on country level compliance. The Paris Agreement, on the other hand, limits the individual country obligations to procedural requirements such as mandatory reporting and transparency that renders the agreement without any 'teeth'. The compliance mechanism for the agreement could be developed to look at 'collective compliance' of the agreed goals within the Paris Agreement. The global stocktake can then provide detailed recommendations that will be collectively agreed upon, promoting international co-operation instead of singling out individual countries. Countries can then take these recommendations into account when taking further domestic action. Third, the recommendations from the global stocktake need to be precise on the actions to be undertaken by the relevant institutions within the UNFCCC. The collective stocktake should not only identify the barriers and gaps in implementation but also come up with concrete plans to address them.

Conclusion

The global stocktake will be critical in enhancing ambition as the implementation of the Paris Agreement begins to take shape. We have argued that in order to be equitable, the process of the global stocktake should look beyond mitigation to other critical elements of the Agreement, such as adaptation and finance.

Moreover, the global stocktake should pay attention to disaggregated data and not just focus on collective progress. This will be key to ensuring equity based on 'common but differentiated responsibilities and respective capabilities, in the light of different national circumstances'. A collective stocktake will also have to look into other elements that potentially address equity like human rights, economic diversification as well as sustainable development goals. Even if these elements are not directly implemented in the global stocktake process, they can provide a good assessment of the quality of implementation of climate action at national level and create synergies for a more comprehensive global climate regime.

8 Stakeholder perceptions of the implementation capacity of the climate change regime

Tim Cadman and Tek Maraseni

Clearly, the effective implementation of the Paris Agreement will be critical to ensuring that the responses developed at COP 21 actually reduce emissions. Public policy theorists look in some detail at the governance issues surrounding implementation, largely on account of their interest in legitimacy (discussed below). Pierre and Peters argue that in order to determine whether a given policy objective has been implemented effectively, it is necessary to trace the final effects of a given policy and its related programmes on society[1]. EU scholars stress the relationship between implementation and compliance[2]. Compliance results from a process of substantive assessment of international rules, in so far as such rules are compatible with existing norms and beliefs; essentially, a rule is complied with if it is considered to be appropriate by stakeholders. A second view places normative influences at a higher level, whereby member states feel obliged to follow EU law, depending on the general culture of compliance within a specific state[3]. Implementation deficits have been alternatively identified as arising from the tensions between on-the-ground learning (open and decentralised) and the need for administrative discipline (hierarchical and centralised)[4]. Zaelke *et al.* define compliance as 'a state of conformity or identity between an actor's behaviour and a specified rule'[5]. They identify compliance as arising from two interrelated, but separate concepts, implementation and effectiveness[6]. Implementation is 'the process of putting... commitments into practice'[7]. In this context, effectiveness

---

1 Pierre, J. and Peters, B.G. (2000), *Governance, Politics and the State*, London: Macmillan Press, p. 31
2 Mastenbroek, E. (2005), "EU Compliance: Still a 'Black Hole'?", *Journal of European Public Policy*, 12 (6), pp. 1103–1120
3 Mastenbroek (note 2), pp. 1111–1112
4 Dryzek, J.S. (1997), *The Politics of the Earth: Environmental Discourses*, Oxford and New York: Oxford University Press, pp. 80–82
5 Raustiala, K. and Slaughter, A.-M. (2005), "International Law, International Relations and Compliance", in *Handbook of International Relations*, Zaelke, D., Kainarum D. and Kružíková, E. (eds.), London: Cameron May, p. 539
6 Zaelke, D., Kainarum D. and Kružíková, E. (2005), "Making Law Work: Environmental Compliance & Sustainable Development", in *Handbook of International Relations*, London: Cameron May, p. 22
7 See Young, O.R. and Levy, M.A. (1999), "The Effectiveness of International Environmental regimes", in *The Effectiveness of International Environmental Regimes: Causal Connections and Behavioural Mechanisms*, Young, O.R. (ed.), Cambridge MA: MIT Press, pp. 3–4

is presented as a measure of the extent to which a policy has been successful in solving the problem it was created to address[8]. Compliance is consequently portrayed 'as a valuable proxy for effectiveness'[9].

Skjærseth *et al.* examine the effective implementation of international environmental agreements[10]. For them, 'an institution is effective if it contributes significantly to solving the problems that motivated its establishment, notably by shaping the behaviour of relevant target groups' (identified as problem-solving effectiveness and behavioural effectiveness)[11]. The relationship between behaviour-change and social-learning within institutions has been linked to transforming the perceptions of participating organisations about how to solve problems[12]. In particular, processes of learning develop an individual's capacity to deal flexibly with new situations[13]. The implications of cultivating such an institutional approach to problem-solving is that governance systems, which incorporate degrees of flexibility, are more resilient in the face of external change and may even benefit from it. Non-resilient systems on the other hand are vulnerable to change[14].

Here, the linkages between behaviour-change, problem-solving and durability are clear. These are explored in the broader evaluation of the governance quality of climate regime included below. An evaluation of stakeholder perceptions in the lead-up to COP 21 provides some insights into which current governance arrangements could present challenges and opportunities in the next period of negotiations, prior to 2020. The remainder of this paper therefore contains a discussion of the research approach adopted, and an assessment of the institutional governance of the regime, including stakeholder observations, followed by analysis and conclusions.

Research approach

The rise of 'new' governance as a means of co-ordinating the processes and structures of governing beyond traditional methods of public administration, has arisen in more recent decades[15]. Governance, as opposed to government, has

---

8 Zaelke (note 6)
9 See Mitchell, R.B. (1996), "Compliance Theory: An Overview", in *Improving Compliance with International Environmental Law*, Cameron, J., Werksman, J. and Roderick, P. (eds.), London: Earthscan, p. 25
10 Skjærseth, J.B., Stokke, O.S. and Wettestad, J. (2006), "Soft Law, Hard Law, and Effective Implementation", *Global Environmental Politics*, 6 (3), pp. 104–120
11 Underdal, A. (1992), "The Concept of Regime Effectiveness", *Cooperation and Conflict*, 27, pp. 227–240.
12 Haas, E.B. (1991), "Collective Learning: Some Theoretical Speculations", in *Learning in United States and Soviet Foreign Policy*, Breslauer, G.W. and Tetlock, P.E. (eds.), Boulder, CO: Westview, p. 63; Haas, E.B. (1990), *When Knowledge is Power*, Berkeley, CA: University of California Press
13 Folke, C., Hahn, T., Olsson, P. and Norberg, J. (2005), "Adaptive Governance of Social-Ecological Systems", *Annual Review of Environment and Resources*, 30, p. 441
14 Folke (note 14), pp. 463–464
15 Rhodes, R. (1997), *Understanding Governance: Policy Networks, Governance, Reflexivity and Accountability*, Buckingham: Open University Press, p. 48. See also Salomon, L. (2002), "The New Governance and the Tools of Public Action: An Introduction", in *The Tools of Government: A Guide to the New Governance*, Saloman, L. (ed.), Oxford: Oxford University Press, pp. 1–41

been identified as a more relevant lens through which to explore state and non-state interactions in contemporary international relations (IR) theory, particularly where these relations intersect with the state, society, and the market[16,17]. Policy-making institutions in this context have been characterised as focusing more on structure and process than command-and-control models of governing, and more concerned with stakeholder participation in decision-making[18]. In the intergovernmental arena, particularly environmental policy regimes (such as UNFCCC), the nature of stakeholder interactions contributes substantively to the effectiveness of governance, and the behaviour of actors one to another, largely determine how regimes are constituted and constructed[19].

For the purposes of understanding the nature of the relationship between the various governance values described below, a framework of principles, criteria and indicators (PC&I) which delineates the relationship between these values is utilized, following the definitive study by Lammerts van Beuren and Blom[20]. PC&I have become an important method for evaluating the effectiveness of sustainable development, popularised through the UN Conference on Environment and Development (UNCED), and its predominantly voluntary, 'soft' approach to norm creation[21]. A *principle* represents a fundamental value to be determined. *Criteria* constitute categories of conditions that make up the principle. *Indicators* may be either qualitative or quantitative measurements, which are aggregated under the relevant criterion, and thereby contributing to the principle, and determination of performance overall. They are placed within a hierarchical framework, from principles, to criteria to indicators, to allow for assessment in a coherent and consistent manner, so as to avoid duplication and/or redundancy[22].

In Table 8.1, the *principle* of meaningful participation is made up of two *criteria*: interest representation and organizational responsibility. This principle is concerned with governance structure. Interest representation is made up of three indicators: inclusiveness, concerning itself with *who* participates in the given institution; equality, demonstrating the *quality* of relationships between participants; and resources, referring to the *capabilities* (financial, technical, institutional, etc.) that a participant can draw on to ensure their interests are represented.

---

16 van Kersbergen, K. and van Waarden, F. (2004), "Governance' as a Bridge Between Disciplines: Cross-disciplinary Inspiration Regarding Shifts in Governance and Problems of Governability, Accountability and Legitimacy", *European Journal of Political Research*, 43, p. 149
17 Kooiman, J. (2000), "Societal Governance: Levels, Models, and Orders of Social-Political Interaction", in *Debating Governance: Authority, Steering and Democracy*, Pierre, J. (ed.), Oxford: Oxford University Press, pp. 138–166
18 Pierre (note 1), p. 14
19 Haas, P. (2002), "UN Conferences and Constructivist Governance of the Environment", *Global Governance*, 8 (1), p. 74
20 Lammerts van Beuren, E.M. and Blom, E.M. (1997), *Hierarchical Framework For The Formulation Of Sustainable Forest Management Standards*, Leiden: The Tropenbos Foundation
21 Rametsteiner, E., Pülzl, H., Alkan-Olsson, J. and Frederiksend, P. (2011), "Sustainability Indicator Development – Science or Political Negotiation?", *Ecological Indicators*, 11 (1), pp. 61–70
22 Lammerts and Blom (note 20), pp. 15–25

Table 8.1 Hierarchical framework for the assessment of governance quality

Principle	Criterion	Indicator
'Meaningful participation'	*Interest representation*	Inclusiveness Equality Resources
	Organisational responsibility	Accountability Transparency
'Productive deliberation'	*Decision-making*	Democracy Agreement Dispute settlement
	Implementation	Behavioural change Problem-solving Durability

Source: Cadman 2011, reproduced with permission from Palgrave Macmillan

Organizational responsibility is made up of two indicators that are often inter-connected: accountability, concerning the degree to which participants' actions can be held to account by other actors (like them, usually delegated representatives of state or non-state organizations) as well as the general public; and transparency, which concerns itself with the visibility of participants' actions, to other actors as well as the public. The principle of productive deliberation consists of two criteria: decision-making and implementation, and addresses the processes of governance. Three indicators are linked to decision-making: democracy, i.e., whether procedures are in place for the exercise of preferences; agreement, addressing how preference selection is applied (voting, or consensus, and so forth); and dispute settlement, which refers to mechanisms for resolving conflicts or taking action when it is not possible to reach agreement. As indicated above, implementation is made up of three indicators: behaviour-change, determining the level to which putting agreements into practice leads to modified conduct; problem-solving, which concerns itself with the degree to which the initial predicament that led to institutional formation is successfully addressed; and durability, comprising longevity, flexibility and adaptability of the problem-solution[23].

Research method

Governance quality was evaluated by means an online survey conducted in 2015, prior to Paris, using an assessment framework of principles, criteria and indicators.

---

23 The terms used here were based on an integrated literature review of over 250 texts in the political science disciplines of comparative politics, public administration, international relations and environmental policy; and from the standards-setting literature. A bibliography is available at: http://eprints.utas.edu.au/9288/

Respondents were selected from state and non-state actors specific to the climate regime (i.e., parties to the Convention, observers, engaged stakeholders and so forth). The intent was to determine their views on the quality of the governance arrangements underpinning the climate regime overall (i.e., perspectives regarding the governance quality of UNFCCC). The subjects were recruited by means of an internet search of documentation containing email addresses of stakeholders active in the field of climate policy and policy processes. Respondents were from both developed countries ('Global North') and developing countries ('Global South'), and could further identify if they were from environmental, social, economic, governmental, academic or regime-specific institutional sectors. Over one hundred respondents participated in the survey. Respondents were also invited to provide specific comments.

Using this framework, participants were asked to select and rate UNFCCC on the basis of the 11 governance indicators of Table 8.1 by means of a five-point Likert scale, using the terms 'very low', 'low', 'medium', 'high' and 'very high'. The questions in the survey, based on the indicators of Table 8.1, are reproduced in Table 8.2 below.

These results are presented using respondents' geopolitical location, and on the basis of sector (environment, government, social, etc.). See Table 8.4.

In the case of sectoral analysis, it should be noted that some cohort sizes were too small to generate representative results (with the exception of Environmental and Academic), and are therefore presented more for their anecdotal value than for representativeness.

Table 8.2 Summary of survey questions

Question	Indicator
Do you think [regime element] is inclusive of your interests?	Inclusiveness
Do you think [regime element] treats all interests equally?	Equality
What level of resources does [regime element] provide for you to participate?	Resources
Do you think [regime element] is accountable?	Accountability
Do you think [regime element] is transparent?	Transparency
Do you consider [regime element] acts in a democratic manner?	Democracy
Do you consider the making of agreements in [regime element] to be effective?	Agreement
Do you consider the settling of disputes in [regime element] to be effective?	Dispute settlement
Do you think [regime element] will contribute to changing the behaviour it was created to address?	Behavioural change
Do you think [regime element] will help solve the problem it was created to address?	Problem-solving
Do you consider [regime element] will be durable?	Durability

Note: explanatory text and introductory materials omitted

Table 8.3 Total number of survey respondents by sector

Sector	Response count
Environmental	42
Social	5
Economic	3
Government	16
Academic	35
Other	6
Total	107

Note: includes those who did not complete the entire survey

Results

Analysis and discussion of results

In the regional survey, the majority of respondents were from the South (57 *cf.* 46). Respondents did not rate UNFCCC very highly, although 'creditable' would be a fair assessment. Southern respondents rated the regime more favourably than those from the North (36.1 out of 55, or 66% *cf.* 31.1 or 57%) – a result that was repeated across both principles. Northern respondents identified decision-making as the weakest criteria overall. At the indicator level, both South and North gave resources the lowest score, but to be fair, the score from Southern respondents was more generous (2.9 out of 5). The highest performing indicator was inclusiveness in both North and South.

In the sectoral results, Government provided the highest score, followed by the Environmental, Academic and Economic sectors. The Social sector consistently provided the lowest scores of all, by which UNFCCC could be said to have 'failed'. All sectors identified resources as the weakest indicator, with the exception of 'other', which selected problem-solving. 'Other' also 'failed' a number of indicators (resources, accountability, democracy, dispute settlement and behavioural change). Caution should be exercised when looking at the results provided by the Social, Economic, 'Other' and Governmental sectors, due to small sample sizes. In the case of Environmental and Academic respondents, however, the sample sizes are sufficiently large to draw some meaningful conclusions. In this case it is interesting to note that both sectors provided similar results overall, and at the principle, criterion and indicator levels; in both cases UNFCCC received a high 'pass' rate, but not a 'credit'.

It is encouraging to note the consistently high score for the indicator of inclusiveness. For those who have attended climate conferences, it is possible to see the broad range of interests that participate. Indeed, that both environmental stakeholder and academic stakeholders should rate inclusiveness as highly as they did in the sector-by-sector survey provides some corroboration of this. Respondents offered some comments on the nature of inclusiveness in the regime. One

Table 8.4 UNFCCC – quality of governance by region (February 2015)

Principle	1. Meaningful participation *Maximum score: 25* *Minimum: 5*					2. Organisational responsibility *Maximum score: 10* *Minimum: 2*				<u>Principle score</u>
Criterion	*1. Interest representation* *Maximum score: 15* *Minimum: 3*				*Criterion score*				*Criterion score*	
Indicator	Inclusiveness	Equality	Resources	Dispute settlement		Agreement	Accountability	Transparency		
North (46)	**3.7**	2.9	**2**	2.5	8.6		3.1	2.8	5.9	14.5
South (57)	4	3.2	2.9	3.1	10.1		3.4	3.4	6.8	<u>17</u>

Principle	2. Productive deliberation *Maximum score: 30* *Minimum: 6*					4. Implementation *Maximum score: 15* *Minimum: 3*				<u>Principle score</u>
Criterion	*3. Decision-making* *Maximum score: 15* *Minimum: 3*				*Criterion score*				*Criterion score*	
Indicator	Democracy	Agreement	Dispute settlement			Behavioural change	Problem-solving	Durability		
North	2.6	2.9	2.5		8	2.8	2.6	3.3	8.6	16.6
South	3.2	3.2	3.1		9.5	3.2	3	3.5	9.7	<u>17.4</u>
Total (out of 55)										
North										31.1
South										36.1

Note: light grey represents the highest scoring indicator by region; dark grey the lowest; numbers in **bold** are below the threshold value of 50%

Northern academic (Canada) noted that the regime was 'a good example of a process that is distanced from the local discourse, even though there is a lot of local discourse on the issues that fall within the scope of the UNFCCC'. Another respondent from 'Other' (Nepal), who identified themselves as 'private sector' thought that it was not so much a question of inclusiveness that affected interest representation within the regime North and South had similar levels of participation, but 'power balance', whereby 'south people rightly could not communicate'. Consequently, they argued that: 'more representation is needed from [the] South'. Another Southern Environmental respondent (India) made the point that 'awareness about the actual provisions of most of the international initiatives is very low at the grassroots, which does not enable [the] poor to participate in these'. According to a second Southern Environmental respondent (Malawi), it was the fact that the application of the programmes of the regime were 'too cumbersome and very expensive for developing countries to manage', leading them to conclude that regime was that programmes were 'mainly geared towards benefiting developed countries'.

In view of the suggested linkages to compliance, and therefore implementation capacity, it is disconcerting to see that resources was the lowest scoring indicator, almost without exception. Providing adequate resources for stakeholders to represent their interests effectively is central to meaningful participation. A good example of this problem in action at the climate talks is the size of delegations. The larger – or richer – the country (such as the US), the greater the number of delegates to cover all the various negotiating streams. In the case of the smaller – or poorer – countries, such as the small island developing states (SIDs), some are lucky to manage even one (in the case of Tuvalu). Inclusiveness by itself is insufficient to ensure adequate interest representation. Financial, institutional and educational capacity-building would go some way to ensuring that the representation of climate-vulnerable states' interests (notably the SIDs) had an impact on climate policy, rather than being merely tokenistic. Having said that, the fact that developing country respondents consistently viewed the regime more favourably than their developed country counterparts is a positive sign. Why this should be is a matter of speculation, but it may be connected to the issue of resources. Southern respondents, as recipient countries of climate finance, rated this indicator considerably higher than their Northern counterparts, who are, in the main, donor countries. Here, there may be some disaffection at play. With the change in dynamics post-Paris regarding differentiation, and with the focus more on common responsibilities, this may change.

There are a few other notes of caution regarding the governance quality of the climate regime, and again, sounded by developed country respondents. Dispute settlement in the UN system, given its make-up of territorial states, is often characterised by coercion and bargaining, rather than genuine consensus-seeking[24]. That Northern respondents should provide only a lukewarm assessment

---

24 Keohane, R.O. (2003), "Global Governance and Democratic Accountability", in *Taming Globalization: Frontiers of Governance*, Held, D. and Koenig-Archibugi, M. (eds.), Cambridge: Polity Press, p. 139

Table 8.5 UNFCCC – quality of governance by sector (February 2015)

Principle	1. Meaningful participation Maximum score: 25 Minimum: 5					2. Organisational responsibility Maximum score: 10 Minimum: 2			
Criterion	1. Interest representation Maximum score: 15 Minimum: 3								Principle score
Indicator	Inclusiveness	Equality	Resources		Criterion score	Accountability	Transparency	Criterion score	
Environment	3.9	3.1	2.9		9.9	3.2	3.2	6.4	16.3
Social	2.8	2	1.4		6.2	2	2	4	10.2
Economic	4.3	1.7	1		7	3.7	3.7	7.4	14.4
Government	4.3	3.4	2.8		10.5	3.7	3.4	7.1	17.6
Academic	3.8	3.1	2.3		9.2	3.5	3.1	6.6	15.8
Other	3.5	3.3	2.4		9.2	2.4	2.6	5	14.2
Avg. of all	3.8	2.8	2.1		8.7	3.1	3	6.1	14.8

Principle	2. Productive deliberation Maximum score: 30 Minimum: 6			4. Implementation Maximum score: 15 Minimum: 3				
Criterion	3. Decision-making Maximum score: 15 Minimum: 3							Principle score
Indicator	Democracy	Agreement	Dispute settlement	Behavioural change	Problem-solving	Durability	Criterion score	
Environment	2.9	3	2.9	3.1	2.9	3.5	9.5	18.3
Social	1.8	1.6	2	2	2	2.4	6.4	11.8
Economic	2.3	2.7	2	3.3	3	3.7	10	17
Government	3.4	3.5	2.9	3.3	3.1	3.4	9.8	19.6
Academic	3.1	3.2	3	3	2.9	3.4	9.3	18.6
Other	2.4	2.8	1.8	2.2	1.5	3	6.7	13.7
Avg. of all	2.7	2.8	2.4	2.8	2.6	3.2	8.6	16.5
Total (out of 55)								
Environment				34.6				
Social				22				
Economic				31.4				
Government				37.2				
Academic				34.4				
Other				27.9				
Average of all				31.3				

Note: light grey represents the highest scoring indicator by sector; dark grey the lowest; numbers in **bold** are below the threshold value of 50%

of the regime's democratic and problem-solving capacities are further indications that more effort is needed in these aspects: especially around problem-solving, given what is at stake. But there are other signs for cautious optimism as well. Southern respondents alike, as well as the Governmental, Academic and Environmental sectors, all provided relatively high ratings for the accountability and transparency of the regime (although again, Northern respondents were not overly enthusiastic regarding transparency – 2.8 out of 5 – even if this indicator did pass). In this regard, it might be said that the climate regime embodies these 'thickish' values.

Looking specifically at implementation and its related indicators of behaviour-change, problem-solving and durability, the differences between Northern and Southern respondents become apparent again, with the South showing a high perception of governance quality (a low 'credit', *cf.* a low 'pass' from the North). This was repeated at the indicator level, with developed countries providing lower ratings for behaviour-change and problem-solving. It is worth noting that there was a near-convergence of views regarding the durability of the regime (3.3 from Northern respondents, *cf.* 3.5 from the South). Apparently, both sets of respondents would seem to be relatively confident that the regime will endure – not surprisingly, given its longevity to date. However, the problem-solving capacity of the regime was identified as the weakest of the three indicators across regions. These trends at the indicator level were also apparent in the sector-based analysis of results, with problem-solving being the weakest of the three, and durability the highest. The difference between environment and government respondents concerning implementation at the criterion level was not especially pronounced (a high 'pass' compared to a low 'credit'), but a note of caution should be sounded regarding the validity of Government respondent results, given the low sample size. Overall, however, it may be concluded that respondents believed that the regime would be durable, but were less optimistic that it would solve the problem it was created to address.

Respondents provided a number of observations regarding implementation, and its associated elements. Speaking at a general level, one Northern Academic (south-eastern Europe) noted that it was 'very difficult to implement initiatives in the country where [there is] government corruption, and where the initiatives [are] decided in a small narrow circle of people'. Another Northern respondent (Switzerland) who identified as 'Other', but with a 'socio-economic' interest suggested that 'need some form of sanctions' was required, 'or at least shaming for not implementing' the agreements made.

Another Northern Academic (USA) commented in some detail:

> UNFCCC has been around for years, yet there is no evidence of improved climate mitigation. Accountability to the objective is lacking. It will be necessary to get to a simpler structure such as an income-neutral carbon tax that actually is effective and supported by some international leaders begging the rest to come along, as it is proven to be effective. Most current support is both sporadic – aimed at options with the least scientific basis. CDM focuses on

incentives without accountability to the impacts because the carbon linkages are so many and interactive, beyond the capability of bureaucratic decision-making.

This individual was of the view that this required 'more science and less politics, a tall order'. Their analysis was that there was little behaviour-change because the 'value of carbon credits' was 'near zero', meaning there was 'no reason to do anything different'; for them, it was 'obvious the program is broken'. Once there was a 'cost on emissions there will be interest and a behaviour-change to do better'. They concluded that carbon trading would 'affect the prices of big utilities if there was an accountable system of what constitutes an offset', and they argued that the 'big utilities' were 'not really the problem'. Rather, they would 'respond to the actions of consumers and policy makers'.

On a more positive note, one Southern Economic respondent (Namibia) provided a more generous assessment: 'In the overall span of human history[,] the idea of global multilateral environmental governance is so recent and starts off from such a low base that its consensus-building achievements to date are nothing short of astonishing.'

How might the climate regime respond to these findings? Given the high rating for inclusiveness, it would be wise for the climate regime to capitalise on multi-stakeholder support to increase the ownership of negotiated outcomes. This could be done by formalising non-state engagement and interest representation. Here, it is worth noting that the Paris Agreement pays some considerable attention to the role of 'non-Party stakeholders'[25]. But in order to be effective, these initiatives need to be accompanied by enhanced role in decision-making. Non-state actors are still locked out of any formal role in reaching agreements under the convention. Nevertheless, this is a positive sign, as is the transition away from the strict developed-developing country delineation between the Annex and Non-annex parties to the Kyoto Protocol, and is now epitomised by the more inclusive, universal approach of combating climate change via nationally determined contributions.

It is a pervasive problem across the UN system that there is never enough money and capacity – whether it is technical, infrastructural or institutional. Better resourcing is required for effective climate finance, stakeholder participation and implementation-related activities. It is therefore encouraging that more recognition has been given to the need for increased resources under the Paris Committee for Capacity-Building (PCCB), although it is disappointing to see that this is a 'should' rather than 'shall' provision for parties[26]. Dispute settlement is another problematic aspect of the regime, and it is important that action is taken to develop clear dispute-settlement mechanisms across the thematic elements of the regime and related sub-institutions. Here there is no change, with the consequence that when countries fail to deliver on their commitments, the negotiations are likely to become bogged down once again.

25 Decisions 122–124 and section V (Decisions 134–137)
26 Decisions 72–84 and Article 11

Nevertheless, the implementation capacity for the new Paris Agreement is an improvement on Kyoto, if ratified by a sufficient number of parties ('at least 55 Parties to the Convention', counting for 'at least an estimated 55 percent of the total global greenhouse gas emissions'[27]). The creation of a new mechanism for implementation and compliance is a step forward. The existing Compliance Committee under Kyoto was not known for taking substantive action. In this regard, it is not entirely positive that it will 'function in a manner that is transparent, non-adversarial and non-punitive'[28].

These observations lead to the conclusion that it is time to think about more consistent standards for the mechanisms/programmes/sub-institutions across the regime. At this stage in the regime's evolution it is vital to move beyond the 'safeguards'-based approach and arrangements for delivering effective and transparent governance developed at Cancun, particularly in the light of the mega-funds that are emerging as the drivers of climate change mitigation and adaptation, and the complex constellations of multi-stakeholders who wish to combat anthropogenic climate change. Without such standards, the potential for competition could lead to a 'race to the bottom', where only the weakest criteria are implemented. This will serve no interests, least of all the investor/donor community, nor those countries most in need of finance for adaptation to the worst impacts of human-caused climate change, which are often the poorest and least developed.

---

27 Paris Agreement
28 Article 15, para. 2

9 Technological ethics, faith and climate control

The misleading rhetoric surrounding the Paris Agreement

Harold P. Sjursen

What is referred to as climate change is an aggregate of problems including the warming of the Earth's atmosphere to a level where many interactive natural processes are challenged. Climate change is thus a term that summarizes many data points from many systems. Our understanding of the constituent systems is not uniform and draws upon sciences that employ variations of method making their integration into a single assessment difficult. This complexity challenges our ability to understand the nature of the threat (if indeed it is a threat) that extreme climate change portends, but it likewise makes knowing what to do all the more difficult. Of course, the question of knowing what to do is highly politicized, with differing national and regional priorities leading to multiple interpretations of the nature and severity of the threat.

With the election of Donald Trump as President of the United States, the political dynamics influencing the global response to how to address the variously interpreted matter of climate change became even more fraught than they were at the time of the Paris Agreement, the approval of which was to a significant degree influenced by an understanding reached between China and the United States under the leadership of President Obama. The assertion by President Trump that the United States will withdraw from the Agreement, from the perspective of scientific assessment, may not make too much difference as new technologies to reduce emissions will continue to be deployed in the United States for economic reasons, but the political turmoil and the sense of what is fair or just is intensified. In various (especially developing) countries, national policies that imply an increase in suffering and cost may now seem exploitative and beneficial to affluent nations at their expense. This reality makes any declaration of a course of action all the more difficult.

Knowing what to do, similarly, is a multi-levelled problem. One level is self-interest where factors of comfort, convenience, safety, economic success and others come into play. One of the apparent consequences of global climate change is the more frequent occurrence of extremely heavy rain storms. The costs incurred in an affected port city like Hong Kong are enormous. Climate change in this example becomes, among other things, a matter of cost-containment necessary to keep the shipping industry viable. Thus one can argue on straightforward business or economic terms for the necessity of controlling the disruptive effects of tumultuous rain. Depending upon cost, this might be in the form of installing better drainage sewers, developing more effective modes of flood abatement, or some other kind of response to improve the infrastructure to manage the increased volume of water. Because it is less immediate and arguably costlier, as well as more ambiguous in its result, to eliminate or significantly abate the source of climate change on this level is a less attractive solution.

The incentive is limited because under no circumstance will the result be the end of climate change nor the return to the predictable and relatively stable climate patterns of an earlier era in the 20th century that many have come to regard as normal. Regardless of whatever measures are put in place, significant and irreversible changes have taken place, and much of the future will have to be taken up with the establishment of new social institutions to respond to the mostly unprecedented circumstances due to new flood-drought cycles, different zones for agriculture, and the consequent shortages of necessary goods and interruptions of essential services. The world may come to look more like a Hobbesian *state of nature* than it ever has before.

Indeed, the environmental activist Bill McKibben argues that our focus now should be on scaling back and building societies that can manage the unprecedented troubles – difficulties that portend the failure of many of the norms of civil society – that will inevitably result from climate change[1]. Whereas McKibben is resigned to the inevitability of climate change-induced crisis, the call to pre-emptive action is still being made.

In a speech on the topic of the Paris climate Agreement, delivered at the Stern School of Business at New York University on 30 May 2017, António Guterres, Secretary-General of the United Nations, stated:

> The moral imperative for action is clear. The people hit first and worst by climate change are the poor, the vulnerable and the marginalized. Women and girls will suffer as they are always the most disproportionately affected by disasters. The nations that will face the most profound consequences are the least responsible for climate change and the least equipped to deal with it. Droughts and floods around the world mean poverty will worsen, famines will spread and people will die. As regions become unlivable, more and more people will be forced to move from degraded lands to cities and to other nations. We see this already across North Africa and the Middle East. That is why there is also a compelling security case for climate action. Around the world, military strategists view climate change as a threat to global peace and security. We are all aware of the political turmoil and societal tensions that have been generated by the mass movement of refugees. Imagine how many people are poised to become climate-displaced when their lands become unlivable. Last year, more than 24 million people in 118 countries and territories were displaced by natural disasters. That is three times as many as were

---

1 McKibben, Bill (2010), *Eaarth*, New York: Henry Holt, *passim*

displaced by conflict. Climate change is also a menace to jobs, to property and to business. With wildfires, floods and other extreme weather events becoming more common, the economic costs are soaring.[2]

Guterres' message is unambiguous: questions of climate science are settled; devastating economic consequences are foreseen; the nations of the world have a moral imperative. By saying that it is a moral imperative, he implies that through the implementation of the Paris Agreement a salutary difference will be made and that we can reasonably hope for a continuation of normalcy for our grandchildren and great-grandchildren.

However, there are other levels to the discussion. One is at no loss to specify the variety of pragmatic approaches to the larger problem that will probably resolve one or several crucial aspects but equally probably not the general problem. Unfortunately, it is not likely that a piecemeal approach, often good engineering practice, will work to fix the problem generally, although the strategy of the Paris Agreement – for real political reasons – appears to be just that.

The problem is that it is global, and the sum of the self-interests does not equal the whole. More than that, it is not so much a problem for the 'now' as it is for the 'future' when both the causal conditions and the world suffering the change may be fundamentally different. There is very little analogy between the kind of problem-solving exercise needed to address flooding in Hong Kong and the sort of thinking required to engage the multiple questions regarding the effects of mutations to the natural environment upon a world beyond the horizons of our imagination. In a sense this is not a 'problem' at all, although certainly economics, natural science, engineering and technology and politics are all in play. But overwhelmingly this is an ethical issue. What must we, citizens of the 21st century, do in order to act responsibly on behalf of a future world community that we do not know and will never meet? What sort of imperative has been given to us? How do we understand our duty to the future? And were we to understand this imperative and duty, would we be able to carry it out? These questions are ethical questions and must be considered as such.

Bjorn Lomborg is well-known as a sceptic when it comes to the approaches to climate change advanced in Kyoto, Copenhagen and now Paris. In an article he himself lauds as 'my peer reviewed research paper', he asserts that the agreement forged in Paris in light of the benefits he has calculated if it is implemented will be 'the costliest in history'. Here is the abstract from his paper:

> This article investigates the temperature reduction impact of major climate policy proposals implemented by 2030, using the standard MAGICC climate model. Even optimistically assuming that promised emission cuts are maintained throughout the century, the impacts are generally small. The impact of the US Clean Power Plan (USCPP) is a reduction in temperature

---

2 Guterres, António (2017), *Vital Speeches of the Day*, 83 (7), p. 197

rise by 0.013°C by 2100. The full US promise for the COP21 climate conference in Paris, its so-called Intended Nationally Determined Contribution (INDC) will reduce temperature rise by 0.031°C. The EU 20-20 policy has an impact of 0.026°C, the EU INDC 0.053°C, and China INDC 0.048°C. All climate policies by the US, China, the EU and the rest of the world, implemented from the early 2000s to 2030 and sustained through the century will likely reduce global temperature rise about 0.17°C in 2100. These impact estimates are robust to different calibrations of climate sensitivity, carbon cycling and different climate scenarios. Current climate policy promises will do little to stabilize the climate and their impact will be undetectable for many decades.[3]

The main critical points Lomborg makes were summarized as follows in a blog posting by Marlo Lewis:

1) Like its predecessor, the Kyoto Protocol, the COP 21 treaty is likely to be a costly exercise in futility – substantial economic pain for no discernible environmental gain.
2) COP 21 is envisioned by its architects as a first step. Decarbonizing of global energy is to be achieved through a succession of more aggressive emission-reduction targets beyond those specified in the current INDCs.
3) Those who say we must adopt the Paris treaty for the sake of our children and grandchildren have not thought things through. Under the global regime envisioned by COP 21, each generation will have to make greater economic sacrifices than their parents did.
4) Since developing country emissions already exceed and are increasing much more rapidly than industrial country emissions[4], the biggest emission cuts under COP 21 and beyond must come from precisely those countries that can least afford to cut emissions.
5) Lomborg says his analysis "clearly indicates that if we want to reduce climate impacts significantly, we will have to find better ways than the ones currently proposed"[5].

Lomborg's position, if his analysis holds up, is fairly simple. It is that unless a solution is economically sustainable its implementation is unlikely. If it is carried out it will not ultimately achieve the intended purpose because of the unintended consequences. His view situates possible solutions to climate problems within the constraints of contemporary capitalism and social/political expectations and does

---

3 Lomborg, Bjorn (2015), "Impact of Current Climate Proposals", *Global Policy*, 7 (1), November, pp. 109–18; doi:10.1111/1758-5899.12295
4 www.globalenergyinstitute.org/european-unions-2050-global-greenhouse-gas-emissions-goal-unrealistic
5 www.globalwarming.org/2015/11/10/lomborg-exposes-paris-accords-climatological-insignificance/

not acknowledge any ethical imperatives that might question either the efficacy or the justice of these constraints or an approach that demands their revision. Whereas Guterres sees a clear moral imperative to arrest an impending global crisis, Lomborg seems to view such an approach as unwarranted, ineffective and ultimately damaging meddling.

Whether or not the facts of climate change propose a moral imperative is an important question, since under that category an entirely different set of actions may be justified. In some cases, an ethical imperative will compel sacrifice or deprivation; that is, under such circumstances, what from an economic point of view is not sustainable as a general practice, may as an exception be compelled by a moral or ethical imperative. The forgiving of loans provides an ordinary example of this principle. Certainly, as a general practice it is not economically sustainable for a financial institution to forgive debts, but under specific circumstances it *should* be done on ethical grounds. That is to say, it should be done even when a consequence might be the serious weakening or even destruction of the financial institution. A question, therefore, is whether the scientifically calculated prospect of climate change and the COP 21 formula for remediation is ethically required even if doing so will weaken global finances in a way that undermines the economic well-being of many.

As a platform on which to debate the ethical issue, specifically the notion that there is an ethical imperative, i.e., duty or obligation, that mandates that we collectively act in ways that may lead to a degree of suffering, and which may also abrogate ordinary notions of fairness, the general proposition of Bjorn Lomborg will be granted, not because it is evidently correct, but because it requires that the ethical questions surrounding global climate change be given full consideration.

In what follows, the criticism of Lomborg will be considered from an ethical perspective that is not limited to the conditions proscribed by economic values he believes to be axiomatic. Indeed, the agreement suggested by the Paris accords does not limit itself to a singular set of economic principles in so far as it explicitly honours that different countries, especially less-developed and wealthy countries, will according to their own various governance models implement the principles of the treaty in various ways. This allowance was the strategy, different in approach from Kyoto or Copenhagen, that makes all signatories responsible for the achievement of the overall goal of the treaty to limit the rise in global temperature to under 2 degrees from pre-industrial levels.

Yet, in Lomborg's view, whether a single economic blueprint were imposed or numerous national approaches to meet the general goals were taken, the result would be that destructive costs and ever-expanding responsibilities would overwhelm whatever benefits resulting from the mitigation of changes to the climate were achieved.

The key question, then, centres on the kind of ethical responsibility that is invoked by the treaty, and more generally to the broad issue of climate change. Lomborg's critique implies either that 1) the approach taken in or implied by the treaty is not as such responsible, or 2) that the responsibilities implied by the treaty cannot be achieved because the economic burdens and escalating costs that will ensue will entail a degree of sacrifice and a burden that is unfair to future generations, particularly in underdeveloped countries. So the treaty will either fail or do significantly more harm than good while obviating economic freedom of choice. This is not an argument to do nothing nor to deny that there may be an ethical imperative; rather his claim would seem to be that the provisions of the Paris Agreement and the strategies for implementation do not express an ethical imperative.

But parsed in this fashion, Lomborg's position is essentially an ethical one resting on ideas of fairness and freedom. He is opposed to the treaty because, framed within his economic assumptions, it would not be a responsible action because it promises what is unattainable and is therefore a kind of lie; if it is forced it will do great harm and/or cause much unnecessary suffering. Lomborg's argument in part is that it is simply wrong for an agreement to deceive or lead to unnecessary suffering. This is an ethical position, although one that accepts *a priori* an economic world order based on certain fundamental principles of global capitalism. Is it possible to step back and understand the ethics of the situation in a manner that is not registered within current political/economic worldviews?

Perhaps some economists might argue that the current system of global capitalism cannot be adjusted, that markets alone are determinative, but if so that means that ethics, law and politics become subordinated to economics. This idea seems incoherent because much of economics is an expression of political will. So the question here becomes, can ethical considerations and priorities demand an adjustment to the economic status quo? Given the consensus that the climate change-induced global warming that is largely the consequence of human behaviour is actual and portends grave destruction to the natural environment that supports the quality of life we know and enjoy, an ethical imperative to do something – to change business as usual – has emerged. Lomborg suggests that the Paris Agreement is not that imperative.

The ethical question

António Guterres' predecessor as United Nations Secretary-General, Ban Ki-moon, in a message for the World Day of Social Justice in February 2011, declared:

> Social justice is more than an ethical imperative; it is a foundation for national stability and global prosperity. Equal opportunity, solidarity and respect for human rights – these are essential to unlocking the full productive potential of nations and peoples.[6]

This formulation finds moral imperatives within the larger category of social justice. In a sense, this is consistent with Lomborg, but makes clear that the

---

6 United Nations (press release): www.un.org/press/en/2011/sgsm13403.doc.htm

economic system should reflect principles of social justice. Let us suggest that social justice implies fairness and responsibility implies freedom and its attendant self-determinism.

Fairness and freedom are both notions that are problematic within the histories of ethics and political theory. John Rawls has famously argued that justice rests upon fairness and tried to reconcile differing views of what constitutes fairness in order to arrive at a universally accepted concept of justice. Despite his grand effort, global justice remains a much disputed idea. However connected, responsibility is perhaps an even more troubled notion. Commonly, the notion of responsibility is associated with accountability. This is insufficient as an explanation because in many instances clear accountability is not possible. Since as a concept responsibility is something like duty or obligation, the first question is the derivation of duties or obligations.

A *prima facie* responsibility to others includes fairness. One cannot exercise responsibility toward others in an unfair way: 'I am responsible for you' or 'My actions on your behalf are based upon my responsibility to you' cannot coherently include exploitive or damaging actions. In the sense that I am responsible for my ward, responsibility minimally demands that my actions be fair. One can say that fairness goes in both directions and thus means that responsibility does not extend to actions that damage me or my interests. Let us consider this a weak or minimalist notion of responsible action, i.e., actions which are mutually fair in their care for another. In short, we can call responsibility in this sense a matter of *fair care*. In fair care actions that affect many, in many-to-many actions for example, all affected must share the burden and responsibility, whether as initiator or recipient. (The objection that may arise in the minds of some, that we do not have a responsibility for the future, except possibly for the immediate future, because we do not interact with it, will be ruled out. We do not interact with the distant or indefinite future directly, but through the extraordinarily increased empowerment of modern technology, but we do – mediated by such technology – affect changes that surely will have a significant impact on this unknown future.)

However, in the case of many-to-many actions that are also long term, and especially those where the consequences may not be recognized until sometime in the fairly distant future, the determination of what would constitute fair care actually becomes nearly impossible. The nature and magnitude of the burdens, financial and otherwise, that the Paris Agreement will create is a function of many interacting factors whose efficacy will change due to undetermined future events. As a general rule this is a problem with technologically driven actions. Modern technology permits, often with great ease for the performer, actions that result in very powerful changes (such as permanent and irreversible alterations to the environment) that become manifest long beyond the horizon of the enabler's awareness. Consider the uses of nuclear energy as an example. This creates a special type of ethical dilemma, one where the consequences of an action are not and cannot be known by the actor.

One could reasonably assume in such cases that some version of the precautionary principle should be applied: because I cannot know reliably whether an action initiated at this time will result in the future in the desired outcome, and because it is possible that it will result in something quite different and undesired – even possibly disastrous – the action should not be performed until such knowledge can be had. The precautionary principle is that, when unaware of what an outcome may be, but given some risk of serious danger, one should not initiate the action. The precautionary principle borders on a principle of non-action.

Of course this strategy would only make sense if precaution didn't default to something about which we do have the knowledge that an undesirable outcome – indeed a disaster – is already in process. This is the case with climate change. Thus, if we accept our ethical responsibility toward the future, regardless of whatever short-term or self-interested actions we take, we cannot for very long adopt the precautionary principle of watchful waiting. But given our ignorance in whatever degree about the future, what action is proper?

There is a tendency, especially during times when ordinary actions and accountability seem inadequate, to embrace *ad hoc* suspensions of ordinary procedures. In times of war or under the threat of terrorism, for example, the principle of habeas corpus may be suspended. Is the crisis of climate change such that a radical approach that abandons normative legal and economic constraints might be justified? And if so, how is the proper course of action to be chosen? These questions will be taken up below.

Technological intervention

This sort of question led Hans Jonas to proclaim that a new technological ethics, an ethics for the future, was needed. In *Das Prinzip Verantwortung* he argues as follows:

> Modern technology, informed by an ever-deeper penetration of nature and propelled by the forces of market and politics, has enhanced human power beyond anything known or even dreamed of before.
>
> The altered nature of human action, with the magnitude and novelty of its works and their impact on man's global future, raises moral issues for which [all] past ethics ... has left us unprepared.
>
> Responsibility is a correlate of power and must be commensurate with [its] scope ... we need lengthened foresight ... a scientific futurology.[7]

His most salient points in this context are that, due to the kind of power technology has granted us, the very nature of human action is different and cannot be measured by traditional standards; all of our great ethical systems rest upon the

---

[7] Jonas, Hans (1984), "The Imperative of Responsibility, In Search of an Ethics for the Technological Age", Preface to the English edition, University of Chicago Press, pp. ix – x

traditional understanding of human action and thus offer only limited guidance; the fate of humanity and the stability of nature are in our unknowing hands and we must learn how to use them responsibly proportionate with their greatly enhanced strength.

In his work Jonas puts environmental issues among the most crucial and vexing ethical problems. There is no question, of course, that environmental issues are pressing and demand response, but his concern is that lacking an adequate ethical-theoretical basis, the response and the decisions taken to initiate action, lacking such grounding, will be short-sighted and expressive of one or another special interest rather than that of the global, common good. Indeed, he thinks that in the fog of unknowing the likely outcome is disaster for the natural world and of course all life including humanity that is dependent upon it.

The issue of climate change leads to this kind of hand-wringing. Not knowing the extent of our power and thus not knowing how to measure its consequences, we wish for a reset, to go back to a *normal* where climate and the natural processes dependent upon it, are restored. The Paris Agreement purports to chart a way to a condition close to that at the time of the rise of industrialization: '...to pursue efforts to limit the temperature increase to 1.5 degrees C above pre-industrial levels, recognizing that this would significantly reduce the risks and impacts of climate change'[8]. Although the document acknowledges that this goal does not represent a restoration, the spirit in which it is encouraged is the same.

Even before the specific problems of technology and climate change are introduced, the question of how to determine responsibility in the sense of fair care is situated at the intersection of knowledge and duty. Simply put, the determination of what one (an individual, agency or collective) *ought* to do requires a complex consideration of what outcome is sought, in this case justice and fairness, what is possible (on the levels of technology, economics and politics), and to what degree. These factors are constrained by the quality and depth of knowledge and when the outcome is at least partly in the distant future by the lack of knowledge.

There is thus an overall issue within the epistemology of ethics. Ethics is not a matter of calculation and although cost-benefit analyses may help to inform ethical judgment, the latter cannot be reduced to the former. In the modern hospital, physicians, their patients and loved ones, routinely face circumstances where it is obvious *both* that the economic cost of a procedure is a factor that cannot be ignored *and* that the basis regarding the decision to use the procedure at all transcends economic considerations, and calls upon fundamental beliefs and values, individual and collective, contemporary and traditional, that cannot be mapped against financial considerations. In the end, however, many life and death choices are made where the influence of these two incommensurable modes of understanding compete in a rush to judgment. The problem of climate change is analogous. Of course projected costs should not be ignored, and certainly accepting the burden of expenses unlikely to be met undermines the very idea

---

8 From the final document of the Paris Agreement

of responsible choice. Yet the dilemma remains. One approach, the one it seems has been taken by Lomborg, is that *if* the procedure is justified on non-economic grounds *then* we will undertake it *only if* the projected costs can be met without undue burden. The argument being presented here is that this approach does not properly acknowledge ethical imperatives. These ethical imperatives rest upon judgments that cannot be made by calculative reasoning. How, then, are they to be made?

Decisions requiring action often appear to be binary. In the case of climate change a series of binary decisions: 1) Is climate change the reality and likely to persist? If yes → 2) Is it (at least partially) the result of human behaviour that can be changed? If yes → 3) Can we (scientifically/technically) do action x? If yes → 4) Will all parties agree? If yes → Commence action. Questions 1) and 2) pertain to the past and thus should permit adequate answers. Questions 3) and 4) are in the future and along with knowledge of the actual outcome and its impacts are beyond complete or certain knowledge. In the realm of ethics or duty, where we are frequently faced with uncertainty about the impact of an action, our choices are formed within categories compatible with cultural traditions and beliefs. These certainly vary, but in the West the pattern tends to be binary, an ethical either/or. Something of the sort seems evident in Lomborg's approach: it's not worth doing because it won't work as anticipated. What is the ethical motivation within such a stark and impractical attitude?

An understanding of this may be found by exploring the kinds of ethical dilemmas that arise in cases where an undefined duty drives one to take an action. In such a case one is required to carry out an action because of a compelling duty which is for the sake of a general good. The precise nature of that general good is not known – it is to preserve a good in a future context which is not disclosed. This is the case with the climate accords. Although the condition of nature (particularly the global climate and all the interacting systems that it comprises) in the indefinite future cannot be known with a high degree of specificity, it is presumed that it will be analogous to what it is in the present. This presumption is based upon inductive reasoning about the history of the globe's climate, the ambiguity of which has led to a number of disputes regarding the cause of climate variations and the attendant arguments, often a matter of ideological bias, about whether these variations are fundamentally *natural* or result from human activity. But regardless of whether one holds climate change to be induced by human activity or not, the duty to respond comes either from a general obligation to nature or out of a sense of responsibility to the future. Neither admits of a calculated or analytical clarification. Thus, our feeling of an imperative to do something to forestall the degradations surely to occur if global warming continues its ascendant path is motivated by a sense of duty that we embrace as an existential commitment.

Divine intervention?

The dilemma posed by the view that there is an ethical imperative to act in the face of incipient disaster on a global level, an apocalyptic scenario overwhelmingly affecting the already disadvantaged, induced by (mostly) post-industrial human activity, leads to a radical kind of ethical thinking that goes beyond ethics. Presented in this way climate change is an existential threat against which rational calculation can offer no more than a temporary and unequally distributed reprieve. That is to say, the reasoned compromise of the Paris Agreement does not answer to the ethical imperative and does not lead to social justice. In the face of the ethical-political crisis one recalls the words of Martin Heidegger in his famous interview with *Der Spiegel*:

> Philosophy will not be able to effect any direct transformation of the present state of the world. This is true not only of philosophy but of any simply human contemplation and striving. Only a god can save us now.[9]

Heidegger is commenting, in the context of modern technology, not only upon the limitation of calculative, techno-scientific problem-solving, but also the human capacity to think, in a deep sense, about the condition of humankind in the world. This sort of fatalistic outlook is one response that appears in the face of ethical crisis of climate change. Alternatively, an existential commitment to action, despite the lack of clear rational grounds, embraces the ethical imperative.

The paradigmatic exposition of an existential commitment to a duty with obvious ethical implications, but where the ethical principle itself is obscure, is Søren Kierkegaard's troubling account of the binding by Abraham of his son Isaac to fulfil a duty he cannot understand[10]. Kierkegaard poses the problem abstractly with the question, *Is a teleological suspension of the ethical possible?* What he means by this formulation is, *can it be that an overriding duty to the highest good demands that one violates the universal-ethical in order to fulfil that duty?* This possible violation of the universal-ethical does not overthrow or refute the ethical but only suspends it to return to it once again. How can this be? He means that the suspension of the ethical does not invalidate the ethical because the suspension is only necessary when the circumstances are beyond the comprehension of the ethical. The argument seems to be that the ethical consciousness itself has limits such that some duties cannot be grasped by ethical reflection or that the rational effort to understand the validating principle of the imperative fails and the demanded duty appears to be wrong. Kierkegaard's discussion of this of course is framed by extremes where the duty (God's instruction to Abraham to sacrifice his son) is utterly inexplicable and even suggests psychological impairment on the part of anyone willing to follow this *duty*. Yet the structure is the same as in Lomborg's reticence to commit to the Agreement. If the duty implied by the ethical imperative of responsibility cannot be justified by ethical norms or

---

9 Heidegger, Martin (1977), "'Only a God Can Save Us Now': An Interview with Martin Heidegger",. Schendler, David (trans.), *Graduate Faculty Philosophy Journal,* 6 (1), pp. 5–27

10 This discussion draws primarily on Kierkegaard's influential text "Fear and Trembling" which was published originally in 1843 under the pseudonym Johannes de Silentio

principles of social justice, one looks with an Abraham-like faith for a redemptive intervention to save the day.

What can be made of this pattern of thinking? Is this another example of the inadequacy of traditional ethics to address the unprecedented dilemmas that result from advanced technology? For whether or not one believes that global warming and the degradation of the natural environment are a consequence of the deployment of too much technological power, it is still the case that our awareness of the likely fate of the planet is expressed in the discourse of science and technology which, unlike an encounter with the God of Abraham and Isaac (or in mythological drama), we must acknowledge it as the stepchild of our own rational endeavours. If we cannot understand our fate, let alone control it, what are our options?

Technology suggests a spirit of optimism. We understand technology to be creative problem-solving, in a sense innovation to save us from ourselves. In the case of climate change the plausible use of technology is ambiguous; our connection to the future is undefined, the specific problem that is to be solved is unknown, the subject of our responsibility unclear.

The impending or already begun crisis of a global climate inhospitable to life on terms that humanity has come to believe is natural may resemble another myth – the myth of the Golem or of Dr. Frankenstein's creation – one where we have through our technology so to speak outsmarted ourselves. Of course we should not take this analogy too far, as we may well be able make the necessary correction. Our understanding of the dynamics of the problem, while certainly not meeting the mostly now abandoned criterion of certainty, on the level of ethics where we are still influenced by the notion of absolute duty, has created an expectation of purity that we are unlikely to realize and which may not serve us well.

Conclusion

The problem of climate change has led us to unusual considerations. As the issue has been moved in its most fundamental form from the arenas of technology, economics and politics and been put forth as a specifically ethical question, the limits of ethical discourse have been reached. The question is not, as Kierkegaard might have put it, one of the ethical within the ethical, but rather one where the limits of ethical reasoning are transcended by what appears to be a duty to act in ways that cannot be justified ethically.

The posing of the problem as one that presents an ethical imperative, and the suggestion that the adoption of the Paris Agreement fulfils that imperative, has obscured the status of the remediation that may be possible. Bill McKibben's argument that we need to adjust to the now inevitable deteriorating natural environment does not seem to have the redemptive purity of an ethical imperative. The Paris Agreement stands for compromise and does not force commitment to a single approach. The often unspoken truth, that the suffering due to climate change that is likely in much of the underdeveloped world will be ameliorated only slightly if at all, undermines the force of declarations of a moral imperative.

Moreover, the notion of an ethical imperative to stop if not reverse the degradation of the natural environment by limiting climate change steers the discussion away from the broader and multifaceted issue of global social justice. It is perhaps this attempt to redirect the prognosis for the world's future toward the phenomenon of climate change that has led critics of the Paris Agreement, like Lomborg, to object.

If Lomborg's analysis is correct then it would be ethically wrong to implement the Paris Agreement. Yet his suggestion is that there are alternative protocols that would not lead to the negative consequences his prognosis anticipates. Although what these would be is not clear, economic accommodation for different levels of development is part of it, otherwise the advantage that developed nations have would place undue burdens on less-advantaged or less-developed nations and still lead to unfair consequences such as those pointed out by António Guterres. To rectify this within the concept of fair care justice would minimally require, given the long-term and dynamic nature of climate change, a kind and degree of knowledge that is not possible. Regarding the impact of climate change in the far distant future, neither the assurance of fair care justice nor the certification that it will not be achieved is possible. Since the precautionary principle in the form of watchful waiting is futile and damaging to short-term interests, it is not ethically justifiable. Thus, from an ethical point of view the nature of the issue changes ground to one of pure duty. What duty or obligation (if any) does humankind have for the well-being of the natural order in an indefinite future we cannot know and in the situation where our actions taken in the present might or might not have the consequences we assume? This situation produces a dilemma where our choice is either to ignore the issue and default to addressing short-term concerns (mostly self-interest) or to take Kierkegaard's 'leap of faith'.

For a transactional and nationalist politician like Donald Trump, the embrace of this notion of duty is simply absurd. According to some, we are faced with an absurd situation and our only hope may be in providence. The concern, on the contrary, is for the loss of the status quo. That this may already have been lost, the position argued by McKibben, is denied by Trump as well as the ethical imperative absolutists.

If we were to believe that a divine imperative enjoined us, for example, to cease using carbon-based energy sources, then we would, despite other concerns that might constrain a rational actor, go forward in the faith that our concerns would somehow be addressed. Lomborg is not willing to take this step. On the contrary he believes that a rational solution can be found that could provide fare care justice in the present and future. Given the epistemological challenges mentioned above, this solution will be elusive.

10 The implementation of the principle of common but differentiated responsibilities within the Paris Agreement

A governance values analysis

Anna Huggins and Rowena Maguire

According to the principle of common but differentiated responsibilities (CBDR), all states have international environmental obligations. However, the manner in which these states meet their obligations varies in relation to states' level of economic development as well as their contribution to the environmental degradation in question[1]. Under the United Nations Framework Convention on Climate Change (UNFCCC)[2] and its Kyoto Protocol[3], the meaning and significance of the CBDR principle was relatively clear-cut. However, differentiation between developed and developing countries is more nuanced and flexible[4] under the Paris Agreement[5] (hereinafter 'the Agreement') as opposed to the UNFCCC.

This chapter analyses the implementation of CBDR under the Paris Agreement from a governance value perspective[6]. While CBDR shapes both mitigation and

---

1 Rajamani, L. (2011), "The Reach and Limits of the Principle of Common But Differentiated Responsibilities and Respective Capabilities in the Climate Change Regime", in *A Handbook of Climate Change and India: Developments, Politics and Governance*, 1st edition, pp. 119–122, London: Earthscan; Hunter, D., Salzman, J. and Zaelke, D. (2011), *International Environmental Law And Policy*, 4th edition, Foundation Press
2 New York, NY (US), 9 May 1992, in force 21 March 1994, available at: http://unfccc.int, at Art. 1. The UNFCCC enjoins parties 'to protect the climate system for the benefit of present and future generations of humankind, on the basis of equity and in accordance with their common but differentiated responsibilities and respective capacities': ibid., at Art. 3(1)
3 Kyoto (Japan), 11 December 1997, in force 16 February 2005, available at: http://unfccc.int/resource/docs/convkp/kpeng.pdf
4 This trend was anticipated in Brunnee, J. and Streck, C. (2013), "The UNFCCC as a Negotiation Forum: Towards Common but More Differentiated Responsibilities", *Climate Policy*, 13 (5), pp. 589–607, at 591. See also Rajamani, L. (2015), "The Devilish Details: Key Legal Issues in the 2015 Climate Negotiations", *Modern Law Review*, 78 (5), pp. 826–53, at 852
5 See Draft Decision -/CP.21, Adoption of the Paris Agreement, Report of the Conference of the Parties (COP) on its Twenty-first Session, held in Paris from 30 November to 11 December 2015, UN Doc. FCCC/CP/2015/L.9/Rev.1, 12 Dec. 2015, Annex, available at: https://unfccc.int/resource/docs/2015/cop21/eng/l09r01.pdf ('Paris Agreement')
6 Breakey, H. and Cadman, T. (2016), "Governance Values and Institutional Integrity", in Cadman, T., Maguire, R. and Sampford, C. (eds.), *Governing the Climate Change Regime: Institutional Integrity and Integrity Systems*, Routledge

adaptation commitments in the Agreement, the present focus is on the interpretation of the principle with respect to legally binding mitigation obligations, as this is where the principle has been most contentious. This chapter argues that the emerging framework for implementing CBDR under the Paris Agreement is flexible and proceduralised, and demonstrates thin governance values[7] through the incorporation of mechanisms for transparency and compliance, which take into account the national circumstances of individual countries. There is also some evidence of accountability mechanisms, which is strengthened by the inclusion of strong negative consequences and sanctions for deficient performance[8]. The multilateral processes for assessing states' progress towards their mitigation goals serve as evidence for governance values, which despite being slightly thicker, deliberative and democratic, are as yet inchoate. The effective operation of these procedurally differentiated oversight mechanisms has the potential to add to the coherence[9] of the Paris Agreement, by which the internal governance arrangements facilitate the pursuit of its 'public institutional justification' (PIJ).[10] The PIJ for the Paris Agreement is reflected in Article 2(1)(a) of the Agreement, which imposes a collective general obligation on all state parties to hold 'the increase in the global average temperature to well below 2°C above pre-industrial levels and to pursue efforts to limit the temperature increase to 1.5°C above pre-industrial levels'[11].

This chapter proceeds as follows. Part two outlines the divergent approaches to differential treatment reflected in the UNFCCC, Kyoto Protocol and Paris Agreement. Part three then focuses on evaluating the governance values evident in the procedural arrangements for implementing the Paris Agreement, with concluding remarks being offered in part four.

Differential obligations under the UNFCCC, Kyoto Protocol and Paris Agreement

Before evaluating the governance values evident in the procedural arrangements for implementing the Paris Agreement, it is useful to explore the evolution and politics surrounding the development of the CBDR principle within the climate regime. The principle of CBDR recognises the difference in capabilities between developed and developing countries when it comes to taking responsibility for

---

7 Breakey and Cadman, "Governance Values and Institutional Integrity", pp. 4–5, see note 6
8 Stewart, R.B. (2014), "Remedying Disregard in Global Regulatory Governance: Accountability, Participation, and Responsiveness", *American Journal of International Law*, 108, pp. 211–70, at 253
9 Breakey, H. and Cadman, T. (2015), "A Comprehensive Framework for Evaluating the Integrity of the Climate Regime Complex" in Breakey, H., Popovski, V. and Maguire, R. (eds.), *Ethical Values and the Integrity of the Climate Change Regime*, Ashgate, pp. 17, 18–19, and Maguire, R. (2015), "Mapping the Integrity of Differential Obligations within the United Nations Framework Convention on Climate Change", pp. 31–42 in the same volume
10 Breakey and Cadman, "Governance Values and Institutional Integrity", p. 8, see note 6
11 Art. 2 Paris Agreement, see note 5

global environmental damage and implementing environmental law reform. The Rio Earth Summit of 1992 gave rise to the modern crystallisation of the principle:

> States shall cooperate in a spirit of global partnership to conserve, protect and restore the health and integrity of the Earth's ecosystem. In view of the different contributions to global environmental degradation, States have common but differentiated responsibilities. The developed countries acknowledge the responsibility that they bear in the international pursuit of sustainable development in view of the pressures their societies place on the global environment and of the technologies and financial resources they command.[12]

This formulation of the principle was acceptable to developed countries mainly because it created obligations for all parties through the inclusion of the words 'global partnership', which necessarily implied co-operation by all nations to address global environmental challenges. Developing countries were satisfied with this statement as well, as it recognised the industrial and colonial practices of developed countries and the role that they play in global environmental destruction, thus placing greater obligations on developed countries to make amends[13]. Furthermore, the principle went some way towards acknowledging poverty alleviation as the priority concern of Southern nations. The incorporation of this principle within the Rio Declaration paved the way for other 'multilateral environmental agreements' (MEAs) to incorporate this principle and create differential obligations for developed and developing countries.

The UNFCCC, which opened for signature in May 1992, and the Kyoto Protocol, which opened for signature in December 1997, strongly reflect the importance given to the CBDR principle in the area of international environmental law at the end of the 20th century. Notably, differential treatment for developed and developing countries is enshrined in the central treaty obligations of both the UNFCCC and its Kyoto Protocol, which is an approach to CBDR that is not replicated amongst other global MEAs[14]. The codification of CBDR under the UNFCCC is found in Article 3(1), which requires that:

> The Parties should protect the climate system for the benefit of present and future generations of human kind on the basis of equity and in accordance with their common but differentiated responsibilities and respective

---

12 United Nations General Assembly, Rio Declaration on Environment and Development (Annex I), A/CONF.151/26 (vol 1), 14 June 1992, www.un.org/documents/ga/confl51/aconf15126-1.htm, principle 7

13 Honkonen, T. (2009), "The Common But Differentiated Responsibility Principle in Multilateral Environmental Agreements: Regulatory and Policy Aspects", Kluwer Law International, p. 40

14 Rajamani, L, Brunnee, J. and Doelle, M. (2012), "Introduction: The Role of Compliance in an Evolving Climate Regime" in Brunnee, J., Doelle, M. and Rajamani, L. (eds.), *Promoting Compliance in an Evolving Climate Regime*, Cambridge University Press, pp. 1–14, p. 3

capabilities. Accordingly, the developed country Parties should take the lead in combating climate change and the adverse effects thereof.

Further guidance as to how differential climate action should be determined can be gleaned from Articles 3(3) and 3(4) of the UNFCCC. Article 3(3) provides that climate policies and measures must take into account the different socio-economic contexts and conditions of states, and Article 3(4) requires that climate policies and measures should be appropriate for the specific conditions of each party while being integrated with their national development programmes, and should take into account that economic development is essential for adopting measures to address climate change. These provisions recognise that economic development is the priority concern for developing countries, and reflect the North/South discourses that permeated the Rio Earth Summit negotiations[15].

The UNFCCC is a framework instrument which merely establishes the architecture of the regime without imposing any legal commitments. Accordingly, the UNFCCC did not specify the interpretation or model of differentiation that should be applied when determining legally binding mitigation commitments. The UNFCCC did, however, create two categories of party states: Annex I parties (essentially developed country parties) and Non-Annex I developing parties. Both groups hold broad commitments under Article 4(1) to publish national inventories of anthropogenic emissions by sources and removals, implement programmes to mitigate climate change, co-operate on technology transfer, promote sustainable management of all sinks, co-operate in preparing for adaptation, take climate change considerations into account in relevant policy development, promote climate research, co-operate in exchange of scientific information, promote public awareness and climate education, and communicate to the COP. Article 4(2) attempts to create more stringent obligations for Annex I parties by requiring the adoption of national mitigation policies along with the communication of detailed information on climate policies and measures. This information is supposed to include projected anthropogenic emissions by sources and removals by sinks based on the best available scientific knowledge. As such, the UNFCCC implements differential treatment by creating more stringent reporting and monitoring practices for Annex I parties. The preamble to the Kyoto Protocol does not provide its own formulation of how CBDR should be construed. Instead, it provides that the Protocol is guided by Article 3 of the UNFCCC. Despite this, the Protocol is renowned for its radical interpretation of CBDR which resulted in legally binding emission reduction obligations being placed on Annex I parties only. Articles 2 and 3 of the Protocol create Quantified Emission Limitation and Reduction Commitments (QELRO) for Annex I parties. Annex I parties are allocated a quota of Assigned Amount Units,

---

15 Atapattu, S. and Gonzalez, C.G. (2015), "The North-South Divide in International Environmental Law: Framing the Issues", in Alam, S., Atapattu, S., Gonzalez, C.G. and Razzaque, J. (eds.), *International Environmental Law and the Global South*, Cambridge University Press, p. 10

which are calculated pursuant to their QELRO. Assigned amount units are the currency used within the regime and represent the carbon dioxide equivalent of all gases covered within the regime (carbon dioxide, methane, nitrous oxide, hydrofluorocarbons, perfluorocarbons and sulphur hexafluoride)[16]. There are two commitment periods for Annex I parties to the the Kyoto Protocol, which place certain obligations on the parties:

- During the first commitment period (2008–2012), 37 industrialised nations and the European Union were required to reduce their overall emissions by at least 5% below 1990 levels[17].
- During the second commitment period (2013–2020), 38 industrialised parties[18] are required to reduce their overall emissions by at least 18% below 1990 levels[19].

Article 10 of the Protocol creates mitigation obligations for Annex I and Non-Annex I parties (i.e., developing countries). These obligations build upon Article 4 of the UNFCCC discussed above, but provide more guidance about the types of policies that should be developed and the modalities for the aforementioned reporting. As such, the Protocol only creates procedural obligations such as reporting and policy formation, rather than imposing substantive mitigation requirements upon non-Annex I parties.

This type of differentiation in substantive obligations between developed and developing countries signifies a high water mark in the influence of the CBDR principle in multilateral environmental agreements. Earlier interpretations of CBDR, for example under the Montreal Protocol on Substances that Deplete the Ozone Layer[20] (Montreal Protocol), allowed for differentiation in timing by giving developing countries delayed compliance schedules[21] and different baseline requirements[22], and creating obligations for developed countries to provide financial and technological assistance[23]. Thus, the differential obligations under the Montreal Protocol 'were designed to assist developing countries in meeting their commitments under the relevant treaty, not to exclude or protect them

---

16 The gases covered by the regime are set in Annex A of the Kyoto Protocol, supra, note 2
17 "The Kyoto Protocol to the United Nations Framework Convention on Climate Change", Art. 3(1)
18 Australia, Austria, Belarus, Belgium, Croatia, Cyprus, Czech Republic, Denmark, Estonia, European Union, Finland, France, Germany, Greece, Hungary, Iceland, Ireland, Italy, Kazakhstan, Latvia, Liechtenstein, Lithuania, Luxembourg, Malta, Monaco, Netherlands, Norway, Poland, Portugal, Romania, Slovakia, Slovenia, Spain, Sweden, Switzerland, Ukraine and United Kingdom of Great Britain and Northern Ireland
19 "Doha Amendment to the Kyoto Protocol", (C.N.718.2012.TREATIES–XXVII.7.C (Depositary Notification), 24 November 2014), para. C
20 Montreal Protocol on Substances That Deplete the Ozone Layer, (1522 UNTS 3; 26 ILM 1550 (1987), 16 September 1987), Preamble
21 Ibid., Art. 5
22 Ibid., Art. 5(3)
23 Ibid., Art. 10(a)

from particular commitments'[24]. This focus on differentiated implementation of commitments contrasts with the Kyoto Protocol's differentiation with respect to substantive mitigation commitments[25].

It is precisely this unique, bifurcated approach to differential treatment that continued to prove highly contentious, and ultimately lead to the failure of the Kyoto Protocol[26]. The lack of binding emission reduction obligations for all parties to the UNFCCC under the Kyoto Protocol caused divisions among developed countries between those willing to abide by this model of differentiation (EU) and those unwilling to accept it (US) unless leading developing countries also assumed responsibility for reducing their greenhouse gas emissions[27]. The Paris Agreement reflects an attempt to reconcile these competing approaches, which entails reconceptualising the meaning and significance of the CBDR principle in the international climate change regime.

The Paris Agreement

The Paris Agreement has generally been viewed as an instrument that blurs the lines between the stark differential treatment model established within the UNFCCC and Kyoto Protocol's top-down approach of codifying differential treatment in central treaty obligations[28]. Given that the second commitment period of the Kyoto Protocol only covered 15% of the global greenhouse gas emissions[29], there was a great need to get all major emitters to pledge legally binding emission reduction commitments. The Paris Agreement does not reference the Annex I and Non-Annex I distinction from the UNFCCC and instead uses the terms of developed and developing countries. There is no definition or criteria within the instrument for identifying a party country as developed or developing. It is due to this lack of a fixed definition that China and India can

---

24 Rajamani, L. (2012), "The Changing Fortunes of Differential Treatment in the Evolution of International Environmental Law", *International Affairs*, 88, p. 608
25 Article 10 of the Protocol seeks to reinforce the obligations created under the UNFCCC and requires both industrial and non-industrialized parties to take mitigation steps in accordance with the principle of CBDR and their specific national and regional development priorities, but it does not require for these actions to be reported upon by non-industrialized parties. For further analysis on the background on the CBDR and its role in shaping differentiation, see Atapattu, S. (2007), *International Law and Development: Emerging Principles of International Environmental Law*, BRILL, 1st edition
26 Rajamani, see note 24, p. 612
27 Maguire, R. and Jiang, X. (2015), "Emerging Powerful Southern Voices: Role of BASIC Nations in Shaping Climate Change Mitigation Commitments", in Alam, S., Atapattu, S., Gonzalez, C.G. and Razzaque, J. (eds.), *International Environmental Law and the Global South*, Cambridge University Press, p. 218
28 Art. 4(2) UNFCCC, and Art. 3 Kyoto Protocol
29 The top ten nations in terms of total CO_2 emissions are China, the US, India, Russia, Japan, Germany, Canada, South Korea, Iran and the UK: see Banerjee, S. (2012), "A Climate for Change? Critical Reflections on the Durban United Nations Climate Change Conference", *Organisation Studies*, 33, no. 12, pp. 1772–73

presumably still be classified as developing, which provides significant strategic benefit to these large and high-emitting economies.

Under the auspices of the Ad Hoc Working Group on the Durban Platform for Enhanced Action (ADP)[30], parties negotiated the framework of the Paris Agreement. Many submissions were made on the principle of CBDR, which primarily concerned the role and status of the CBDR principle and different methods of applying differential treatment. Brazil, China and India's submissions under the ADP process emphasised that the Paris Agreement would sit beneath the Convention and as such the existing Annex I and Non-Annex I categorisation should stay in place[31]. The US submission argued that the Paris Agreement should further the Convention's objective, which left scope for reinterpreting the differentiation model applied under the UNFCCC and Kyoto Protocol[32]. The final text of the Paris Agreement does not reference the Annex I and Non-Annex I distinction, suggesting that Brazil, China, and India conceded that the existing model of differentiation would not apply. This compromise by Brazil, India and China meant that the nature of the legally binding commitments for all parties would necessarily need to be flexible and driven by a bottom-up pledge-and-review process.

The principle of CBDR is explicitly recognised in the Paris Agreement in Article 2(2), which states 'this agreement will be implemented to reflect equity and the principle of common but differentiated responsibilities and respective capabilities, in the light of different national circumstances'. The interpretation of CBDR within the Paris Agreement is, however, radically changed by Article 3, which requires 'all parties to undertake and communicate ambitious efforts [...] while recognising the need to support developing countries for the effective implementation of this agreement'. Furthermore, Article 4(9) requires all parties to contribute a nationally determined amount every five years, thus effectively binding all parties to ongoing five-year commitment periods with increasingly stringent commitments. The more nuanced and flexible instantiation of the CBDR principle in the Paris Agreement seems to be the result of a political compromise designed to ensure the participation of major emitters including the US, China and India, which did not have binding commitments under the Kyoto Protocol. The inclusion of the qualifying clause 'in the light of different national circumstances' introduces a dynamic element to the interpretation of the CBDR principle as it recognises that as countries' circumstances change, so too will the responsibilities of state parties[33]. It is perhaps in recognition of states' evolving

---

30 Decision 2/CP.17, Establishment of an Ad Hoc Working Group on the Durban Platform for Enhanced Action, UN Doc FCCC/CP/2011/9/Add.1, 15 March 2012
31 For further information on this see Maguire, R. (2014), "The Role of Common but Differentiated Responsibility in the 2020 Climate Regime", *Carbon and Climate Law Review*, p. 1
32 Ibid.
33 Rajamani, L. (2016), "Ambition and Differentiation in the 2015 Paris Agreement: Interpretative Possibilities and Underlying Politics", *International and Comparative Law Quarterly*, 65 (2), pp. 493–514, at 508

national circumstances that there is no definition of developed and developing countries included in the Paris Agreement, though a more cynical interpretation might suggest that the difficulties associated with developing criteria or guidelines to define developed or developing nations would have resulted in failure to reach consensus on the Paris Agreement.

Submissions made under the ADP process generated a range of methods to differentiate mitigation commitments and many of these methods can be seen within the Paris Agreement. There are four methods of differentiating mitigation commitments within the Paris Agreement: 1) self-differentiation (bottom-up pledges); 2) economy-wide versus sector-specific obligations; 3) delayed compliance for developing countries; and 4) financial and technological support for developing countries to meet obligations. First, the Paris Agreement enshrines a new paradigm of bottom-up 'self-differentiation'[34], as parties have the leeway to decide their own mitigation targets[35]. This paradigm shift is reflected in the Agreement's emphasis on nationally determined contributions (NDCs), rather than centrally imposed targets that differentiate between developed and developing state parties. A distinctive feature of the mitigation obligations under the Paris Agreement is their collective nature. As previously noted, the Paris Agreement imposes a collective general obligation on all state parties to hold 'the increase in the global average temperature to well below 2°C above pre-industrial levels'[36], which contrasts with the Kyoto Protocol's imposition of legally binding emissions targets on industrialised countries only. In order to achieve its aim, the Paris Agreement requires each state to produce successive and progressively strengthened NDCs[37].

Secondly, Article 4(4) states that developed countries should continue to take the lead through economy-wide absolute emission reduction targets. Developing countries, too, are expected to make mitigation efforts and are encouraged to adopt economy-wide emission reduction targets in the future in the light of national circumstances[38]. Thus, the collective obligations on all state parties are qualified by the provisions which recognise the differing national circumstances of developed and developing countries.

Thirdly, Article 4(1) implements a model of delayed compliance by requiring parties to aim to reach global peaking of greenhouse gas emissions as soon as possible and recognising that peaking will take longer for developing country parties. The Montreal Protocol has successfully utilised a delayed compliance model, though the success of differential standards by time was contingent upon the development of specific emission profile reductions and commitments

---

34 Rajamani, L. (2015), "The Devilish Details: Key Legal Issues in the 2015 Climate Negotiations", *Modern Law Review*, 78 (5), pp. 826–53, at 852
35 Brunnee, J. and Streck, C. (2013), "The UNFCCC as a Negotiation Forum: Towards Common but More Differentiated Responsibilities", *Climate Policy*, 13 (5), pp. 589–607, at 591
36 Art. 2 Paris Agreement
37 Arts. 3 and 4(2) Paris Agreement
38 Art. 4(4) Paris Agreement

for developing countries. As the Paris Agreement does not contain specific detail on the peaking of emissions, it will be vital for future COP decisions to set more specific limits in order to prevent warming beyond 2° Celsius.

Fourthly, differential treatment is provided for by Article 4(5) which calls for financial, technological and capacity-building support for developing countries to ensure implementation of nationally determined contributions. A distinctive feature of the Paris Agreement is the emphasis it places on states' procedural steps towards achieving their mitigation obligations, rather than their substantive fulfilment[39]. Unlike the substantive mitigation obligations imposed upon developed states under the Kyoto Protocol, under the Paris Agreement states' obligations are binding with respect to fulfilling procedural requirements to prepare, communicate, maintain and periodically report national contributions, and pursue domestic mitigation measures[40]. To assist developing states to meet their largely procedurally orientated commitments, developed states are expected to demonstrate leadership through the provision of finance[41], technology[42] and capacity-building[43] support. Article 4(6) goes some way towards recognising the broad range of capacities within the 'developing country' grouping by singling out least- developed countries and small island developing states, which may need more flexibility in the procedural requirements of preparing and communicating their climate strategies and plans than other developed and developing states. Thus, despite the absence of bifurcated differentiation between developed and developing states with respect to substantive mitigation commitments, there is nonetheless evidence of procedurally orientated differentiation in the Paris Agreement[44].

Governance values and CBDR under the Paris Agreement

This section argues that the emerging procedural framework for implementing CBDR under the Paris Agreement demonstrates thin governance values. Breakey

---

39 Huggins, A. and Karim, S. (2016), "Differential Treatment and Substantive and Procedural Regard in the International Climate Change Regime", *Transnational Environmental Law*, 5 (forthcoming)
40 Ibid., at Art. 4(2) and 4(3). Under Art. 4(3), parties are required to communicate their contributions every five years
41 Under the Agreement, developed countries are obligated to provide financial resources to assist developing country parties in fulfilling their obligations in continuation of their existing obligations under the UNFCCC: Paris Agreement, Art. 9(1). In addition, developed states are expected to take the lead in mobilizing, and progressively increasing funds for, climate finance: Paris Agreement, Art. 9(3)
42 The Agreement provides for a technology framework to facilitate enhanced action on technology development and transfer through the Convention's Technology Mechanism: ibid., at Art. 10(4)
43 Developed countries are urged to 'enhance support for capacity-building actions in developing country Parties': ibid., Art. 11(3)
44 Huggins, A. and Karim, S. (2016), "Differential Treatment and Substantive and Procedural Regard in the International Climate Change Regime", *Transnational Environmental Law*, 5 (forthcoming).

A governance values analysis 173

and Cadman argue that thin governance values encompass accountability, transparency, and rule-compliance[45], which provide *indicia* for the following analysis. There is sufficient evidence of the existence mechanisms to achieve each of these governance values, which take into account the national circumstances of developing states. There is also some potential for thicker deliberative and democratic governance values, however these are incipient at this stage. There is significant scope to develop these existing oversight mechanisms to ensure they are robust and comprehensive in the future development of the institutional apparatus for the Paris Agreement. This analysis is carried out with an understanding of the political environment in which the Paris Agreement was construed and makes suggestions for reform on the basis of the necessary adjustments needed to ensure more effective implementation of the Agreement.

Transparency

Transparency is an emerging norm in global environmental governance[46], and holds a prominent place in the Paris Agreement. Transparency refers to the 'governance of information, including demands for active transparency and access to information, but also demands for confidentiality and privacy, and for legal or political controls on the gathering and use of policy-shaping information'[47]. In MEAs such as the Paris Agreement, there are both internal and external dimensions to transparency. Internal transparency relates to transparency between regime members whereas external transparency refers to information that is more widely available, including to the public[48].

The Paris Agreement's well-developed transparency arrangements are exemplified in the requirements for state reporting of progress towards national goals, and expert review of state reporting on mitigation and finance. Every two years, each party is required to provide a national inventory report of greenhouse gas emissions and removals[49], information necessary to monitor progress towards implementing and achieving NDCs[50], and information on the impact of climate change and adaptation[51]. Further, developed country parties are required to provide information on the financial, technology-transfer

---

45 Breakey and Cadman, "Governance Values and Institutional Integrity", pp. 4–5, see note 6
46 Hunter, D.B. (2014), "The Emerging Norm of Transparency in International Environmental Governance", in P. Ala'i and R.G. Vaughn, *Research Handbook on Transparency*, Edward Elgar Publishing, p. 343
47 Kingsbury, B. and Casini, L. (2009), "Global Administrative Law Dimensions of International Organizations Law", *International Organizations Law Review*, 6, p. 325
48 Stewart, R.B., Oppenheimer, M. and Rudyk, B. (2013), "Building Blocks for Global Climate Protection", Stanford Environmental Law Journal, 32, pp. 385–86; Huggins, A. (2015), "The Desirability of Administrative Proceduralisation: Compliance Rules and Decisions in Multilateral Environmental Agreements", Ph.D. thesis, University of New South Wales (Australia), October 2015, pp. 56–57
49 Paris Agreement, Art. 13(7)(a)
50 Paris Agreement, Art. 13(7)(b)
51 Paris Agreement, Art. 13(7)(b)

and capacity-building support they provide to developing countries, and developing countries are urged to provide information on the support they need and have received[52]. State reporting of performance in relation to mitigation and support goals is thus a key way in which transparency is fostered under the Paris Agreement.

A fraction of these types of reporting is subject to expert review, which further enhances transparency in relation to states' progress towards their goals under the Agreement. All parties' mitigation information, and developed countries' provision of support to developing countries, will undergo a technical expert review[53]. Expert review is thus a second key component of the 'transparency framework for action and support' in Article 13 of the Paris Agreement. Differentiation is evident with respect to the procedural requirements in the transparency framework, which are intended to be implemented 'flexibly' in the light of parties' different capacities[54]. Article 13(14) states that '[s]upport shall be provided to developing countries for the implementation of this Article', however this provision fails to specify who should be providing this support. Thus, transparency in relation to states' mitigation and support is actively promoted through reporting and review requirements, which will be implemented flexibly in view of states' capacity constraints.

Accountability

There is widespread agreement on the importance of accountability in global governance, however understandings of its meaning diverge[55]. Both broad[56] and narrow[57] conceptualisations of accountability are evident in the literature. For the purposes of analysing the Paris Agreement in this chapter, a narrow definition of accountability is to be taken. Richard Stewart proposed a narrow definition encompassing three structural elements: 1) a specified accounter, who is

---

52 Paris Agreement, Art. 13(9) and 13(10)
53 Paris Agreement, Art. 13(11); Decision 1/CP.21, Adoption of the Paris Agreement, Report of the COP on its Twenty-first Session, held in Paris from 30 November to 11 December 2015, UN Doc. FCCC/CP/2015/L.9/Rev.1, 12 December 2015, at paras. 97 and 98
54 Paris Agreement, Art. 13(1)
55 Koppell, J. (2005), "Pathologies of Accountability: ICANN and the Challenge of 'Multiple Accountability Disorder'", *Public Administration Review*, 65 (1), p. 94
56 See, e.g., Grant, R.W. and Keohane, R.O. (2005), "Accountability and Abuses of Power in World Politics", *American Political Science Review*, 99 (1), pp. 29–43, at p. 36 (identifying hierarchical, supervisory, fiscal, legal, market, peer reputational, and public reputational accountability mechanisms); and Mashaw, J.L. (2005), "Structuring a 'Dense Complexity': Accountability and the Project of Administrative Law", *Issues in Legal Scholarship*, 5 (1), pp. 1–38, at p. 27 (identifying political, administrative, legal, product market, labour market, financial market, family, professional and team accountability)
57 See, eg, Stewart (2014), pp. 244–55, see note 8; Bovens, M. (2007), "Analysing and Assessing Accountability: A Conceptual Framework", *European Law Journal*, 13, pp. 449–50; Black, J. (2008), "Constructing and Contesting Legitimacy and Accountability in Polycentric Regulatory Regimes", *Regulation and Governance*, 2, p. 150

A governance values analysis 175

subject to being called to provide account for his conduct; 2) a specified account holder who can require the accounter to render account; and 3) the ability and authority of the account holder to impose sanctions or other remedies for deficient performance[58].

This definition sees accountability as a distinct procedural tool, and provides *indicia* for evaluating the achievement of accountability for states' mitigation commitments under the Paris Agreement. Within the UNFCCC, these functions are carried out by the state parties acting as the specified accounters, and the COP serving as the player able to render an account of climate action. Under the Kyoto Protocol, states would account to the Enforcement Branch of the compliance committee for their progress towards achieving their commitments under the Protocol, and could be deprived of treaty privileges if they were found to be in non-compliance with key commitments[59].

The accountability mechanisms evident in the Paris Agreement require states to account for their progress towards, *inter alia*, their emissions goals in their NDCs, however, there is a paucity of sanctions or other remedies for deficient performance, which is a key element of accountability[60]. There are three key oversight mechanisms provided for in the Paris Agreement: 'multilateral consideration of progress'[61], 'global stocktakes'[62] and non-compliance processes[63]. Article 13(11) specifies that each party shall (i.e., must) participate in a facilitative, multilateral consideration of progress with respect to both implementation and achievement of mitigation goals, and developed states' provision of climate finance. The practical details of this process remain unclear, as 'how these processes will be conducted, who will they be conducted by, what the outputs will be, and how these outputs will feed into the global stocktake' are not yet specified[64]. The multilateral nature of these processes suggests the presence of slightly thicker governance values pertaining to deliberation and participation by those states affected by decisions, in accordance with Breakey and Cadman's framework[65]. However, in line with the emphasis on facilitation in the rest of the Agreement, there are no consequences specified for deficient performance, drawing into question the robustness of the Paris Agreement's accountability arrangements in this regard.

---

58 Stewart (2014), p. 253, see note 8
59 In the Kyoto Protocol's first commitment period (2008–2012), the suspension of parties' eligibility to participate in the Protocol's flexibility mechanisms – the clean development mechanism, joint implementation and emissions trading – was frequently recommended in practice as a consequence of non-compliance
60 Huggins, A. and Karim, S. (2016), "Differential Treatment and Substantive and Procedural Regard in the International Climate Change Regime", *Transnational Environmental Law* (forthcoming)
61 Paris Agreement, Art. 13(11)
62 Ibid., Art. 14
63 Ibid., Art. 15
64 Rajamani, "Ambition and Differentiation in the 2015 Paris Agreement", p. 503, see note 33
65 Breakey and Cadman, "Governance Values and Institutional Integrity", pp. 6–7, see note 6

Similarly, there are gaps with respect to the accountability mechanisms associated with the global stocktake. This stocktake will be held every five years with the aim of assessing 'collective progress towards achieving the purpose of this Agreement and its long term goals'[66]. The stocktake will focus on mitigation, adaptation, and the means of implementation and support, and shall be conducted in a facilitative manner 'in the light of equity and the best available science'[67]. Given the emphasis on collective progress, there appears to be little scope for states to be held accountable individually for failure to meet national goals[68]. This is compounded by the absence of quantifiable targets in relation to mitigation, finance and technology support, and capacity-building. It is as yet unclear what meaning 'in the light of equity' has in the context of Article 14, however the inclusion of this qualifier creates an opening for further dialogue on 'equitable burden sharing'[69]. Thus, there are significant limitations associated with both the multilateral consideration of progress and the global stocktake in terms of their creation of robust accountability mechanisms to hold states to account for their national emissions reduction goals.

Compliance

The third element of Breakey and Cadman's schema for thin governance values is rule-compliance[70], which in this instance relates to states' compliance with their binding obligations under the Paris Agreement. Under international law, states have a binding obligation to comply with rules of law in treaties to which they are a party, and this binding quality is usually made explicitly in treaty texts via mandatory language[71]. The primary 'hard' mitigation obligations under the Paris Agreement that use mandatory language (e.g., 'shall') and are addressed to individual state parties, as opposed to parties collectively, are in Articles 4(2) and 4(9) of the Agreement[72]. These provisions relate to obligations to 'prepare, communicate, and maintain successive nationally determined contributions', pursue domestic mitigation measures[73], and communicate an NDC every five years[74]. As previously noted, the obligation to hold the global average temperature to well below 2°C above pre-industrial levels is a collective obligation, which means that individual states cannot be held accountable for compliance with a communal

---

66 Ibid., Art. 14(1)
67 Ibid, Art. 14(1)
68 Rajamani, "Ambition and Differentiation in the 2015 Paris Agreement", p. 504, see note 33
69 Ibid, p. 504
70 Breakey and Cadman, "Governance Values and Institutional Integrity", p. 11, see note 6
71 Zahar, A. (2015), *International Climate Change Law and State Compliance*, Routledge, p. 165
72 Rajamani, L. (2016), "The 2015 Paris Agreement: Interplay Between Hard, Soft and Non-Obligations", *Journal of Environmental Law*, 28, pp. 337, 344. Rajamani also notes that Art. 4(17) specifies that each party to an agreement to act jointly to achieve the requirements in Art. 4(2) shall be responsible for its emission level as set out in the terms of its joint fulfilment agreement: ibid
73 Paris Agreement, Art. 4(2)
74 Paris Agreement, Art. 4(9)

A governance values analysis 177

obligation[75]. In contrast, the Kyoto Protocol imposed an obligation on Annex I parties to 'individually or jointly' ensure that their emissions did not exceed their assigned amounts, calculated pursuant to their QELRO, with a view to reducing their overall emissions by at least 5% below 1990 levels by 2012[76]. This requirement created scope for individual states to be in non-compliance with their obligations.

In addition to identifying the rules of law that are binding upon states, issues arise as to how compliance with multilateral environmental agreements is best achieved. Traditionally in international environmental law scholarship, debates regarding compliance have been dominated by advocates of the managerial model[77] and the enforcement model[78]. The managerial school of thought is premised on 'managing' the causes of non-compliance, which arguably stem from ambiguous and indeterminate treaty norms and states' capacity limitations, and taking measures to facilitate states' return to compliance[79]. In contrast, the 'enforcement model' approach focuses on the use of sanction measures that create costs or remove benefits, which, proponents argue, are especially important when there are strong incentives for states not to comply with their international commitments[80]. In the Paris Agreement, the former school of thought appears to have prevailed as Article 15 provides for the establishment of a compliance mechanism to 'facilitate implementation' and 'promote compliance' with the provisions of the Agreement. As part of this expert-based, facilitative approach, the compliance committee is required to 'pay particular attention to the respective national capabilities and circumstances of Parties'[81], thus ensuring the CBDR principle is embedded in the procedural framework for non-compliance.

Therefore, unlike the Enforcement Branch of the Kyoto Protocol's compliance committee, under which parties could be deprived of treaty privileges if they were found to be in non-compliance with key commitments, the Paris Agreement's compliance framework reflects elements of a managerial, facilitative approach only. The detailed modalities and procedures for this mechanism will be adopted by the conference of the parties serving as the meeting of the parties to the Paris Agreement in 2016[82], and may potentially strengthen the current accountability arrangements. However, present indications suggest that non-compliant state parties are likely to face limited concrete consequences, except

---

75 Zahar, A, (2015), *International Climate Change Law and State Compliance*, Routledge, p. 166
76 Kyoto Protocol, Art. 3(1)
77 See, e.g., Chayes, A. and Handler Chayes, A. (1998), *The New Sovereignty: Compliance with International Regulatory Agreements*, Harvard University Press
78 Downs, G.W., Rocke, D.M. and Barsoom, P.N. (1996), "Is the Good News About Compliance Good News About Cooperation?", *International Organization*, 50 (3), p. 379
79 Chayes and Handley Chayes, *The New Sovereignty*, at pp. 10–5, 22–5, see note 77
80 Downs, Rocke and Barsoom, "Is the Good News About Compliance Good News About Cooperation?", see note 78
81 Paris Agreement, at Art. 15(2)
82 Paris Agreement, at Art. 15(3)

perhaps public 'naming and shaming'[83]. This is significant because, as noted above, an important aspect of accountability is the imposition of 'sanctions or other remedies for deficient performance'[84], raising questions about the robustness and efficacy of a purely 'facilitative', 'non-adversarial' and 'non-punitive'[85] approach to non-compliance[86].

Conclusion

The Paris Agreement provides a framework for the development of an institutional apparatus that is attuned to the national circumstances of state parties – and developing states in particular – in regime decision-making. This nuanced, proceduralised manifestation of the CBDR principle reflects a departure from the way this principle was reflected in the Paris Agreement's predecessors, the UNFCCC and the Kyoto Protocol. Whilst the Kyoto Protocol adopted a top-down, compliance-backed approach to implementation, which had greater power to compel states to take action, the Paris Agreement's approach is bottom-up and places increasing reliance on procedural rather than substantive commitments, with the aim of encouraging widespread participation and self-directed action from states. Within the framework created by the Paris Agreement, there is evidence of thin governance values pertaining to transparency and rule-compliance, however these arrangements are undermined by accountability mechanisms that are not as yet buttressed by robust consequences for non-compliance. The coherence integrity of the Paris Agreement, and the achievement of its public institutional justification, would be significantly enhanced by the inclusion of stronger accountability mechanisms in the future development of the Agreement's procedural apparatus.

---

83 Oberthur, S. (2014), "Options for a Compliance Mechanism in a 2015 Climate Agreement", *Climate Law*, 4, pp. 30–49, at 43
84 Stewart, "Remedying Disregard in Global Regulatory Governance", at p. 253, see note 8
85 Paris Agreement, at Art. 15(2)
86 Huggins, A. and Karim, S. (2016), "Differential Treatment and Substantive and Procedural Regard in the International Climate Change Regime", *Transnational Environmental Law* (forthcoming)

11 After Paris

Do we need an international agreement on green compulsory licensing?

*Dong Qin**

In December of 2015, the Paris Agreement was passed by the 21st conference of parties to the UNFCCC. The Paris Agreement was opened for signature at the United Nations Headquarters in New York on 22 April 2016, and 175 parties (174 countries and the European Union) signed the Agreement, with 15 states depositing instruments of ratification[1].

Before the Paris Agreement, the focus of international climate co-operation was primarily on how parties to the UNFCCC could be encouraged to make ambitious promises. However, now that the Paris Agreement is in force, the focus will be turned towards helping state parties implement their promises.

One of the primary differences between the Paris Agreement and the Kyoto Protocol is that, under the latter, no developing country had to make qualified carbon reduction or control promises, whereas under the new agreement almost all developing countries are to make such promises.

Naturally, helping developing countries fulfil their commitments has become a very imminent issue, one which the Paris Agreement aims to realize. What developing countries need urgently for fulfilling their promises are green technologies, which are also called climate technologies or environment sound technologies (EST).

In fact, international transfer of green technology is capable of becoming a form of 'development dividend' under the Paris Agreement, because it can both reduce greenhouse gas emissions and contribute to local social and economic development[2].

However, there are barriers that prevent green technology from being transferred to developing countries from developed countries – as a result, many developing countries keep calling on the parties of UNFCCC to develop an

---

* Dong Qin, Doctor of Law, Associate Professor of Nanjing University of Information Science & Technology, China
1 List of 175 signatories to the Paris Agreement: http://newsroom.unfccc.int/paris-agreement/175-states-sign-paris-agreement/.
2 See Forsyth, T., "Promoting the 'Development Dividend" of Climate Technology Transfer: Can Cross-sector Partnerships Help?" at http://personal.lse.ac.uk/FORSYTHT/WD_CPS_proof.pdf

international agreement on compulsory licensing for promoting international transfer of green technologies. *Compulsory licensing* is a legal system that allows courts or patent administrations to permit someone to use the patents without the permission of their owners. In fact, the effort on developing a green compulsory licensing agreement has not thus far progressed smoothly, because there are some very strong adverse opinions from some developed countries.

This article will discuss international agreements on compulsory licensing of green technologies, and whether such an agreement is needed. The remainder of this article proceeds as follows: part one attempts to figure out if international transfer of green technology is necessary for developing countries to fulfil their commitments under the Paris Agreement. Part two discusses whether patent suppression constitutes a major obstacle for developing countries to apply green technologies to reduce greenhouse gases. Next, part three discusses whether the international intellectual property rights system should be improved to solve the problem of green patent suppression. Finally, part four discusses whether developing a green compulsory licensing agreement is necessary for achieving the goals of the Paris Agreement.

Necessity for international transfer of green technologies

Article 2 of the Paris Agreement states that the Agreement aims to strengthen the global response to the threat of climate change by holding the increase in the global average temperature to well below 2°C above pre-industrial levels and pursuing efforts to limit the temperature increase to 1.5°C above pre-industrial levels, and also that all response to the threat of climate change shall be in the context of sustainable development and efforts to eradicate poverty.

It is challenging to achieve both of the above goals of the Paris Agreement simultaneously. This challenge mainly arises due to the incapability of human beings to reduce greenhouse gas emissions. The COP 21 Decision of Adoption of the Paris Agreement by UNFCCC notes with concern that the estimated aggregate greenhouse gas emission levels in 2025 and 2030, resulting from the intended nationally determined contributions, is projected to be 55 gigatonnes in 2030, while the necessary emissions reduction target should be no more than 40 gigatonnes if the parties of the Paris Agreement want to hold the increase in the global average temperature to below 2°C above pre-industrial levels[3]. Undoubtedly, the gap between the goal and reality is quite significant.

Similarly, the task for achieving sustainable development and poverty eradication is also tough. The world becomes increasingly unequal owing to the ever-widening gap between rich countries and poor countries. In 2015, Ban Ki-moon, the Secretary-General of the United Nations, said in one of his reports that inequalities were growing in all societies and the poorest of the poor were

---

3 FCCC/CP/2015/L.9/Rev.1, http://unfccc.int/resource/docs/2015/cop21/eng/l09r01.pdf (accessed 1 March 2016)

being left farther behind[4]. Considering the task of safeguarding the security of the global climate system and to promote sustainable development and poverty eradication, the Secretary-General stressed the urgency of moving quickly down a lower-carbon pathway in this report. He warned all nations that there was no time to waste, and made an urgent appeal that action had to be accelerated at every level and all countries had to be part of the solution if we were to stay within the global temperature rise threshold of 2°C[5].

If we do realize the urgency of moving quickly down a lower-carbon pathway and decide to do something about that, the action for promotion, development and diffusion of green technologies must be accelerated. During the negotiation of the Paris Agreement, both developed countries and developing countries recognized that the planet could find its way out of the dilemma only with the help of green technologies. Therefore, Article 10 of the Paris Agreement states that accelerating, encouraging and enabling innovation is critical for an effective, long-term global response to climate change and promoting economic growth and sustainable development, and such effort shall be supported by the Technology Mechanism.

Technologies will need to be 'transferred' and made accessible, since most innovation takes place in the developed countries and private corporations in those countries are the main owners of the intellectual property (IP) rights covering the majority of green technology[6]. The goal of the Paris Agreement may never be achieved if the developed countries hesitate to share their green technologies with developing countries. The developing countries have the economic systems with the greatest potential in reducing greenhouse gas emissions. Developing countries, especially low-income ones with relatively low rates of electricity usage, may be able to 'leapfrog' into electricity generation based on renewable forms of primary energy[7]. However, if we keep waiting instead of acting, more and more developing countries will be locked into the use of non-green technologies. If so, the misfortune belongs not only to developing countries but also to developed countries, because the insecure climate system does not recognise any national boundaries.

The importance of international technology transfer has been eloquently illustrated by the IPCC in its report titled *Methodological and Technological Issues in Technology Transfer*. The report states:

> Sustaining development globally will require radical technological and related changes in both developed and developing countries. Economic development

---

4 See Ban Ki-moon, *Report of the Secretary-General on the work of the Organization*, A/70/1, at p. 4, www.un.org/en/ga/search/view_doc.asp?symbol=A/70/1
5 Ibid. at p. 12
6 See Department of Economic and Social Affairs, *World Economic and Social Survey 2011: The Great Green Technological Transformation*, E/2011/50/Rev. 1, ST/ESA/333, at ix, United Nations publication Sales No. E.11.II.C.1.
7 Ibid.

is most rapid in developing countries, but it will not be sustainable if these countries simply follow the historic polluting trends of industrialized countries. Rapid development with modern knowledge offers many opportunities to avoid bad past practices and move more rapidly towards better technologies, techniques and associated institutions.[8]

The importance of green technologies transfer has also been proved by the Intended Nationally Determined Contributions (INDCs) of many developing countries. According to Brazil's INDC, the nation intends to commit to reducing greenhouse gas emissions by 37% below 2005 levels in 2025, and to reduce greenhouse gas emissions by 43% below 2005 levels in 2030[9]. Meanwhile, Brazil emphasizes the fact that it will strive for a transition towards energy systems based on renewable sources and the decarbonization of the global economy by the end of the century, only in the context of access to the technological means necessary for this transition[10]. What is more, Brazil also insists that technological development is one of the preconditions of additional actions[11]. According to South Africa's INDC, technological support is really important to achieve its goal for 2030[12]. South Africa explains that it only submits its INDC because it assumes that the Paris Agreement will make affordable technology support be available[13]. According to the INDC of Mexico, there are not only unconditional greenhouse gas (GHG) emissions reduction plans, but also conditional reduction plans, with the condition for the latter being international technical co-operation[14].

According to the INDC of Argentina, its unconditional goal is to reduce GHG emissions by 15% in 2030 with respect to projected business-as-usual emissions for that year[15]. However, Argentina could increase its reduction goal under the conditions including support for transfer, innovation and technology development[16]. Under these conditions, a reduction of 30% GHG emissions could be

---

8 See IPCC, *Methodological and Technological Issues in Technology Transfer (Summary for Policymakers)* (2000), at p. 3, http://www.ipcc.ch/ipccreports/sres/tectran/index.php?idp=0
9 See Federative Republic of Brazil Intended Nationally Determined Contribution Towards Achieving the Objective of the United Nations Framework Convention on Climate Change, at pp.1–2, www4.unfccc.int/submissions/INDC/Published%20Documents/Brazil/1/BRAZIL%20iNDC%20english%20FINAL.pdf
10 Ibid. at p. 1
11 Ibid. at p. 4
12 See South Africa's Intended Nationally Determined Contribution, at p. 1, www4.unfccc.int/submissions/INDC/Published%20Documents/South%20Africa/1/South%20Africa.pdf
13 Ibid. at p. 3
14 See Intended Nationally Determined Contribution of Mexico, at p. 2, www4.unfccc.int/submissions/INDC/Published%20Documents/Mexico/1/MEXICO%20INDC%2003.30.2015.pdf
15 See Argentine Republic Intended Nationally Determined Contribution (INDC), at p. 7, www4.unfccc.int/submissions/INDC/Published%20Documents/Argentina/1/Argentina%20INDC%20Non-Official%20Translation.pdf
16 Ibid. at p. 7

achieved by 2030 compared to projected business-as-usual emissions in the same year[17]. The INDC of Egypt emphasized that Egyptian national efforts alone would not be able to fulfil the state's aspirations in contributing to the international climate change abatement efforts. Thus, the INDC of Egypt stated that transfer of technology was needed and Article 4 of the UNFCCC, which states that developed parties shall provide support to developing countries in applying their liabilities, should be enacted[18].

Patent suppression: a major obstacle for green technology transfer

Of the registered green technology patents, 40–90% have never been used[19]. There are mainly two kinds of patent non-use: one is that some patentees have neither capability for commercializing their patents by themselves nor want to sell their patents at unreasonably low prices; the other is that some patentees want their patents never to be used, no matter whether by themselves or by someone else. The former is a kind of normal market behaviour, because the real purposes of these patentees is trying to bargain for a reasonable price. However, the latter is abnormal, because the only purpose of the patentees' research on new technologies is to suppress them, so that their old technologies cannot be supplanted and they can continue to make money from them.

This patent suppression behaviour has many negative impacts on technology research, development and diffusion. For example, many patentees build *patent thickets*, which are thick patent webs consisting of various related and overlapping patents, so that their competitors will have much more trouble researching and developing new technologies. Facing patent thickets, firms can require access to dozens, hundreds or even thousands of patents to produce just one commercial product[20]. The most troublesome quality of a thicket is the risk that one may not be able to conclusively determine that all of the patents have already been read on a product or service[21]. Relevant patents can pop up and catch even sophisticated manufacturers by surprise[22]. Addressing this awkward situation, the Secretary-General of the United Nations pointed out that the rise of strategic patenting and a series of legislative changes to expand monopoly rights had led to a very

---

17 Ibid.
18 See The Arab Republic of Egypt Intended Nationally Determined Contributions as per United Nation Framework Convention on Climate Change, at p. 13, www4.unfccc.int/submissions/INDC/Published%20Documents/Egypt/1/Egyptian%20INDC.pdf
19 See Saunders, K.M. (2002), "Patent Nonuse and the Role of Public Interest as a Deterrent to Technology Suppression", *Harvard Journal of Law & Technology*, 15, p. 389
20 See Federal Trade Commissionn, *To Promote Innovation: The Proper Balance Of Competition And Patent Law And Policy*, at p. 9, www.ftc.gov/sites/default/files/documents/reports/promote-innovation-proper-balance-competition-and-patent-law-and-policy/innovationrpt.pdf
21 See Cahoy, D.R. and Glenna, L. (2009), "Private Ordering and Public Energy Innovation Policy", *Florida State University Law Review*, 36 (3), pp. 415–458
22 Ibid.

complex system of patents, which was increasingly geared to support the rights of incumbent large firms over new, smaller, innovative firms[23]. Additionally, the system in many countries had moved from its original objective of stimulating innovation through the provision of incentives to innovators, to preventing new domestic and foreign market entrants[24].

In many green industries, core technologies have already been monopolized by a few large companies. For example, the technologies in hybrid vehicles are very important for developing countries in reducing greenhouse gases under the Paris Agreement. However, more than 90% of patents in hybrid vehicles belong to companies in the United States, Germany and Japan[25]. It is very difficult for developing countries to get access to these technologies at affordable prices. In the field of LED, a kind of low-carbon light, some companies in developed countries monopolize most of the core technologies and never permit companies in developing countries to use their patents.

Because of patent suppression, the technology gap between developing countries and developed countries keeps widening. On the one hand, patenting rates for clean energy technologies have increased faster than for other sectors, at a rate of about 20% per year since the adoption of the Kyoto Protocol by the United Nations Framework Convention on Climate Change, in 1997[26]. On the other hand, most green technology patents continue to be controlled by only a few developed countries. According to statistics provided by the Secretary-General of the UN, six developed countries, including Japan, the United States, Germany, the Republic of Korea, the UK and France, account for almost 80% of all patent applications in clean energy technology[27]. Some other statistics show that developing countries own too few high-value inventions in the field of climate change technology. Taking China and Brazil as examples, the former owns only 2.3% high-value inventions in the field of climate change technology and the latter owns only 0.2%[28].

Necessity for IP system improvement

Although green patent suppression is now very serious and has become an important barrier to technology transfer, it is not right to jump to the

---

23 *Options for a Facilitation Mechanism that Promotes the Development, Transfer and Dissemination of Clean and Environmentally Sound Technologies*, Report of the Secretary-General, A/67/348, 4 September 2012, at p. 9, www.un.org/zh/documents/view_doc.asp?symbol=A/67/348&referer=http://www.un.org/zh/documents/&Lang=E.
24 Ibid.
25 Lara, A., Parra, G. and Chávez, A. (2013), *The Evolution of Patent Thicket in Hybrid Vehicles, Commoners and the Changing Commons: Livelihoods, Environmental Security, and Shared Knowledge*, the Fourteenth Biennial Conference of the International Association for the Study of the Commons, 3–7 June 2013, at p. 7
26 *Options for a Facilitation Mechanism that Promotes the Development, Transfer and Dissemination of Clean and Environmentally Sound Technologies, supra* note 23, at p. 9
27 Ibid.
28 Dechezleprêtre, A. *et al.* (2011), "Invention and transfer of climate change-mitigation technologies: a global analysis", *Review of Environmental Economics and Policy*, 5 (1), pp. 115–117

conclusion that the governments of parties to the UNFCCC are devoid of political willingness to deal with it. On the contrary, these governments have already shown some resolve on removing barriers to the international transfer of green technology.

Article 4, para. 5, of the UNFCCC states that the developed countries shall take all practicable steps to promote, facilitate and finance the transfer of environmentally sound technologies to other parties, particularly developing countries, to enable them to implement the provisions of the Convention. Article 5 of the UNFCCC also states that the parties shall support international and intergovernmental efforts to strengthen national technical research capacities and capabilities, particularly in developing countries. Moreover, Article 10 of the Kyoto Protocol also rules that all parties shall take all practicable steps to promote, facilitate and finance the transfer of environmentally sound technologies pertinent to climate change, in particular to developing countries.

The parties of the UNFCCC tried to develop more detailed plans to promote the international transfer of green technologies after the signing of the Kyoto Protocol in 1997. For example, the Conference of the Parties, on its seventh session held in Marrakesh from 29 October to 10 November 2001, made the decision on development and transfer of technologies (Decision 4/CP.7)[29]. According to this decision, the parties would establish an expert group on technology transfer, the objective of which was enhancing the implementation of Article 4, para. 5, of the Convention, including, *inter alia*, by analysing and identifying ways to facilitate and advance technology-transfer activities. The decision also decided to urge developed country parties to provide technical assistance through existing bilateral and multilateral co-operative programmes. The decision even provided a framework for meaningful and effective actions to enhance the implementation of Article 4, para. 5, of the Convention[30]. According to the framework, all parties of the UNFCCC were urged to improve the enabling environments for technology transfer, which focused on government actions, such as fair-trade policies, removal of technical, legal and administrative barriers to technology transfer, sound economic policy, regulatory frameworks and transparency.

Although many efforts have been made by the international community to promote international transfer of green technologies, the results are quite disappointing. For example, the Kyoto Protocol created the Clean Development Mechanism (CDM) to help developing countries to contribute to the ultimate objective of UNFCCC. According to Article 12 of the Kyoto Protocol, developing countries will benefit from CDM project activities resulting in certified emission reductions. Other countries that have qualified greenhouse gas reduction obligations may use the certified emission reductions accruing from such

---

29 *Report of the Conference of the Parties on its Seventh Session, Held at Marrakesh from 29 October to 10 November 2001*, FCCC/CP/2001/13/Add.1, 21 January 2002, at p. 22, http://unfccc.int/resource/docs/cop7/13a01.pdf
30 Ibid. at p. 24

project activities to contribute to compliance with part of their own quantified emission limitation and reduction commitment. When the Clean Development Mechanism was designed during the negotiations of the Kyoto Protocol, almost all parties of the UNFCCC expected the mechanism to be a helpful tool in promoting green technology transfer between developed countries and developing countries. In fact, it was estimated that about 26% of the projects in relation to the CDM would involve at least some kind of technology transfer[31]. However, the results have proved very frustrating. Statistics shows that only 0.6% of projects involved technology transfer and the contribution of the CDM to technology transfer can at best be regarded as minimal[32]. Of course, the reasons for the frustrating results are many, but undoubtedly one of them is that some entities who own advanced green technologies have strong IP protection tactics, including building patent thickets, so that others have little opportunity to get technologies relating to their CDM projects.

Yet another important reason why many efforts of the parties of the UNFCCC have been frustrated is that they only aim to regulate the behaviour of governments rather than the behaviour of patentees. However, the fact is that patentees, rather than governments, have the final say in green technology transfer. The right of patentees to refuse to share their patents with other people is strictly protected by the international intellectual property rights system. According to Article 28 of the Agreement on Trade-Related Aspects of Intellectual Property Rights (TRIPS), where the subject matter of a patent is a product, the owner of the patent has exclusive rights to prevent third parties from the acts of making, using, offering for sale, selling or importing for these purposes that product unless they have the consent of the owner. Where the subject matter of a patent is a process, the owner of the patent has exclusive rights to prevent third parties from the act of using the process unless they have the consent of the owner. Accordingly, the problem of green patent suppression can never be solved if the parties of UNFCCC cannot manage to improve the current IP system.

Necessity for green compulsory licensing agreement

If the owners of green technologies neither use their technologies nor permit others to use their technologies to reduce greenhouse gases, the goal of the Paris Agreement can never be fulfilled. If we want to make the Earth, which is becoming warmer and warmer, safer for us to live, attention should be paid not

---

31 *Options for a Facilitation Mechanism that Promotes the Development, Transfer and Dissemination of Clean and Environmentally Sound Technologies*, supra note 23, at p. 15
32 See *Options for a Facilitation Mechanism that Promotes the Development, Transfer and Dissemination of Clean and Environmentally Sound Technologies*, supra note 23, at p. 15; Das, K., "Technology transfer under the clean development mechanism: an empirical study of 1000 CDM projects", *The Governance of Clean Development, Working Paper Series, No. 14*, Economic and Social Research Council and University of East Anglia, July 2011, at p. 28, www.indiaenvironmentportal.org.in/files/file/gcd_workingpaper014.pdf

only to the protection of the private interests of patentees, but also to the protection of public interests.

Compulsory licensing is an important legal tool to protect public interests by preventing patent-right holders from abusing their rights. However, the most important and influential international agreement on intellectual property, TRIPS, has extremely strict regulations on compulsory licensing.

According to Article 31 of TRIPS, there are more than ten kinds of limitations on compulsory licensing. For example, authorization of compulsory licensing shall be considered on its individual merits; compulsory licensing may only be permitted if the proposed user has made efforts to obtain authorization from the right holder on reasonable commercial terms and conditions and that such efforts have not been successful within a reasonable period of time; the scope and duration of compulsory licensing shall be limited to the purpose for which it was authorized; compulsory licensing shall be non-exclusive; compulsory licensing shall be non-assignable except with that part of the enterprise; compulsory licensing shall be authorized predominantly for the supply of the domestic market of the country authorizing it; authorization for compulsory licensing shall be terminated if and when the circumstances which led to it cease to exist and are unlikely to recur; the right holder shall be paid adequate remuneration in the circumstances of each case taking into account the economic value of the authorization; the legal validity of any decision relating to compulsory licensing shall be subject to judicial review or other independent review by a distinct higher authority in the country authorizing it; any decision relating to the remuneration provided in respect of compulsory licensing shall be subject to judicial review or other independent review by a distinct higher authority; where compulsory licensing is authorized to permit the exploitation of a patent ('the second patent') which cannot be exploited without infringing another patent ('the first patent'), the invention claimed in the second patent shall involve an important technical advance of considerable economic significance in relation to the invention claimed in the first patent.

What cannot be ignored is that the main parties of the UNFCCC are also the parties of TRIPS. Therefore, these parties are not allowed to implement less strict limitations on green patent compulsory licensing because the regulations of TRIPS are the minimum standards for intellectual property rights protection. If there is no new international agreement on green patent compulsory licensing to modify those strict limitations under TRIPS, the parties of the UNFCCC can do almost nothing to solve the problem of green patent suppression.

What is also worth mentioning is that there already exists an important precedent for the international community that successfully managed to modify strict limitations under TRIPS with a new international agreement. In 2001, World Trade Organization (WTO) members adopted a special Ministerial Declaration at the WTO Ministerial Conference in Doha, to clarify ambiguities between the need for governments to apply the principles of public health and the terms of TRIPS[33].

The Declaration responds to the concerns of developing countries about the obstacles they faced when seeking to implement measures to promote access to affordable medicines in the interest of public health[34]. The Doha Declaration refers to several aspects of TRIPS, including the right to grant compulsory licences[35]. According to paragraph 4 of the Doha Declaration, the TRIPS Agreement does not and should not prevent its members from taking measures to protect public health. On the contrary, the TRIPS Agreement can and should be interpreted and implemented in a manner supportive of right of WTO members to protect public health. The Doha Declaration affirms in paragraph 5 that each member has the right to grant compulsory licences and the freedom to determine the grounds upon which compulsory licences are granted. More importantly, the Doha Declaration makes it clear in paragraph 5 that each member of TRIPS has the right to determine what constitutes a national emergency or other circumstances of extreme urgency and that public health crises, including those relating to HIV/Aids, tuberculosis, malaria and other epidemics, can represent a national emergency or other circumstances of extreme urgency.

According to Article 31 of TRIPS, compulsory licensing may only be permitted if, prior to it, the proposed user has made efforts to obtain authorization from the right holder on reasonable commercial terms and conditions and that such efforts have not been successful within a reasonable period of time. However, Article 31 of TRIPS regulates that this requirement may be waived by a member of TRIPS in the case of a national emergency or other circumstances of extreme urgency. Obviously, this is helpful for the members of TRIPS to be able to grant compulsory licences to address public health crises, because the Doha Declaration clearly regards public health crises as national emergencies.

Climate change is a serious challenge to all members of TRIPS as well. It is at the very least as grim or urgent as many public health crises, if not more so. The IPCC Fifth Assessment Report stated that each of the last three decades had been successively warmer at the Earth's surface than any preceding decade since 1850, and the period from 1983 to 2012 was likely the warmest 30-year period of the last 1,400 years in the northern hemisphere[36]. The report emphasized that changes in climate have caused impacts on natural and human systems on all continents and across the oceans, and that glaciers continued to shrink almost worldwide due to climate change[37]. The Decision on Adoption

---

33 The Doha Declaration on the TRIPS Agreement and Public Health: www.who.int/medicines/areas/policy/doha_declaration/en/.
34 Ibid.
35 Ibid.
36 IPCC, *Climate Change 2014: Synthesis Report (Summary for Policymakers)*, at p. 2: www.ipcc.ch/pdf/assessment-report/ar5/syr/AR5_SYR_FINAL_SPM.pdf
37 IPCC, *Summary for Policymakers. In: Climate Change 2014: Impacts, Adaptation, and Vulnerability*, at p. 4: http://ipcc-wg2.gov/AR5/images/uploads/WG2AR5_SPM_FINAL.pdf

Green compulsory licensing 189

of the Paris Agreement by COP 21 of the UNFCCC clearly recognized that climate change represented an urgent and potentially irreversible threat to human societies and the planet[38]. Ban Ki-moon, then Secretary-General of the UN, pointed out in his statement on World Meteorological Day on 23 March 2016 that climate change was accelerating at an alarming rate and the window of opportunity for limiting global temperature rise to well below 2° Celsius – the threshold agreed by world governments in Paris – is narrow and rapidly shrinking[39].

Many developing countries have been calling for the development of an international agreement on compulsory licensing of green technologies. The 2007 Joint Position Paper of Brazil, China, India, Mexico and South Africa participating in the G-8 summit, stated that agreement on transfer of technologies at affordable costs was needed for accelerated mitigation efforts in developing countries[40]. On 7 June 2008, the G-77/China highlighted equal treatment for mitigation and adaptation technologies, and emphasized the need to establish a technology-transfer mechanism under the UNFCCC in a meeting convened by the Ad Hoc Working Group on Long-term Cooperative Action under the Convention (AWG-LCA)[41]. What is more, the African Group directly identified intellectual property rights as a major barrier to the international transfer of green technologies, and Pakistan stressed the necessity of compulsory licensing in this meeting[42]. On 6 February 2009, China provided its views on the fulfilment of the Bali Action Plan and stated:

> The existing IPR system does not match the increasing needs for accelerating D&T&D of ESTs to meet challenges of climate change. Compulsory licensing related patented ESTs and specific legal and regulatory arrangement to curb negative effects of monopoly powers shall be put in place as part of the efforts to implement the UNFCCC.[43]

Prompted by developing countries, the AWG-LCA prepared a negotiation text and listed compulsory licensing for specific patented technologies as one of

---

38 Adoption of the Paris Agreement, Decision 1/CP.21, 29 January 2016: http://unfccc.int/resource/docs/2015/cop21/eng/10a01.pdf
39 Secretary-General's message on World Meteorological Day, 23 March 2016: www.un.org/sg/statements/index.asp?nid=9559
40 See Fair, R. (2009), "Does Climate Change Justify Compulsory Licensing of Green Technology?", *Brigham Young University International Law & Management Review*, 6, p. 23
41 *Earth Negotiation Bulletin*, 12 (375), 16 June 2008, p.4: www.iisd.ca/download/pdf/enb12375e.pdf.
42 Ibid.
43 See UN Framework Convention on Climate Change, Ad Hoc Working Group on Long-Term Cooperative Action Under the Convention (2009), "China's Views on the Fulfillment of the Bali Action Plan and the Components of the Agreed Outcome To Be Adopted by the Conference of the Parties at Its 15th Session", U.N. Doc.FCCC/AWGLCA/2009/MISC.1at p. 23; http://unfccc.int/resource/docs/2009/awglca5/eng/misc01.pdf.

the optional measures to address intellectual property right problems on 9 May 2009[44].

However, the negotiation of compulsory licensing agreements has not progressed smoothly, because there are huge divergences between some developed countries and developing countries. For example, in the first meeting of the Ad Hoc Working Group on Further Commitments for Annex I parties under the Kyoto Protocol and the Ad Hoc Working Group on Long-term Cooperative Action under the Convention held in Bangkok in April 2008, some developed countries, including Australia and the US, affirmed their belief that IP was not a barrier to green technologies international transfer, but a catalyst for technology transfer[45].

As a matter of fact, the above viewpoint of the negotiators of the US and Australia reflects a common understanding on IP protection. Many judges in the United States do not think patent suppression is illegal and believe the monopoly right which a patentee receives does not need further explanation, for it has been the judgment of Congress from the beginning that the sciences and the useful arts could be best advanced by giving an exclusive right to an inventor[46]. Therefore, they support the system that patentees have the absolute right to dispose of their patents, including suppressing them, because patents are owned by them just as they own their other properties such as houses, cars and computers. In *Continental Paper Bag Co.* v. *Eastern Paper Bag Co.*[47], the plaintiff did not use the patent because it required a substantial investment in machines that could not be improved or replaced without great expense[48]. Moreover, the plaintiff declined to license the patent to any of its competitors[49]. As a defence to infringement, the defendant asserted that the plaintiff should be denied injunctive relief because it was holding the patent in non-use[50]. The Supreme Court rejected the defendant's argument and stated that competitors being excluded from the use of the new patent was the very essence of the right conferred by the patent, as it is the privilege of any owner of property to use or not use it, without question of motive[51].

However, it cannot always be just and fair to say that the patentees have the right to not use their patents and exclude others from the use of the same. It is common legal knowledge that it is necessary to limit the monopoly right

---

44 Negotiating text prepared by the Chair of the Ad Hoc Working Group on Long-term Cooperative Action under the Convention (AWG-LCA), FCCC/AWGLCA/2009/8, 9 May 2009, at p. 48: http://unfccc.int/resource/docs/2009/awglca6/eng/08.pdf
45 International Centre for Trade and Sustainable Development (ICTSD), *Climate Change, Technology Transfer and Intellectual Property Rights*, August 2008, at p. 4: www.ictsd.org/downloads/2008/11/climate-change-technology-transfer-and-intellectual-property_ictsd-2008-2.pdf
46 Continental Paper Bag Co. v. Eastern Paper Bag Co., 210 U.S. 405 (1908), at p. 429
47 Continental Paper Bag Co. v. Eastern Paper Bag Co., 210 U.S. 405 (1908)
48 See Saunders, "Patent Nonuse and the Role of Public Interest as a Deterrent to Technology Suppression", note 19
49 Ibid.
50 Ibid.
51 Continental Paper Bag Co. v. Eastern Paper Bag Co., 210 U.S. 405 (1908), at p. 429

of owners of properties for public interests. In 1926, the Permanent Court of International Justice took the position that 'expropriation for reasons of public utility, judicial liquidation and similar measures' was lawful[52]. The 1962 United Nations General Assembly Resolution on Permanent Sovereignty over Natural Resources stated that the private properties of owners could be expropriated if public utility, security or the national interest were recognized as overriding purely individual or private interests[53]. Recently many bilateral investment agreements (BIT) recognize that the property rights of their legal owners can be limited for protection of public interest. In 2007, a document of the United Nations Conference on Trade and Development pointed out that, increasingly, bilateral investment agreements 'not only emphasize the objectives of investment promotion and protection, but also underline that this goal must not be pursued at the expense of other public interests, such as health, safety, environment and labour'[54].

For example, the BIT between France and Uganda (2002) regulated that it would be an exception for contracting parties to take any measures of expropriation or nationalization or any other measures having the effect of dispossession, direct or indirect, of nationals or companies of the other contracting party of their investments on its territory and in its maritime area[55]. The protocol of the BIT between Germany and Mexico (1998) regulated that the measures taken by reason of national security, public interest, public health or morality should not be considered as a 'less favourable treatment'[56]. Article 15 of the BIT between Australia and India (1999) regulated that: 'Nothing in this Agreement precludes the host Contracting Party from taking, in accordance with its laws applied reasonably and on a nondiscriminatory basis, measures necessary for the protection of its own essential security interests or for the prevention of diseases or pests'[57].

Similarly, Article XVI of the BIT between Mozambique and the United States (1998) stated: 'This Treaty shall not preclude a Party from applying measures that it considers necessary for the fulfilment of its obligations with respect to the maintenance or restoration of international peace or security, or the protection of its own essential security interests.'[58] Accordingly, if patents can be protected as other properties such as lands, houses, cars, etc, why cannot the rights of patent

---

52 See Treeger, C., "Legal analysis of farmland expropriation in Namibia", at pp. 2–3: www.kas.de/wf/de/21.38/wf/doc/kas_4800-544-2-30.pdf
53 Permanent sovereignty over natural resources, General Assembly Resolution 1803 (XVII), para. 4, 1962: www.un.org/en/ga/search/view_doc.asp?symbol=A/RES/1803(XVII)&referer=http://www.un.org/depts/dhl/resguide/r17_resolutions_table_eng.htm&Lang=E
54 United Nations Conference on Trade and Development, *Bilateral Investment Treaties 1995–2006: Trends in Investment Rulemaking*, 2007, at p. xi, http://unctad.org/en/Docs/iteiia20065_en.pdf
55 Ibid., p. 46
56 Ibid., p. 86
57 Ibid., p. 85
58 Ibid.

owners be limited as are the rights of other property owners? Besides, if we do not regard global climate change as an issue concerning public interests, why do we need the Paris Agreement? Hence, it is easy to draw a conclusion that it is necessary and reasonable to take some measures including green compulsory licensing to achieve the goals of the Paris Agreement.

Another reason for some developed countries to refuse to develop a compulsory licensing agreement is that they are afraid that compulsory licensing will weaken the incentive of green technology research and development. However, the fact is that the aim of green compulsory licensing is not to deprive reasonable profits of patentees. Instead, the aim is to prevent patentees from suppressing their green technologies for unreasonable monopoly profits. If the international agreement regulates that a green compulsory licence can only be granted when a patentee refuses to permit someone else to use his patent who already offers a price obviously higher than the market price or the price suggested by neutral assessment centres, the incentive of green technology research and development can be effectively protected.

As long as the system of compulsory licensing is well designed, it will have next to no negative influence on technology research and development. For example, Professor Frederic M. Scherer from Harvard University and his research team found from interviews, mail survey responses and statistical analyses that compulsory licensing decrees had little or no unfavourable impact on research and development decisions[59].

Conclusion

We have to face the fact that we do not have a guarantee for the safety of our planet, which is becoming warmer and warmer, even though we have the Paris Agreement. If we do want the goals of the Paris Agreement to be achieved, the conditions for the parties cannot be ignored, and a very important condition of them is that developing countries urgently need more green technologies. However, some larger companies will never allow their green technologies to be used to reduce carbon emissions, because more carbon emissions means less cost to their business.

Large companies have little regard about the safety of the ecological systems of the Earth and they operate with the sole aim of making more monopoly profits with their high-carbon technologies. It will be a shame for everyone on the planet if necessary measures cannot be adopted to change the situation. However, the very strict limitations on compulsory licensing in TRIPS make it very difficult for parties of the UNFCCC to adopt measures to solve the problems of green patent suppression. Although it is the privilege of any owner of property, including any owner of a patent, to use or not use it without question of motive, there

---

59 Scherer, F.M., *Political Economy of Patent Policy Reform in the United States*, at p. 6: www.hks.harvard.edu/m-rcbg/papers/scherer/PATPOLIC.pdf

should be exceptions for the protection of public interest. Green compulsory licensing is an effective legal system to balance the private interests of patentees and public interests of human beings, and it will have little or no unfavourable impact on green technology research and development. Therefore, it is necessary to develop an international agreement on green compulsory licensing to promote the international transfer of green technologies for achieving the goals of the Paris Agreement.

12 Low-carbon market opportunities and a brief discussion on lessons learned from the Adaptation Fund

Andrea Ferraz Young

Introduction

Nowadays, adaptation to climate change represents a significant challenge to structure of governance at all scales, and cross-scale dynamics in human-environment systems[1]. It involves social perceptions of climate risk, environmental conservancy, reducing of potential damages, recognition of opportunities[2].

Adaptation consists of a wide variety of actions by an individual, community, or organization such as: 1) improvement of water use efficiency and additional water storage capacity; 2) protection of river banks; 3) implementation of early warning systems and emergency response to changes in the frequency, duration and intensity of extreme weather events; 4) increase in energy efficiency; and 5) promotion of the use of alternative energy sources[3].

A growing number of countries (e.g., India, Zambia, Colombia, Indonesia, China, Mexico and Thailand) have prepared their institutional organizations to protect people and infrastructure from climate change impacts. In China and India, for example, adaptation measures focus on solar energy policies to support solar power development[4]. In Zambia, the adaptation measures focus on disaster risk reduction through the establishment of an Interim Inter-Ministerial Climate

---

1 O'Brien, K., Pelling, M., Patwardhan, A., Hallegatte, S., Maskrey, A., Oki, T., Oswald-Spring, U., Wilbanks, T. and Yanda, P.Z. (2012), "Toward a sustainable and resilient future", in *Managing the Risks of Extreme Events and Disasters to Advance Climate Change Adaptation Field*, Barros, V., Stocker, T.F., Qin, D., Dokken, D.J., Ebi, K.L., Mastrandrea, M.D., ... Midgley, A.M. (eds.), Cambridge, UK & New York, USA: Cambridge University Press, pp. 437–486
2 Adger, W.N., Agrawala, S., Mirza, M.M.Q., Conde, D., O'Brien, K., Pulhin, J., Pulwarty, R., Smit, B. and Takahashi, K. (2007), "Assessment of adaptation practices, options, constraints and capacity", in *Climate Change 2007: Impacts, Adaptation, and Vulnerability*, Canziani, O.F., Palutikof, J.P., van der Linden, P.J. and Hanson, C.E. (eds.), Cambridge, UK: Cambridge University Press, pp. 717–743
3 USAID (2013), "Addressing Climate Change Impacts on Infrastructure preparing for Change", Engility Corporation Washington DC report prepared for the United States Agency for International Development (USAID), under the Climate Change Resilient Development Task Order No. AID-OAA-TO-11-00040 Integrated Water and Coastal Resources Management Indefinite Quantity Contract (WATER IQC II)
4 International Energy Agency (2016), Energy Efficiency Market Report, Paris, France, p. 142

Change Secretariat (IIMCCS) attached to the Ministry of Finance – which is also responsible for national development planning[5].

On the other hand, the adaptation has been implemented through the design of a national climate change system in order to improve the resilience in Colombia by Ministry of Environment and Sustainable Development through the National System of Climate Change. In Indonesia, the efforts to respond to climate change are achieved by the State Ministry of National Development Planning, which seeks to mobilize, manage and allocate funding in alignment with Indonesian development priorities in order to implement greenhouse gas (GHG) emissions mitigation and adaptation measures to climate change initiatives[6].

The increase in demand for action has required significant levels of private and public investment and, despite considerable efforts, one of the greatest challenges has been to ensure increased investment in creating a productive, equitable and resilient environment through climate finance. In this sense, a close dialogue between the authorities and the industry will be the key to reap the rewards of innovation[7].

A particular focus has been on understanding goals and instruments in a more cost-effective way over time, managing some systemic risks and enhancing a pull of factors involved by encouraging parties to establish predictable, transparent and responsive actions, which could include: 1) alignment of climate finance interventions with national development goals; 2) adjustment of the goals and policies aimed at achieving a low-carbon resilient economy; and 3) evaluation of the results[8].

This paper addresses the following question: How can actions on adaptation be stepped up? What lessons can be drawn from implementation of climate policies at national and sub-national levels?

Inspired by these questions, this document is organized into four sections and covers the progress made to date and lessons learned with the operationalization of procedures for the Adaptation Fund:

In section one, we present some concepts that represent the fundamental characteristics of adaptation to climate change and analyse some aspects of the relationship between adaptation and social justice.

In section two, we describe the provision of financial resources, including the potential diversification of revenue streams to fund concrete adaptation projects

---

5 Ministry of National Development Planning (2017), Zambia Integrated Forest Landscape Project Environmental and Social Management Framework Report, p. 137
6 Nakhooda, S. and Jha, V. (2014), "Getting it together. Institutional arrangements for coordination and stakeholder engagement in climate finance", p. 28; www.giz.de/fachexpertise/downloads/giz2014-en-climate-finance-coordination-study.pdf
7 World Economic Forum (2017), "Balancing Financial Stability, Innovation, and Economic Growth", p. 20
8 Kato, T., Ellis, J. and Clapp, C. (2014), "The Role of the 2015 agreement in mobilizing Climate Finance", Organisation for Economic Co-operation and Development, p. 49

and programmes that has been driven based on the needs, views and priorities of each country. Then, in section three, we describe the institutional linkages and relations between the Adaptation Fund and Implementing Entities. We conclude by describing the lessons learned from the application of the access modalities of the Adaptation Fund.

We believe that it is important to highlight some efforts in which the actions on adaptation could improve the mobilization of further climate finance both directly (e.g., by implementation of clean energy) and indirectly (e.g., by legal rules for reducing the effects of climate change). There are some important issues that need to be considered to ensure sustainability, predictability and adequacy of financial resources.

Concepts and approaches

Adaptation

Adaptation to climate change implies actions to reduce the risks of disasters and improving the range of opportunities associated with global climate change[9]. The emphasis on the actions and use of evidence-based policy as part of results based on adaptation management often indicates a gap between the concepts and the complex nature of interventions. For example, a biopolitical reading offers a critical alternative to this field, however it also suggest that the emergence of vulnerability and resilience approaches provide progressively more complex solutions[10].

Besides, there is no single approach for adaptation assessing, planning and implementation[11]. In this way, we could define adaptation as a range of options considering socio-economic impacts of environmental stresses associated with global environmental change[12].

The climate change impacts and vulnerability assessment vary widely, depending on the situation (e.g., natural resources, industrial production, agriculture and economy); time-frame (e.g., near-term consistent with annual crop-planning or a longer time-frame comparable to the design of a road transport system); region (e.g., a transboundary watershed or a single site); and purpose of the assessments (e.g., technical design of required infrastructure)[13].

Ideally, adaptive negotiations should be supported by a clear identification and description of the environmental problems involved. However, it is difficult to

---

9 Füssel, H.M. (2007), "Adaptation planning for climate change: concepts, assessment approaches, and key lessons", Integrated Research System for *Sustainability Science* and Springer, 11pp.
10 Grove, K. (2014), "Biopolitics and Adaptation: Governing Socio-Ecological Contingency Through Climate Change and Disaster Studies", *Geography Compass*, 8 (3), pp. 198–210
11 Füssel (note 9)
12 Adger (note 2)
13 UNFCCC (2016), "Methodologies for assessing adaptation needs with a view to assisting developing country Parties, without placing an undue burden on them", Tenth meeting of the Adaptation Committee Bonn, Germany, 13-16 September; AC/2016/13

realize such an ideal situation in climate negotiations, and sometimes, large-scale multilateral negotiations have shown signs of strain[14].

For this reason, adaptation has become an important policy priority in the international agreements, mainly in terms of reducing global inequities and resource distribution[15].

Accordingly, assessment for adaptation has been discussed by parties with focus on the preparation of National Adaptation Programs of Action (NAPA), where adaptation is a process that requires decisions from public, non-governmental, to private sectors[16].

The political dynamics around adaptive negotiations is the result of multiple intersecting factors[17]. For this reason, adaptation to environmental change is best formulated under governance perspectives, adaptive capacity, and under robustness of response strategies[18].

As part of this strategy, the federal government must be one of the key agents that should provide incentives for local and state authorities, guiding across jurisdictions, sharing lessons learned, and supporting the scientific research in order to expand knowledge about the impacts and adaptive solutions[19].

Actually, the technical solutions run into lack of solid information on the costs, benefits and effectiveness of the solutions, caused by uncertainty on the impacts and lack of co-ordination in many countries[20].

'Whereas recognition of the dignity and of the equal and inalienable rights of all members of the human family is the foundation of freedom, justice and peace in the world', full participation of society is required for legitimacy of the agreements, turning social justice into reality. In this sense, it is fundamental not to ignore the origin of people and the difference between goods for distinct persons[21].

14 O'Brien, E. and Gowan, R. (2012), "What Makes International Agreements Work: Defining Factors for Success", Center of International Cooperation, p. 38
15 Lahsen, M., Sanchez-Rodriguez, R., Romero Lankao, P., Dube, P., Leemans, R., Gafney, O., Mirza, M., Pinho, P., Osman-Elasha, B. and Smith, M.S. (2010), "Impacts, adaptation and vulnerability to global environmental change: challenges and pathways for an action-oriented research agenda for middle-income and low-income countries", *Environmental Sustainability*, 2, pp. 364–374
16 UNFCCC (2014), "Institutional arrangements for national adaptation", Planning and Implementation Adaptation Committee Thematic Report, p. 44
17 O'Brien and Gowan (note 14)
18 Nelson, D.R., Adger, W.N. and Brown, K. (2007), "Adaptation to environmental change: contributions of a resilience framework", *Annual Review of Environment and Resources*, 32, pp. 395–419
19 National Research Council (2010), *Adapting to the Impacts of Climate Change*, Washington DC: The National Academies Press
20 Parry, M., Arnell, N., Berry, P., Dodman, D., Fankhauser, S., Hope, C., Kovats, S., Nicholls, R., Satterthwaite, D., Tiffin, R., and Wheeler, T. (2009), *Assessing the Costs of Adaptation to Climate Change: A Review of the UNFCCC and Other Recent Estimates*, London, UK: International Institute for Environment and Development and Grantham Institute for Climate Change
21 O'Brien and Gowen (note 14)

Environmental justice and adaptation

Environmental justice has been defined in many different ways, depending on how the context was defined, delimited and even reconstituted in order to attend specific, and usually dominant, social interests[22]. Most understandings of environmental justice refer to social equity, or the distribution of environmental ills and benefit[23].

Ideologically, environmental justice refers to a safe, clean, healthy, productive and sustainable environment for all living beings. In this case, environment is considered in its totality as a system, which includes the ecological (biological), physical (natural and built), social, political, aesthetic and economic aspects[24].

It should be highlighted that poorer segments of the population disproportionately live in environmentally degraded conditions, even in industrialized countries. Therefore, environmental justice refers to a set of conditions that should support the fulfilment of human rights through economic activities, and political and legal instruments[25].

Clearly, we could consider that a comprehensive definition would extend beyond the traditional perspective of human rights. It must be understood within a larger social and historical context, since it is about equitable distribution, or in other words, a process by which efforts are made to ensure equal opportunities for all[26].

Effective implementation of legislation, regulations, executive demands, policy directives and programmes can serve as key tools to advance environmental justice. However, it is necessary to understand that environmental justice is not universally defined. It is based in place, time and different local perspectives; therefore it has different meanings to various communities and institutions[27].

Adaptation poses significant governance challenges at the international, national and local levels. Governance structures and decisions affect the distribution of environmental costs and often perpetuate rather than alleviate environmental injustices. For this reason, risk governance indicates when policy process and institutional structure restrain the activities of a group to regulate, reduce or control general risk problems[28].

---

22 Castree, N. and Braun, B. (2001), *Social Nature Theory, Practice, and Politics*, Massachusetts, USA: Blackwell Publishers, p. 263
23 Schlosberg, D. (2004), "Reconceiving Environmental Justice: Global Movements And Political Theories", *Journal of Environmental Politics*, 13 (3), pp. 517–540
24 Tompkins, E.L. and Adger, W.N. (2004), "Does adaptive management of natural resources enhance resilience to climate change?", *Ecology and Society*, 9 (10); www.ecologyandsociety.org/vol9/iss2/art10 (accessed July 2017)
25 Hawkins, C.A. (2010), "Sustainability, human rights, and environmental justice: Critical connections for contemporary social work", *Critical Social Work* 11 (3), pp. 67–81
26 O'Brien and Gowan (note 14)
27 EPA (2015), "Environmental Justice Implementation Progress Report", US Department of Health and Human Services, Washington, DC, p. 34
28 Renn, O., Klinke, A. and van Asselt, M. (2011), "Coping with Complexity, Uncertainty and Ambiguity in Risk Governance: A Synthesis", *Ambio*, 40 (2), pp. 231–246

This examination of governance challenges in promoting environmental justice finds that existing state and international governance institutions are insufficient mechanisms for securing environmental justice; a multisector, multi-level governance approach that integrates civil society and social movement actors is needed[29].

The dynamics, structures and functionality of risk governance processes require a comprehensive understanding of procedural mechanisms and structural configurations with emphasis on opportunities and benefits of adaptation. In this sense, environmental justice must go deeper because environmental justice deals with the fundamental manner that political decisions are regulated[30].

Even with a comprehensive perspective of environmental justice, governance and procedural mechanisms, legal and institutional frameworks are essential for establishing the roles and responsibilities of different actors in co-ordination with administrative policies. For this reason, it is essential to examine the conditions of organizational networks for action effectiveness of social protection[31].

Social and environmental justice is not only an income issue or a sociological/operational concept. Some prerequisites are necessary for proper operation. For example, the connections between social economic sustainability, human rights and environmental justice, need to be more clearly articulated through a democratic culture based on balanced and alive discussions on the one hand and appropriated social structure on the other[32].

In the contemporary world, we still live in battlegrounds that result from conflicts of interests and dissenting beliefs about justice. Indeed, consensus and conflict characterize our world. Society is not an entity homogenized through the integration of values[33].

Given the relevance of this, a number of general questionings about social justice are still recommended for increasing procedural justice in society[34] and environmental decision-making processes. An understanding of these separate but closely linked concepts is required to effectively pursue the purpose of making the world more just[35].

Besides, the presence of resilience in competing grounds of society has strong implications on environmental justice that can be observed through the application of inequality evaluation in general terms, and in particular income distribution assessment[36].

---

29 Ibid.
30 Ibid.
31 Provan, K.G. and Kenis, P. (2008), "Modes of Network Governance: Structure, Management, and Effectiveness", *Journal of Public Administration Research and Theory*, 18 (2), pp. 229–252
32 Verwiebe, R. and Wegener, B. (2000), "Social Inequality and the Perceived Income Justice Gap", *International Social Justice Project*, 13 (2), p. 28
33 Adger, W.N., Hughes, T.P., Folke, C., Carpenter, S.R. and Rockström, J. (2005), "Social-Ecological Resilience to Coastal Disasters", *Science* 309 (5737), pp. 1036–39
34 Ibid.
35 Hawkins (note 25)
36 Sen (note 33)

A critical component of legitimate resilience is the recognition of the role of governance. Governance mechanisms were design to include the public perspective because the decisions normally affect all members of society and therefore must promote involvement of people who have been deeply impacted by them. Every community must have access to information on the potential benefits of decisions, equal opportunities and knowledge on environmental risks[37].

The involvement of stakeholders must extend beyond voicing opinions and include some actual decision-making power and ability to influence outcomes established by formal roles and relationships between corporations and community groups, such as agreements that supplement government regulations and allow increased local participation in managing relationships through collaborative problem-solving[38].

Examining policies through the lens of environmental justice and governance suggests that the complex interplay of inequities in terms of distribution should be formalized through the recognition that participation must be better understood to promote investments in long-term education, fair tax collection with coherent directions, equal rights (procedural mechanisms), fair participation, and consequently better employment opportunities[39].

Socio-economic evidence, such as public finances, economic efficiency and macro-economic stability, shows that without extended investments in education, there will be no better opportunities. Therefore, rapid and sustained growth will be rather critical without education[40], since the productive activities of individuals are considered an important aspect of social environmental justice. They represent forms of movement to achieve social and environmental justice on the ground[41].

Socio-environmental justice is more complex than the distribution of income for the poorest and environmental risks recognition[42]. No doubt that both are important aspects of inclusive growth. Notwithstanding, education associated to appropriate legal and institutional frameworks are expressions of paramount importance[43].

37 Anand, A., Milne, F. and Purda, L. (2006), *Voluntary Adoption of Corporate Governance Mechanisms,* Quebec, Canada: Department of Economics Queen's University, p. 40
38 Kahane, D., Loptson, K., Herriman, J. and Hardy, M. (2013), "Stakeholder and Citizen Roles in Public Deliberation", *Journal of Public Deliberation,* 9 (2), p. 37
39 Bulkeley, H., Gareth, A.S., Edward, G.A.S and Fuller, S. (2014), "Contesting climate justice in the city: Examining politics and practice in urban climate change experiments", *Global Environmental Change,* 25, pp. 31–40
40 Reynolds, D., Teddlie, C., Chapman, C. and Stringfield, S. (2015), "Effective school processes", *The Routledge International Handbook of Educational Effectiveness and Improvement, Research, Policy, and Practice;* www.routledgehandbooks.com/doi/10.4324/9781315679488.ch3 (accessed July 2017)
41 DESA-UN (2006), "The International Forum for Social Development – Social Justice in an Open World. The Role of the United Nations", Department of Economics and Social Affairs, Division for Social Policy and Development. ST/ESA/305. 157pp
42 Bulkeley (note 39)
43 Reynolds (note 40)

In this sense, it is important to persevere and insist on the establishment of an appropriate legal framework by providing a statement of clear entitlements with a coherent set of strategies and rules[44].

What is required?

Social justice may be broadly understood as the fair distribution of the outcomes of economic growth. However, it is necessary to understand that tackling environmental change will require substantial financial and investment flows to support mitigation and adaptation measures[45].

The concept of social justice must integrate these dimensions, starting with the right of all human beings to benefit from a safe and pleasant environment. However, overcoming this challenge will largely depend on the efficiency of both fiscal transfers and economic transactions, although existing financing instruments have clear limits and inefficiencies[46].

As most organizations and governments are only beginning to implement their actions for climate adaptation, the scale of the financing gaps, the diversity of needs, and the differences in terms of national emplacement require a broad range of analysis and instruments[47].

Financing needs are linked to the scope and timing of international agreements on climate change (e.g., the Adaptation Fund). The Adaptation Fund was established to finance concrete adaptation projects and programmes in developing countries that are particularly vulnerable to the adverse effects of climate change[48]. In order to increase climate resilience in 40 countries around the world, the fund has dedicated more than $232 million between 2011 and 2014[49].

The systematization of finance process should involve education, access to technology and capacity-building, but the developing countries must shift to a lower-carbon development path as defined in the agreement between the parties[50].

Whatever the legal sources of climate finance, it is vital to ensure that adequate and reliable climate finance reaches the vulnerable people, where the impacts can be clearly evaluated and monitored, and the social-environmental justice can be safeguarded[51].

44 DESA-UN (note 41)
45 UNFCCC (note 16)
46 DESA-UN (note 41)
47 UNFCCC (note 16)
48 UNFCCC (2016), "Subsidiary Body for Scientific and Technological Advice", Report of the Adaptation Committee, United Nations, FCCC/SB/2016/2. 25pp
49 Parker, C., Keenlyside, P. and Conway, D. (2014), *Early Experiences In Adaptation Finance: Lessons From The Four Multilateral Climate Change Adaptation Funds,* Amsterdam, the Netherlands: Climate Focus, p. 100
50 UNFCCC (note 16); UNFCCC (note 48)
51 UNFCCC (note 48)

Co-operative agreement initiative for environmental justice

Given the relevance of environmental justice goals and financing needs, the first efforts to assist different countries in achieving adaptation measures should begin with grants for programme pilot projects, which should be developed and awarded through an evaluation conducted by international advice[52].

In this context, key elements should be considered for a co-operative agreement initiative such as:

1) State strategies, programmes and local activities for identifying, developing, planning and working on local environmental issues with society, for building consensus, and setting community priorities;
2) Collaboration with other stakeholders (e.g., community-based organizations, environmental groups, businesses, industry, federal and local governments, and academic institutions) to realize their goals and objectives;
3) Achieve measurable and meaningful environmental results in society;
4) Build broad and robust, results-orientated partnerships, particularly with community organizations within affected areas;
5) Projects in communities that create models which can be expanded or replicated in other geographic areas;
6) Strengthen the development and implementation of specific approaches to achieve environmental justice.

The purpose of co-operative agreement initiatives is specifically to support and produce specific activities targeting climate solutions that lead to adaptation measures and public environmental results in different strata of society burdened by environmental harms and risks by leveraging or utilizing the existing resources or assets of state agencies[53].

Eligible applicants

Another important aspect related to efforts to assist different countries is the political consistency which is essential for creating an attractive investment environment. A lack of consistency, for example, when entrepreneurs agree to participate regardless of the opinion of some alienated politicians, undermines investor confidence, which takes time to rebuild. Investors are generally operating on a longer time-scale than politicians, so changing politics can have a huge impact. Cross-parties consensus is vital, as is cross-government alignment[54].

---

52 UNFCCC (2012), "Report of the Global Environment Facility to the Conference of the Parties", United Nations FCCC/CP/2012/6, 124pp
53 Ibid.
54 Ibid.

Low-carbon market opportunities 203

Furthermore, institutions are important but they are not a guarantee of successful engagement or effective climate-related investment. They need consistent political support as well as a positive behaviour across organizations[55].

Under the UNFCCC, various processes have been established to support parties in their planning efforts on adaptation. The National Adaptation Plan of Action (NAPA) enables least developed countries (LDCs) to identify and prioritize urgent and immediate needs with regard to adaptation to the adverse effects of climate change.

Under the UNFCCC (2010), the Cancun Adaptation Framework supports parties to implement adaptation actions, including two focus areas: 1) the formulation and implementation of national adaptation plans; and 2) a work programme to consider approaches to address loss and damage associated to climate change in vulnerable developing countries[56].

Many activities have been carried out towards the implementation of adaptation actions under the Nairobi work programme, including NAPAs[57].

In this sense, a large number of activities have been carried out by parties and by partner organizations under the Nairobi work programme in relation to: 1) climate-related risks and extreme events; 2) socio-economic information; 3) adaptation planning and practices; and 4) economic diversification, including activities in Action Pledges.

Ultimately, all these approaches rely on the definition of eligible applicants. An eligible applicant has to be a designated authority (DA) which is represented by government officials who act as points of contact for the Adaptation Fund[58].

On behalf of their national governments, the designated authorities endorse the accreditation applications of National or Regional Implementing Entities before they are sent to the fund's secretariat for assessment and/or proposals by National, Regional, or Multilateral Implementing Entities for adaptation projects and programmes in the DA's country[59].

This designation requirement is often seen as a difficulty. But, actually, some barriers to implementation of adaptation include limited funding, policy and legal impediments, lack of proposals, and absence of reliable recognition to become a DA. There is no magic bullet for application to the adaptation fund, but there are some similarities in the processes[60].

The most common challenges faced by Entities (e.g., Ministry) during the Adaptation Fund (AF) accreditation process range from a lack of understanding of fiduciary standards and limited competencies in some areas, to an underestimation

---

55 Ibid.; UNFCCC (note 16)
56 UNFCCC (note 48)
57 UNFCCC (2017), "Nairobi work program on impacts, vulnerability and adaptation to climate change", FCCC/SBSTA/2017/L.7, 3pp
58 UNFCCC (note 16)
59 Ibid.
60 Ibid.

of the workload involved and the importance of involving designated staff and directors during the process. The Entity's willingness to actively drive the accreditation process is also decisive[61].

It is important to realize that the Entities should demonstrate experience of using their own monitoring and evaluation frameworks, and must demonstrate commitment to zero tolerance for fraud, financial mismanagement and other malpractices at the highest level in the organization. In addition, policies and procedures such as a code of conduct, whistleblower protection, and measures to address conflicts of interest and individual complaints, should be clear and must contain a track record in applying those policies and procedures[62].

Building upon its experience so far, the AF experts recommend that National Implementing Entities (NIEs) establish an independent internal audit service and demonstrate its effectiveness. In addition, they should demonstrate the internal control framework with documented roles and responsibilities, and appropriate procurement policies, and provide the Accreditation Panel with tangible evidence and recent documentation (less than ten years)[63].

In the past couple of years, many new multilateral and bilateral climate funds have been established in order to develop channels of international climate financing. Hence, being clear, multilateral climate funds are those which receive contributions from different countries, such as the Global Environment Facility (GEF) under the financial mechanism of the UNFCCC and the Climate Investment Funds (CIFs) created by the World Bank and other regional Multilateral Development Banks (MDBs).

In the next section, selected case studies reveal some elements of incentive associated with interconnections between climate adaptation to extreme weather events and climate financing.

Case studies

India

Over recent decades, the climate adaptation community in India has made important contributions. For example, the Indian energy market transformation is accelerating under Energy Ministry leadership and solar power is a fast-growing industry. However, in general terms, the climate finance in India is highly fragmented among central government, states, private sectors and civil society[64].

More recent efforts to address well-defined policies in the solar energy and energy efficiency markets have encouraged climate financing through a variety of domestic and international sources, both public and private[65]. 'In December

---

61 Parker (note 49)
62 Ibid.
63 UNFCCC (note 16)
64 Jha, V. (2014), "The coordination of climate finance in India", Centre for Policy Research, p. 43
65 Ibid.

2016, India released its 10-year Draft National Electricity Plan, aiming for the installation of a cumulative 275 GW of renewable energy capacity by 2027.[66]

The main institutional response of the Government of India on climate finance has been to establish a Climate Change Finance Unit within the Department of Economic Affairs in the Ministry of Finance. Both of these efforts focus on accessing international climate funds by the Ministry of Environment and Forests that has led selection and oversight of projects, while the Ministry of Finance has been the nodal department for receiving financial assistance from multilateral and bilateral funds[67].

In India, a critical consequence of such challenges has been the need for a coherent strategy on climate finance based on the integration of ongoing efforts on mitigation and adaptation with the emerging domestic and international financial arrangements[68].

A key challenge to achieving greater implementation of adaptation initiatives is the fact that there is no formal co-ordination mechanism around climate finance. Therefore, an important component of stakeholders, at the national and sub-national levels, could be the development of a clearer sense of opportunities and priorities using both domestic and international finance[69].

Despite their limitations, the initiatives have helped to build awareness and understanding about solar energy investments. In 2013, when India provided a new energy-policy roadmap, any plan would have been considered impracticable. 'Nowadays, it still looks ambitious but absolutely feasible. The results of solar growth make the US$200–300bn capital investment requirement commercially viable. The capital inflows into India are the ultimate endorsement by global financial markets.' In recent years, the cost of renewable energy in India has seen an unexpectedly rapid decline, some 65% over the last three years[70].

Indonesia

One of the central concerns in Indonesia has been the intensive process of policy and capacity-building to respond to climate change. The adoption of mitigation and adaptation strategies has resulted in the recent calls for the integration of national institutions through a framework of actions[71].

Key institutions involved in efforts to implement these measures include the National Council on Climate Change, established by the former president, the Ministry of Finance, the planning Ministry BAPPENAS (Indonesian Ministry of

66 Buckley. T. (2017), "India's Electricity-Sector Transformation Is Happening Now", *IEEFA Asia;* http://ieefa.org/ieefa-asia-indias-electricity-sector-transformation-happening-now/ (accessed July 2017)
67 Jha (note 64)
68 Ibid.
69 Ibid.
70 Buckley (note 66)
71 Maulidia, M. and Halimanjaya, A. (2014), "The coordination of climate finance in Indonesia", Centre for Policy Research, p. 39

National Development Planning) and its Indonesian Climate Change Trust Fund (ICCTF), and the REDD+ Agency[72].

Economic regulation is required to create the necessary financial viability and allow finance sector leaders to become more proactive[73]. Financial regulators have encouraged investments, and there are increasing efforts to engage the private sector in order to make the investments environmentally and socially beneficial. The financial viability of projects is one of the prerequisites[74].

Unequal investment capabilities among different economic sectors have long been a basic feature of market[75]. For example, in practice the ICCTF is one of the smallest institutions in the domestic climate finance framework, in part because of its modest levels of capitalization, but also because the arrangements for the fund have not defined a clear role for important agents involved[76].

Despite this, the ICCTF has also struggled to meet international fiduciary standards. It is evident that financial sector regulation could potentially further incentivize financial institutions to supply capital to domestic assets. But, in the first years of attempt, existing international climate funds have been docked in one of the key ministries involved[77].

Nowadays, in the fast-evolving regulatory framework of the Indonesian power sector, it is hard to say for sure how changes will affect developers, investors and lenders. Nonetheless, the trends are at least instructive. In short, the regulatory changes make it financially riskier to develop and finance coal-fired power projects in Indonesia[78].

Actually, all of these regulatory changes represent opportunity for renewable energy, which is already competitive with coal-fired electricity. Besides, there is a recognized opportunity for new climate funds such as the Green Climate Fund (GCF) that work in collaboration with the national designated authority (NDA) to take a more proactive approach involving diverse stakeholders, and putting in practice new operational processes in order to foster progress in achieving mitigation and adaptation actions[79].

Colombia

The analysis presented here grounds its interpretation based on the new politics made by Colombia. This country has made important advances in the

---

72 Ibid.
73 Chung, Y. (2017), "IEEFA Indonesia: Shifting Regulatory Landscape Makes Coal-Fired Plants Riskier to Finance. Rule Changes That Undermine Traditional Guarantees to Developers", IEEFA; http://ieefa.org/ieefa-indonesia-shifting-regulatory-landscape-makes-coal-fired-plants-riskier-finance/ (accessed July 2017)
74 Maulidia (note 71)
75 Chung (note 73)
76 Maulidia (note 71)
77 Ibid.
78 Chung (note 73)
79 Maulidia (note 71)

co-ordination of national climate change action through the design of a national climate change system recognized as 'Sistema Nacional de Cambio Climático' (SISCLIMA)[80].

The process has been started, but much work is still to be done. We see here an interesting system which brings together national and international agents developing work on climate change, but that has to date been spread widely with few inter-linkages[81].

For example, one of the most important facts is that, according to the Institute for Energy Economics and Financial Analysis, the coal industry has entered a phase of terminal and rapid decline. IEEFA projects a 25% drop in global demand for thermal coal by the end of the decade – a crash of a quarter in the next years. In 2016, it was estimated that the US coal market would decline (by an additional 11%), and this decline would be expected to be around 3% in 2017[82].

Moreover, India and China would be importing less coal for economic and security reasons. They would be considering some negative fiscal and monetary consequences such as those caused by excessive pollution. Therefore, China's coal use dropped 2.9% in 2014, 4% in 2015 and 6.8% to this point in 2016[83]. Asia would be accounting for near zero consumption because both the metallurgical and thermal markets were going down and getting worse. Coal plants are being mothballed or closed from South Australia to Queensland.[84] On the other hand, solar pricing is becoming more competitive over time[85].

Following the same path, the European coal sector was deeply affected by new air quality mandates and one-third of existing capacity must retrofit or close. Actually, in May 2017, more than 100 separate plants, representing one-third of Europe's large-scale coal-fired power plant capacity, face costly air quality upgrades or closure as a result of new European Union emissions limits[86].

With regard to government co-ordination, we could say that this kind of support has being developing through an informal work with SISCLIMA and has acquired propulsion with international processes supporting climate change actions. However, there is no doubt that the regional development

---

80 Jaramilo, M. (2014), "The coordination of climate finance in Colombia", Centre for Policy Research, p. 33
81 Ibid.
82 Silverstein, K. (2016), "SNL: U.S. Coal Exports Fall Again in 2015 as China, India Are Becoming Self-Sufficient", IEEFA; http://ieefa.org/snl-u-s-coal-exports-fall-2015-china-india-becoming-self-sufficient/ (accessed July 2017)
83 The Green Institute (2016) The end of coal: How should the next government respond? www.greeninstitute.org.au
84 Ibid.
85 Ibid.
86 IEEFA (2017), "IEEFA Report: European Coal Sector Woes Deepen With New Air Quality Mandate. One-Third of Existing Capacity Must Retrofit or Close", http://ieefa.org/ieefa-report-european-coal-sector-woes-deepen-new-air-quality-mandate-requiring-one-third-existing-capacity-retrofit-close/ (accessed July 2017)

banks of Colombia could play a more prominent role in developing solutions with financing and implementing actions through programmatic approaches[87].

Certainly, the Inter-American Development Bank and the policy-based loans provided to Colombia have played a fundamental role in the development of SISCLIMA. However, dealing with conflicting government priorities remains a big challenge for effectiveness of climate change actions[88].

The identification of mutual benefits is now under way, aiming to tackle this issue. Nevertheless, stronger stakeholder engagement of civil society, private sectors, sub-national entities and law-makers is required to increase awareness and understanding of climate change vulnerabilities and opportunities[89].

Besides, improvements in terms of transparency of finance flows for climate-related activities are also required because this could help identify financing gaps. In addition, international institutions such as the Green Climate Fund (GCF) could allow stronger stakeholder engagement. Existing national and sub-national entities could support implementation, measurement, reporting and verification of processes. But, certainly, institutional strengthening, clear mandates and improved capacities are still required[90].

Zambia

The focus on place helps us to elaborate and specify the meaning of key concepts in each country, and how financing processes are occurring. In Zambia, considerable attention has been given to the institutional arrangements (1990s) in order to effectively co-ordinate the environmental policy agenda. The country has received support from successive donor-funded programmes since 1997[91].

However, the separation and the overcharging of institutions, in terms of co-ordination of the national development and climate change agendas, has undermined the effectiveness of past arrangements[92].

Nowadays, the establishment of an Interim Inter-Ministerial Climate Change Secretariat (IIMCCS) attached to the Ministry of Finance, which is also responsible for national development planning in Zambia, represents a real opportunity to harmonize and integrate these agendas[93].

It is important to highlight that uptake of utility-scale solar power also made a movement in Africa in 2016[94]. In 2015, the Industrial Development Corporation (IDC) of Zambia signed an agreement with the International

---

87 Jaramilo (note 80)
88 Ibid.
89 Ibid.
90 Ibid.
91 van Rooij, J. (2014), "The coordination of climate finance in Zambia", Centre for Policy Research, p. 36
92 Ibid.
93 Ibid.
94 Buckley, H. and Nicholas, S. (2016), "2016 Year in Review: Three trends highlighting the accelerating global energy market transformation", IEEFA Report, p. 37

Finance Corporation (World Bank) to explore the development of two large-scale solar projects through a Scaling Solar programme. A competitive operation was organized through this programme, which attracted 48 solar power-developers, seven of whom with final proposals that yielded the lowest solar power tariffs in Africa to date[95].

Thereby, Zambia set a new African low-price record (excluding South Africa) of just $60/MWh fixed for 25 years under the Scaling Solar programme, helping to change the perception that low renewable energy costs are unattainable in poor countries with underdeveloped institutions[96]. It is expected that, given this kind of programme, capacity-building support from a greater diversity of multi- and bilateral programme or funds is likely to occur.

Conclusion: lessons from experiences

In the first part of this chapter, we have argued that, in addition to the conventional international scale, it is necessary to examine how environmental justice is being pursued at the national and regional scales. We can realize that the principles of environmental justice need to be better understood in order to attend to the multiple scales and diverse forms of social organization involved in responding to climate change.

In essence, the principles of human rights do not consider the distinct origin of people, assuming that everyone is equal. Considering this aspect, the concept of socio-environmental justice is pivotal for supporting and dismantling unjust structures, or these can perpetuate this inequitable system for a long time, more than we can estimate. It is obvious that socio-environmental justice, considering the reality of each country, should involve more than only distribution of income.

In order to become Designated Entities, a national institution must demonstrate experience with their own monitoring and evaluation frameworks, with zero tolerance for frauds as much as for financial negligence and other illegal practices at the highest level in these institutions.

More and more, new environmental opportunities are being consolidated in the global market. These elements can interact in a myriad of ways to influence overall well-being and also constitute a chance for important social justice change. The rapid expansion of renewable energy combined with its increasing demand and the beginning of the end of coal is an opportunity for driving to a structural decline in thermal coal markets and consequently reduction of CO_2 emissions.

Besides strategies for improving adaptation practices and local enhancement through adaptation funds, this expansion of renewable energy also includes the installation of new plants provided by increasing solar energy-efficiency impacts, weak coal demand, and policy initiatives that continue to move the world energy markets toward an inevitably lower carbon market in the future[97].

95 World Bank Group (2017), "Unlocking Private Investment in Emerging Market Solar Power"; www.scalingsolar.org/ (accessed July 2017)
96 Buckley (note 93)
97 IEEFA (note 85)

This signals a change in perception about renewable energy costs, and there are some linkages between adaptation investments (the Adaptation Fund) and these 'new' energy sources. Globally, this position is reinforced by 'a decline of more than 50 percent in coal prices and about 80 to 90 percent of value of most listed coal companies in the last four years, an unprecedented underperformance against the equity market overall'.[98]

Given the global economic difficulties, it appears crucial for entities to have a comprehensive understanding about risk assessment regarding lack of social environmental justice to deal with the financing of projects. More and more, the financing, and consequently the implementation, of projects will depend on Environmental Social Management Plans (ESMP). It will be fundamental to demonstrate clear linkages between the projects, programme, budgets and social environmental safeguarding measures.

In the past, some Designated Entities have delineated inadequate internal controls and audit systems, which impeded a smooth process. On their side, there are some misunderstandings about the accreditation process and its requirements[99]. Meeting Adaptation Fund standards for monitoring, evaluation and risk management appeared to be challenging in some cases[100], given the relatively small size of some entities. Meeting international fiduciary standards can be challenging, especially if the Implementing Entities (IE) follow national standards that are not compatible with international ones.

It is important to realize that entities that have been successfully accredited adopted strategies that were particularly useful to overcome such barriers[101]. They highlighted the importance of building close relations and interactions with the Adaptation Fund Secretariat, and to network and forge partnerships with other Designated Entities, including accredited ones.

In many cases, field visits from Accreditation Panel experts of the Adaptation Fund Board and representatives from the Secretariat have greatly helped in clarifying some elements of the applications. Similarly, getting institutional buy-ins and ensuring that top management and other relevant stakeholders are on board is a crucial cornerstone during the process[102].

In this respect, it is important to ensure that the process of getting accredited aligns with institutional priorities, and that senior management fully understands what the process means for their institution, as it requires sustained efforts.

Consequently, accreditation must be on the institutional agenda and included in work plans, budget and performance measurement processes. Delegating social environmental responsibility for the accreditation process to a willing expert with

---

98 Ibid.
99 UNFCCC (note 16)
100 van Rooij (note 90); Jaramilo (note 86); Jha (note 64)
101 UNFCCC (note 16)
102 Ibid.

a strong sense of responsibility and commitment has also helped some nations to be accredited. Another key recommendation from accredited entities is to anticipate and start gathering documentation that will most likely be requested during the process as early as possible. Finally, establishing and maintaining a professional working and suitable relation between the Adaptation Fund Secretariat and Designated Authorities appears crucial.

13 Understanding the relationship between global and national climate regimes and local realities in India

Arnab Bose[1] and Seema Sharma[2]

Climate change has become a global environmental problem caused by the build-up of greenhouse gases, particularly carbon dioxide and methane, in the Earth's atmosphere[3]. The impacts of this problem threaten a range of issues, stretching from threats to biodiversity all the way to national security. Given the multidimensional nature of the climate change problem, there is a need to explore systems and mechanisms that can translate ideas into ground realities. As Popovski *et al.* (2015)[4] have pointed out:

> As we move towards the post-Paris climate regime, understanding the complex and multi-faceted structure of integrity systems can help us construct agreements and mechanisms capable of fulfilling the roles we need them to play.

Thus, a need to identify suitable arrangements and mechanisms can be observed. In this paper we shall precisely describe a mechanism that is capable of fulfilling the aspirations of the global climate regimes (primarily bestowed with the UNFCCC, the United Nations Framework Convention on Climate Change, and its adjoining complex involving governments/policies at multiple levels). The espoused mechanism also ties in with the national climate regime, primarily the NAPCC (National Action Plan on Climate Change) in India.

Part of the national regime addresses the global concern, and part of it is more geared towards internal concerns. We should know that while the two regimes (represented by UNFCCC and NAPCC) are not in conflict with each other, there are considerable diplomatic, negotiation-centric disconnects. However, this paper is not the forum for that discussion. Instead, the emphasis of this paper lies in identifying the disconnect between higher (global and national) climate regimes and local realities. We demonstrate that (see Figure 13.1):

---

1 Resilience Relations
2 Delhi University/Resilience Relations
3 Swain, A. (2015), *Climate Change: Threat to National Security, Encyclopedia of Public Administration and Public Policy*, 3rd edition; doi: 10.1081/E-EPAP3-120053262
4 Breakey, H., Popovski, V. and Maguire, R. (eds.) (2015), *Ethical Values and Integrity of the Climate Change Regime*, Ashgate, p. 27

Global regimes, local realities in India 213

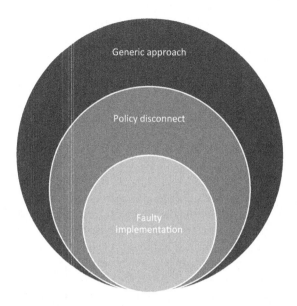

Figure 13.1 Causal mapping of global climate policy at community level

1) *There is a policy disconnect:* while there is an understanding of sustainability issues at the global and national level, there is a considerable lack of understanding of those issues at a community level.
2) *The implications of this disconnect:* we highlight the existing disconnect in the fact that, while individuals are aware of what needs to be done (individual level), and policies/plans (including urban plans) are made keeping sustainability issues in mind (global/national level), things go awry when it comes to implementing these urban plans at the community level.
3) *Key takeaway message:* we identify the key takeaway from the exercise, which is that instead of general information asymmetry, a more granular implementation approach that specifies details on how a problem needs to be solved is the real challenge. To meet this challenge, we recommend the treatment of various localities separately, and enable them to take part in the decision-making process on their own. Empowering decentralized decision-making is key, along with the understanding that every locality is culturally different and the agents that bring about change have different textures.

The inferences given above are drawn from the work done by the authors of the paper, starting from 11 January 2015 to date, with the help of various entities, most importantly Delhi University under the aegis of Resilience Center Global Network (RCGN), which is demonstrated in the subsequent sections of the chapter. Currently, RCGN is implementing and assessing the effects of the 3i (Inform, Inspire, Implement) framework/mechanism for instilling resilience in a region across ten localities in Delhi, West Bengal and Bihar (the 3i framework is described later). Moreover, with significant assistance from Vivekananda College (Delhi University), the authors have started the Resilience Center Vivekananda College Chapter (RCVNC), which works primarily with localities adjacent to the college campus. RCVNC has conducted numerous community engagement activities utilizing an internally devised framework and has touched 2,000 or more lives, from local electricians to elected representatives (includes local councillors, members of parliament, district magistrates, residents' welfare associations, unskilled work people, etc.).

Direction for action

To understand the need for specific, directed action, let us examine two directions for action emanating at the global (originating from the UNFCCC) and the national level (originating from the NAPCC). Table 13.1 gives us an overview of the priority areas of the Green Climate Fund (GCF), which is borne out of the processes of the UNFCCC. Although the fund amount might not be substantial enough to fight climate change, it is a first move that may well create the climate of market/ economy and set the rules of the game. Figure 13.2 elucidates the eight original missions and nodal ministries and departments under the NAPCC. These missions, and four new additions, give the texture of climate/sustainability action described at the national stage.

A scan of Table 13.1 reveals important points. Even as the layouts help to identify and understand the directions in which action will flow, there is silence on the granularity of the action: this disconnect clearly highlights 'the disconnect', consequent to which there will be improper implementation. There are limitations to the depths to which global action can be planned, and therefore there is an inability to identify the role of community action. These have repeatedly been pointed out as the chief weaknesses of past global climate mitigation and adaptation approaches. Similar expectations had been voiced in 2015, where civil society had clearly stated a wish to enable people to overcome challenges posed by climate change in the immediate run (which holds true even in the long run); they wanted this to be done in a manner that empowered them, and that built on their knowledge and ideas: this would require a paradigm shift from the GCF compared to standard practice at international financial institutions (IFIs)[5]. Even as the GCF continues to evolve to address high community-level expectations, one notices that projects approved for funding from GCF are of a small ticket-size (up to $50 million)[6], a repeat of its earlier approvals in 2015, barring one project. Such project sizes are ideal for direct community empowerment for execution of the projects, but the institutional capacity at the relevant level seems to be absent.

5 http://cdkn.org/2015/03/opinion-paradigm-shift-want-green-climate-fund/?loclang=en_gb
6 www.greenclimate.fund/documents/20182/226888/GCF_B.13_16_Add.12-Consideration_of_funding_proposals_-_Independent_Technical_Advisory_Panel_s_assessment.pdf/b98da11f-1c32-421e-8aa8-356f9cf1fdba?version=1.0

Table 13.1 Overview of priority result areas for the Green Climate Fund

Mitigation	Option M1:	Reducing energy use from buildings and appliances
	Option M2:	Enabling reduction in the emission intensity of industrial production
	Option M3:	Increasing access to transportation with low-carbon fuels
	Option M4:	Providing households with access to low-carbon, modern energy
	Option M5:	Supporting the development, transfer and deployment at scale of low-carbon power generation
	Option M6:	Reducing emissions from agriculture and related land use management
	Option M7:	Supporting implementation of the phased approach to REDD+
Cross-sectoral	Option C1:	Facilitating design and planning of sustainable cities
	Option C2:	Joint mitigation and adaptation approaches for the integral and sustainable management of forests
Adaptation	Option A1:	Support across the full range of adaptation result areas
	Option A2:	Support for a selective set of sectoral result areas
	Option A3:	Support for selected themes cutting across result areas ('flagships')
	Option A4:	Facilitating capacity for programmatic and transformative activities
	Option A5:	Facilitating scaling-up of effective community-based adaptation (CBA) actions
	Option A6:	Supporting co-ordination of public goods such as 'knowledge hubs'

Source: Green Climate Fund (2013)
Green Climate Fund Business Model Framework (2016): www.greenclimate.fund/documents/20182/24934/GCF_B.04_07_-_Business_Model_Framework__Private_Sector_Facility.pdf/fb909f84-1c95-42bd-973f-54bc9bcada8f?version=1.0 (accessed 7 March 2017)

Focusing on the national level, one can see the considerable amount of detail present in the National Action Plan on Climate Change, released in 2008, in order to guide India's climate policy (Figure 13.2). However, the policy fails to give voice to what local communities want; this is evidenced by the absence of community considerations from many of the specific missions that the policy envisages. Such a strategy tends to fail good intentions at multiple levels. Totin et al. (2015)[7] have identified the following as barriers to policy development and

---

[7] Totin, E., Traoré, P.S., Zougmoré, R., Homann-Kee, S.T., Tabo, R., and Schubert, C. (2015), "Barriers to effective climate change policy development and implementation in West Africa: Findings from a qualitative study in Mali, Ghana and Senegal", Info Note, Research Program on Climate Change, Agriculture and Food Security, October

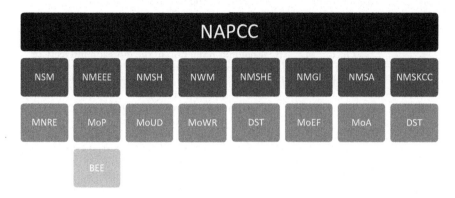

Figure 13.2 The original eight missions and nodal ministries and departments under the NAPCC

effective implementation in the countries of their study: a lack of operational capacity at lower administrative levels, little involvement from stakeholders, lack of awareness and funding. They observed that lack of information flows between national and local levels around existing climate policy processes prove to be a major hindrance in the effective implementation of the policy. Totin *et al.* (2015) also noticed that supervised knowledge-sharing platforms for national, regional and local policy-makers and other stakeholders can offer the advantage of strengthening information flows and support policy development and implementation. Similar observations have been made by Salon, Murphy and Sciara (2014)[8], who stated that,

> Important factors that enable action include strong local champions, supportive residents, and state and national policies and actions. Important barriers to action include lack of local-government staff time and financial resources.

Thus, both the cases clearly enable the reader to easily predict a departure of resultant actions from the intended purpose, without making any significant contributions towards mitigation of the adverse effects of climate change.

The 3i 'Inform, Inspire, Implement' mechanism/framework: the case of Vivekananda College, Delhi University (DU)

Community action can succeed only if the community can be pushed in the right direction, which implies engaging with its members using an integrated approach. To that end, a three-stage approach has been put forward by us

---

8 Salon, D., Murphy, S. and Sciara, G.C. (2014), "Local climate action: motives, enabling factors and barriers", *Carbon Management*, 5 (1), pp. 67–79

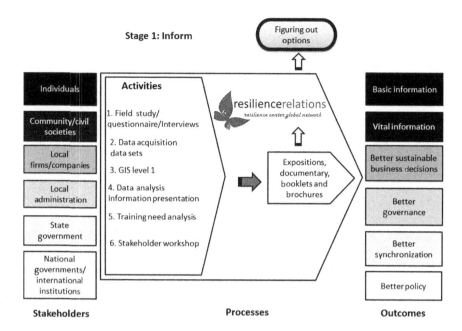

Figure 13.3 Stage 1: information collection and dissipation
Source: the authors

(Vivek Vihar Atlas, 2016). The three stages involved in this process, which are imperative in engaging with the local community, are: Inform, Inspire and Implement (hereinafter, referred to as 3i). The channels thus created work both ways – appropriate information channels can be generated to help quickly reframe approaches, if necessary, and the public at large can learn about the projects being undertaken, and participate in the decision making process in an effective manner.

Stage 1 – Inform (information collection and dissipation): The simplest way to achieve the objective of this stage is to collect the information directed towards identifying strengths, weaknesses, opportunities and threats (SWOT) of the locality; the training needs and other basic, relevant information. Pursuant to this, the information is dissipated in a condensed, yet simple format amongst the concerned stakeholders. The information processes are contextually aligned to cultural characteristics of the local community.

Stage 2 – Inspire (creating a set of solutions or actions that 'may' be taken): It is at this stage that human resources are trained, and capacities are built, keeping in mind an option or set of options. This is considered the crux in the 'keep options open/alive' method of robust decision-making.

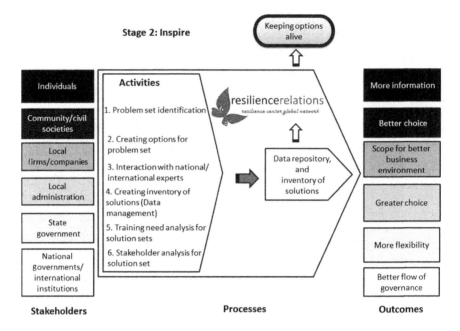

Figure 13.4 Stage 2: creation of solutions/options inventory
Source: the authors

Stage 3 – Implement (project implementation and maintenance stage): In this stage, if one of the options is chosen as a part of the multi-stakeholder interaction then it is implemented.

In order to instil concepts and values of urban resilience in its surrounding areas, the faculty of Vivekananda College collaborated with the Resilience Center Global Network (RCGN) in a project mode. The project was named 'Resilience Project', with the motive of meeting the challenges of emerging India.

The hosting of the project by the university is a demonstration of 'Context-Integrity', as defined by Breakey *et al.* (2015)[9], as the external environment facilitates the public institutional justification (PIJ) of RCGN. The hosting facilitates and empowers the agent integrity, despite the original PIJ of the university being rather limited and distinct from that of RCGN and its project.

For the successful implementation of the strategic approach, the need to develop a replicable methodology of implementation was formulated. It was identified that the steps to be undertaken within the method adopted had to be consistent with the framework's strategic approach. To that end, a sub-plot of the gamut of activities against the three stages was drawn up (Table 13.1).

---

9 Breakey, H., Popovski, V. and Maguire, R. (eds.) (2015), *Ethical Values and Integrity of the Climate Change Regime,* Ashgate

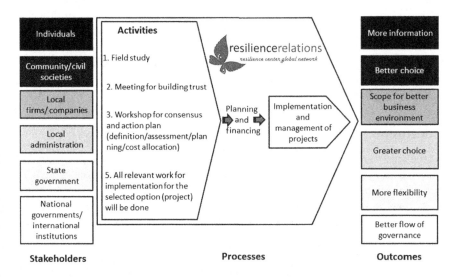

Figure 13.5 Stage 3: project implementation and maintenance stage
Source: the authors

In the first stage of the framework's pilot, the authors of the study first conducted an engagement activity. The activity essentially entailed inviting community members, through the college forum, to participate in a discussion and a walkabout. They were encouraged to identify the shortcomings and positive achievements within the neighbourhood area of Vivek Vihar with respect to public infrastructure and the related activities, in an open-house discussion following the walkabout. This led to the identification of a preliminary set of problems (Figure 13.2 that were immediately identified by the community residents to which they sought redress in some form or the other.

Similarly, students of the college's Environment Society were taken for the walkabout to understand the framework's implementation, which in turn would be utilized to instil resilience into the community.

In the second stage, the inspiration-inducing activity was taken up. A public toilet that was clean and pointed out by the community was noted, and the responsible staff-member of the urban local body directly involved in that operation was traced. The sanitation worker was handed a certificate of appreciation to acknowledge his contribution towards the efforts of keeping the Vivek Vihar community clean.

In the third stage, workshops and open-forum discussions were organized on the themes that were identified during the first stage of the framework's implementation. Foremost among them was the workshop on health and sanitation from women's perspective, wherein issues of health, sanitation, menstruation-management and the safe disposal of sanitary protection were discussed frankly. An industrial organization working in this area was brought in to provide more information on the subject; they even carried out a distribution of free healthcare samples. Similarly, given that financial literacy and livelihoods are major challenges for the community's women, an interactive session was organized on the occasion of Women's Day, 2015. The session was on the subject of female empowerment and plausible livelihood options as a way forward, with the help of the bank that had been previously identified.

Conclusion

At the level of the community or locality, particularly for India, there seems to be a strong disconnect between the goals of higher global or national policy regimes and ground realities.

Poverty alone cannot explain the failure, as even in affluent parts of India there is a failure of governance and non-aligned stakeholders. Very often this failure is attributed to non-effective institutions and corruption. However, from our experience, this was very far from the truth.

Drawing from our experiences, we state that information flows are often incorrect; for instance, very few people were aware who their elected representatives were, let alone what their roles were, or how these roles were different, from the district magistrate, to who was accountable for footpaths, parks, the children's swings, etc. The narrative is quite long, but what 3i essentially strived to do was create information flows which are accurate, whereby the district magistrate can take up issues which s/he can resolve, or that the member of parliament or local councillors could address.

Our final concluding thoughts are:

- We have been successful in developing an interface between academia, community, industry and policy, with different community co-ordinators who have voluntarily agreed to work to keep a common dialogue going, where information channels are becoming clearer.
- The 3i approach has been beneficial in taking up local issues very often interconnected with each other, and if treated in silos will lead to a temporary solution.
- Cultural aspects play an important part in addressing 'the disconnect'; different localities have different textures and strategic actors; solutions of one size fits all lead very often to mal-development.
- Participatory approaches will find academic institutions a great, effective and robust partner when there is correct guidance and leadership; appropriate understanding of culture and adult learning methods shall be realized too.
- Resilience is about relations, and academic institutions are a great place to build the trust needed to bridge the relationship gap and instil resilience in India.

- Academic institutions play a pivotal role in climate regimes, in connecting the higher-level thinking to local realities; therefore, in the concept of context-integrity which includes 'nested' relationships with larger institutions, nesting a smaller resilience centre to work for the development of an adjoining locality by an academic institution is noteworthy.

14 Paris Agreement and climate change in India

To be or not to be?

Aditya Ramji[1]

India and climate action: the past, present and future

As the Paris Agreement underwent negotiations in December 2015, the opening remarks of various world leaders conveyed a message – a call for breaking all shackles and coming together for a unified cause with one purpose.

As the world leaders gathered together to discuss and debate the way forward to tackle the impacts of climate change, the coastal city of Chennai in southern India witnessed unprecedented rains, while the capital city of New Delhi was enveloped in dense smog. The question remains as to whether such instances are to be written off as one bad phase in time, or whether we must take notice, heed the warning signs and take immediate action.

Although India intends to reduce its emissions intensity of GDP by 33–35% from 2005 levels by 2030, the Paris Agreement, as many say, is just the beginning. India's main goal will be to invest in an economy of the future, an economy that is built on the principles of equity and justice, which will primarily strengthen India's stand at the UNFCCC negotiations.

While the Cancun climate talks (Conference of Parties or COP 16) in the year 2010 set forth a comprehensive package on climate action, the Durban Platform for Enhanced Action or COP 17 was a critical point in international climate negotiations. The Durban talks saw countries agreeing to work towards an agreement that would be in force by 2015 and implemented by 2020, with the principle of equity forming the bedrock of the agreement. It also kept open the possibility of a legal outcome. Developing countries including India fought hard to ensure that equity remains the principle for any agreement, the ambiguity around a 'legally binding' agreement left any 'notional victory' as just partial relief[2].

---

1 Aditya Ramji is a Programme Lead with the Council on Energy, Environment and Water. The views expressed in this paper are those of the author and do not necessarily reflect the views of the Council on Energy, Environment and Water. The author can be contacted at aditya.ramji@ceew.in

2 CSE (2011), "The final outcome of the Durban Conference on Climate Change", Centre for Science and Environment, December (available at www.cseindia.org/content/final-outcome-durban-conference-climate-change); International Institute for Sustainable Development (2011), "Summary of the Durban Climate Change Conference", *Earth Negotiations Bulletin*, 12 (534), December (available at www.iisd.ca/download/pdf/enb12534e.pdf); UNFCCC (2012),

From Durban to Doha (COP 18), the developing world continued its stand as initially proposed in Durban, slowly emerging as a strong bloc of countries as compared to earlier climate talks. Doha was important for India as it ensured the principles of equity, and common but differentiated responsibilities (CBDR), stayed on the negotiating table, with the Indian government making the enhancement of its pledge of reducing emissions intensity conditional on differentiation[3].

Countries debated the historic responsibilities of developed nations, with the developing world expected to take actions commensurate to their level of development. The Warsaw COP (COP 19) in 2013 saw a discussion around 'loss and damage' – a demand for the developed world to compensate the less-developed countries (LDCs) for the losses caused due to the already existing levels of GHG emissions, which the existing or future adaptation and mitigation efforts would not correct. India was seen supporting this demand, although the developed nations suggested that such compensation could be raised through the market or insurance mechanisms currently available. Ultimately, it was only in the Paris Agreement of 2015 that this suggestion was completely left out of the negotiation text.

Once again, with negotiations not seeming to proceed in the desired direction, the Warsaw round of talks ended with yet another important amendment to the draft agreement. The word *'commitments'* for nationally determined GHG emissions reduction was replaced by *'contributions'* – considered to be a weaker term, from the perspective of moving forward on an agreement that could be legally binding, but seen as a means to build greater consensus. This set the stage for the talks at Lima (COP 20), which ended with an agreement popularly known as the Lima Call for Climate Action. The agreement at Lima set the ground for 'nationally determined contributions' for emissions cuts and adaptation measures that each nation was to submit before the next round of talks at Paris (COP 21), and retained the principles of CBDR and equity at the core of the agreement. By this time, India had emerged as a key player in the negotiations, with smaller South Asian nations and other developing countries looking towards India to take the lead[4].

Thus, what does the Paris Agreement mean for India going forward? The Agreement is binding to the extent that all countries are bound by the commitments they make as part of their intended nationally determined contributions (INDCs),

---

"Decisions adopted by the Conference of the Parties", United Nations Framework Convention on Climate Change (available at http://unfccc.int/resource/docs/2011/cop17/eng/09a01.pdf)

3 International Institute for Sustainable Development (2011), "Summary of the Doha Climate Change Conference", *Earth Negotiations Bulletin*, 12 (567) (available at www.iisd.ca/download/pdf/enb12567e.pdf); MoEF (2013), "Outcome of Doha Climate Change Conference 2012", Ministry of Environment and Forests, Government of India (available at http://pib.nic.in/newsite/PrintRelease.aspx?relid=93042)

4 CEEW (2016), "Getting a Deal: CEEW Climate Research, Engagements and Contributions to COP21 Negotiations", Council on Energy, Environment and Water (available at http://tinyurl.com/pt938de)

but at the same time their domestic actions are not subject to international law. But does that mean the Paris Agreement has not achieved much since Durban?

What the Paris declaration has achieved is very important. First, it brought an actionable point of intersection between a top-down approach and a bottom-up approach by asking countries to determine the limits of their own actions towards tackling climate change through the INDCs. Second, it put in place a mechanism which involves a five-yearly review of country pledges, a technical review, transparency features and a global stock-take. The third, and probably the most important point, is the hope that this mechanism could be replicated at the domestic level, so as to drive domestic action towards climate change and ensure that after every five-year review there would be an opportunity to significantly raise the bar for emissions reductions[5].

The key question for India is where it should draw the line with regards to its climate policy – should there be a trade-off between what is equitable and what is legally binding? In spite of India's strong beliefs in equity and CBDR, its intended contributions are ambitious and disproportionate to its fair share. While India has always been committed to combating climate change, its stand at international negotiations is based on its domestic priorities with regards to development, poverty eradication, food security, clean energy, sustainable cities and other welfare goals.

Irrespective of what happens in the post-Paris global regime, undoubtedly domestic action is critical to tackle the impacts of climate change. The case for domestic action in India is all the more urgent, with the Minister for Environment, Forests and Climate Change stating that India's INDCs would be a domestic action plan and not be contingent on the nature of any international agreement[6]. The mechanism that emerged from the Paris Agreement provided a golden opportunity for India, as a way forward, to create a mechanism or legislative framework that fosters collective action among states to tackle climate change with accountability and assistance from the centre, subject to certain conditions.

Reviewing domestic action on climate change in India

The Indian economy saw a growth rate of 7.6% for the year 2015–16, with the last quarter observing the growth of the manufacturing sector at about 9.3% while the agriculture sector grew at 2.3%[7]. It is important to note that this growth rate has been achieved in spite of nominal global demand, and two consecutive monsoons

---

5 Dubash, K.N. (2015), "A climate for congenial for India", *The Hindu* (available at www.thehindu.com/opinion/op-ed/cop-21-and-paris-agreement-a-climate-more-congenial-to-india/article7992802.ece); UNFCCC (2015), "Adoption of the Paris Agreement", United Nations Framework Convention on Climate Change (available at https://unfccc.int/resource/docs/2015/cop21/eng/l09r01.pdf)

6 MoEF (2014), "Lima COP: India's Priority", Ministry of Environment and Forests, Government of India (available at http://pib.nic.in/newsite/PrintRelease.aspx?relid=112749)

7 Economic Times (2016), "India's growth at 7.6% in 2015–16 fastest in five years" (available at http://tinyurl.com/zzy3bb5)

that were below-normal, thus impacting agricultural output and productivity. India is one of the world's fastest emerging economies, and with its programmes such as Make-in-India, which aims at increasing the manufacturing base, the need for conscious and strategic policy-planning is imperative. In the context of climate change, India needs to contextualize climate impacts and uncertainty within its plans for the dual goals of economic growth and development.

India has taken a suite of measures addressing sectorial issues to improve resource efficiency, enhance resource security and pursue a sustainable development strategy. Some of the recent notable initiatives of the Government of India include[8]:

1) Upscaling of installed solar power capacity from 20,000 MW to 100,000 MW by 2021–22;
2) Increasing the Clean Energy Cess (renamed the Clean Environment Cess) from INR 200/tonne to INR 400/tonne on coal to fund research and development on clean energy technologies and for cross-subsidy on solar and other renewables;
3) Setting up of Ultra Mega Solar Power Projects in the states of Tamil Nadu, Rajasthan, Gujarat, Andhra Pradesh and Jammu & Kashmir (Ladakh region);
4) Setting up of the National Adaptation Fund for Climate Change in 2015–16;
5) Integrated Ganga Conservation Mission;
6) Launching the National Air Quality Index;
7) Unlocking INR 40,000 crore of funds for the 'Green India' initiative through the Compensatory Afforestation Funds (CAF) Bill 2015. Under CAF, funds would be made available to states to take up afforestation programmes, and to increase density of existing forests to substantially boost tree cover in the country;
8) Controlling vehicular pollution by moving to Bharat VI emission norms by April 2020.

While India had earlier voluntarily pledged to reduce the emissions intensity of its GDP by 20–25% of its 2005 levels by 2020, the INDC commitment made by India pledges 33–35% reduction in emissions intensity by 2030. As a result of its continuous efforts on the domestic front, India's emission intensity of GDP has decreased by 12% between 2005 and 2010, as per India's first Biennial Report to the UN submitted in 2015[9]. Figure 14.1 shows sector-wise contribution to India's total emissions, dominated by the energy and agriculture sectors.

In the year 2008, India announced its National Action Plan on Climate Change (NAPCC) which stressed the need for a high growth rate while at the same time reducing the vulnerability of the Indian population to climate change.

---

8 Ministry of Power (2016), "India to ratify Global Climate Agreement", Government of India (available at http://pib.nic.in/newsite/PrintRelease.aspx?relid=138511)
9 MoEFCC (2015), "India's First Biennial Report to the UNFCCC", Ministry of Environment, Forests and Climate Change, Government of India

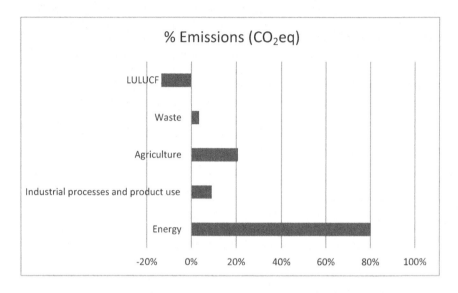

Figure 14.1 India's sector-wise emissions (CO_2 eq), 2010
Source: Ministry of Environment, Forests and Climate Change, Government of India (2015)

The NAPCC outlined a national strategy to tackle climate change with eight flagship missions. Subsequent to the NAPCC, all states in India were mandated to develop their respective State Action Plans on Climate Change (SAPCC), which were to outline their climate action plan at the state level, including budgetary outlay.

Various states submitted their SAPCCs in the period 2010–2015, in the post-Paris regime, and these plans would have to be assessed with a fine lens to understand whether the magnitude of all actions combined would meet the targets set by India as part of its INDC commitments. It should be noted that states have given indicative budgetary figures in their SAPCC documents of expected spending over a period of five to ten years. An analysis of the SAPCC documents of a few select states indicates that agriculture, forests and water resources have been allocated a significant share of the indicated budgetary allocations. States such as Andhra Pradesh (before its division) and Assam have allocated 90% and 57% of their resource allocation under the SAPCC towards agriculture, respectively, with almost all of the allocation to agriculture in Andhra Pradesh going to improve credit access for farmers and crop insurance[10]. In Bihar, animal husbandry (37%), forests and biodiversity (30%) and water resources (21%) constitute

10 EPTRI (2012), "State Action Plan on Climate Change for Andhra Pradesh", Environment Protection Training and Research Institute, Government of Assam, 2015; "Assam State Action Plan on Climate Change", Department of Environment, Government of Assam

close to 90% of the resource allocation under the SAPCC[11]. Gujarat has allocated close to 80% towards water resources while Uttarakhand has allocated a similar share (71%) of its SAPCC resources towards building roads infrastructure[12,13]. Such varying importance given to different sectors could be taken as an indication as to what each state identifies as the vulnerable sectors within its economy. The key question is, how have the strategies as outlined in the state action plans been arrived at? Are these assessments accurate or is there a need to revisit these strategies?

Understanding the Paris Agreement is critical for India before it takes the next steps in preparing an operational strategy to realize its ambitious climate action plan. In Article 11.2 of the Paris Agreement, it states that capacity of countries for tackling climate change should take into consideration their national needs and *'foster country ownership at the national, subnational and local levels'*[14].

The Paris Agreement outlines very clearly in Article 14 that all countries will be subject to a periodic global stocktake which would review each country's actions and assess whether collective progress is on the right track. The Agreement also expects each country to share detailed information on its GHG emissions inventory, a tracking framework to measure the progress made in achieving its INDCs, as well as detailed documentation of its adaptation efforts and plans.

The Council on Energy, Environment and Water (CEEW), based on a detailed assessment of industrial emissions in India, found that there is a clear need to ensure a streamlined reporting framework at a disaggregated level to help evaluate sub-sectorial emissions trends more accurately[15].

While all those who signed the Paris Agreement pledged to keep the global temperature rise under the 2°C mark, a recent study estimates that as many as 800 million people living across nearly 450 districts in India are currently experiencing significant increases in annual mean temperature going beyond the 2°C warming pathway. The study further estimates that the costs of adaptation for India may increase to about $360 million by 2030. In such scenarios, securing the livelihoods of over a billion people and minimizing the risk towards development outcomes due to climate change is all the more important[16].

11 Government of Bihar (2015), "Bihar State Action Plan on Climate Change", Government of Bihar
12 Government of Gujarat (2014), "Gujarat State Action Plan on Climate Change", Climate Change Department, Government of Gujarat
13 Government of Uttarakhand (2014), "Uttarakhand Action Plan on Climate Change", Government of Uttarakhand
14 UNFCCC (2015), "Adoption of the Paris Agreement", United Nations Framework Convention on Climate Change (available at https://unfccc.int/resource/docs/2015/cop21/eng/l09r01.pdf)
15 www.business-standard.com/article/opinion/arunabha-ghosh-karthik-ganesan-shining-the-light-on-climate-action-116051601188_1.html
16 Garg, A., Mishra, V. and Dholakia, H. (2015), "Climate Change and India: Adaptation Gap (2015) – A Preliminary Assessment", Working paper of Indian Institute of Management Ahmedabad (IIMA) W.P. No. 2015-11-01

Both national and state-level plans need to be reassessed and reviewed to ensure that climate action plans are integrated with development plans, allowing India to move forward in the right direction. As we go ahead, states need to review their climate action plans and revisit the processes followed in carrying out such assessments. Studies and reviews have revealed that state action plans are most often seen as similar to sustainable development planning, leading to the exclusion of climate resilience as a core outcome of the plan document. Further, gaps in knowledge and data on climate impacts have also been a hindering factor[17,18]. Thus, strong mechanisms and processes need to be put in place to make data on climatic and development indicators available at a disaggregated level, to enable evidence-based decision-making.

Last but not least, for India to meet its ambitious commitments of 40% of non-fossil fuel power capacity by 2030, a 33–35% reduction in emissions intensity of its GDP by 2030 and an increase in forest cover to serve as a carbon sink, access to finance is critical[19]. In the last decade, between 2003–04 to 2014–15, India's budget outlay for development and climate action has increased five times, with fiscal spending on adaptation around 2% of GDP (~ INR 2130 billion or about $31 billion). In addition, state governments spent another INR 3100 billion (~$46 billion), through their respective budgets[20]. In a recent study, it was estimated that, despite such spending, the adaptation gap could be around $1 trillion, if other non-monetised risks were to be valued, by the year 2030[21]. India will have to strategically leverage international finance to meet its goals of climate action.

Moreover, many states within India have indicated the adaptation and mitigation measures in their plans clearly state that financial resources would be required from either the central government or external agencies such as the Green Climate Fund, World Bank or other such multilaterals. However, for this to happen, there is an urgent need to translate the international climate negotiations to all state governments in India and their respective legislators in terms of how they must prioritize their actions, as adaptation and mitigation measures, and what mechanisms are available to them to raise fiscal resources. Often, current international finance is targeted towards mitigation actions, but the lack of resources to adaptation also lead to cross-border impacts such as migration. Thus, what constitutes for international co-operation and domestic action needs

---

17 Dubash, K.N. and Jogesh, A. (2014), "From Margins to Mainstream? Climate Change Planning in India as a 'Door Opener' to a Sustainable future", Centre for Policy Research (CPR), Climate Initiative, Research Report
18 Jogesh, A. (2013), "Time to Dust Off the Climate Plan". *Economic and Political Weekly*, XLVIII (48)
19 MoEFCC (2015), "India's Intended Nationally Determined Contributions", Ministry of Environment, Forests and Climate Change, Government of India
20 MoEFCC (2015), "India's First Biennial Report to the UNFCCC", Ministry of Environment, Forests and Climate Change, Government of India
21 Garg, A., Mishra, V. and Dholakia, H. (2015), "Climate Change and India: Adaptation Gap (2015) – A Preliminary Assessment", Working paper of Indian Institute of Management Ahmedabad (IIMA) W.P. No. 2015–11–01

to be identified, with strategic flows of finance resulting in co-benefits such as reduced risk of trans-boundary consequences of climate change.

Institutional frameworks and legislation for furthering climate action in India

In independent India, environmental legislation came into prominence in the 1970s, with the Wildlife (Protection) Act 1972, the Water (Prevention and Control of Pollution) Act 1974, the Forest (Conservation) Act 1980 and the Air (Prevention and Control of Pollution) Act 1981 being some of the notable legislations early on. Since then, India has come a long way in enacting legislation on various aspects of environmental protection and abatement of pollution. As the impacts of climate change and the risks it poses rise, the challenge is to ensure that legislations are amended as new and reliable information on environmental impacts and climate change is made available. Given the trans-boundary nature of climate impacts and the associated risks, it is critical for India that different states begin what could be called 'climate co-operation' to mitigate and adapt to the potential risks in a collective effort.

In the context of concerted domestic action, this brings to the fore the key question of who does what and how much. It cannot be denied that it is cumulative and co-ordinated action that will get us through. For this, there are a few critical questions that need to be answered over the coming months on a priority basis. These are given below.

Should all states set targets and be held accountable to them? How does one set such targets?

The starting-point for this would have to be a detailed revision of the existing SAPCC documents prepared by all states, with efforts to identify potential climate risks and then prepare a strategy which accounts for climate impacts within the larger policy framework of the state. It is critical to assess the existing actions outlined in the respective state action plans and quantify each intended intervention to a common metric, so as to understand the cumulative impact that can be achieved, and thus decide whether revisions are required. All states must report on an annual basis their progress with respect to their action plan in a common reporting framework, and make periodic (upward) revisions to their implementation strategy as the need arises. For this, a dynamic evaluation framework is critical to both build accountability and ensure efficient use of resources. For this, having a robust monitoring-reporting-verification (MRV) framework domestically is essential for India, with those institutions involved in reporting emissions information and those verifying them working under a transparent mechanism[22]. The

---

22 Ghosh, A. and Ganesan, K. (2016), "Shining the light on climate action", *Business Standard*, 16 May (available at www.business-standard.com/article/opinion/arunabha-ghosh-karthik-ganesan-shining-the-light-on-climate-action-116051601188_1.html)

common reporting framework and its dynamic nature will facilitate a smoother transition for India towards a low-carbon economy and ensure that India is also prepared for the periodic review as stated in the Paris Agreement.

The methodology of target-setting may be based on an assessment of available resources and future emissions of the state, and can be a useful way of leveraging the domestic resources available as part of the National Clean Energy Fund (NCEF) or the recently set-up National Adaptation Fund for Climate Change (NAFCC). The challenge in evaluating adaptation actions in the context of climate change is the attribution of interventions and contextualizing them in the long-term view of climate change. Thus, to begin with, a practical evaluation framework should probably be based on an indicator approach, which aims to provide a trend analysis, to enable better planning. An indicator approach involves setting specific metrics which are direct indicators or suitable proxy indicators, which can be quantified.

Germany has developed its indicator system based on the DPSIR approach (Drivers-Pressures-States-Impacts-Responses) with emphasis on impact and response indicators[23,24]. Evaluation frameworks ranging from pyramid-based approaches which address various levels of monitoring, to evaluations based on indicators developed according to thematic areas have been developed, a combination of which could be a potential way forward for India[25,26].

Could we have a negotiations forum for all states to come together to discuss their intended actions?

Next, as states prepare their strategic action plans integrating climate action within the larger ambit of development policy and programmes, it is important that they have a common platform to discuss their intended actions and take it a step further, so as to integrate any trans-boundary concerns and challenges that may arise due to climate change. The objective of *this* forum will not be to raise expectations of different states to leave certain efforts to their neighbours, nor to create any form of political conflict; such discussions are essential to move forward. The National Development Council (NDC), set up in 1952 by an executive order of the Government of India, is constituted by the Prime Minister, all Union Cabinet Ministers, Chief Ministers of all States and Union Territories and members of the NITI Aayog (erstwhile Planning Commission of India). The key objective of a body like the NDC was to promote co-ordination and consultation between the centre and state governments, and review national plans and key issues affecting social and

---

23 Schönthaler, K., von Andrian-Werburg, S., Wulfert, K., Luthardt, V., Kreinsen, B., Schultz-Sternberg, R. and Hommel, R. (2010), "Establishment of an Indicator Concept for the German Strategy on Adaptation to Climate Change", German Federal Environment Agency
24 OECD (2013), "National Level Monitoring and Evaluation of Climate Change Adaptation in Germany", Organisation for Economic Cooperation and Development
25 UNFCCC (2011), "Assessing Climate Change Impacts and Vulnerability: Making informed adaptation decisions", United Nations Framework Convention on Climate Change
26 GIZ (2014), "A Framework for Climate Change Vulnerability Assessments", German Development Cooperation, Federal Government of Germany

economic policy in the country. There have also been deliberations on dissolving the NDC and transferring its powers to the new planning body, the NITI Aayog, which comprises almost the same members as the NDC[27]. In either context, such a forum could be a potential institution to set the ball rolling in the right direction, in terms of planning and action on climate change and development.

From inaction to incentives: the way forward

To take this objective forward, policy-makers and legislators representing various states will have to join hands to accept the fact that environmental stress exists, and is only increasing as economic activities go unchecked, is well established and understood to a large extent by legislators as well. Their inaction is probably due to a lack of incentives, lack of accountability and a general but mistaken perception that either some foreign intervention will solve their problems, or the central government will take responsibility.

It is understandable that the Union Government has an important role to play, however it cannot bear all the costs of climate action. State governments will have to allocate fiscal resources in a more focused manner to tackle climate challenges. As experts say, perhaps the most important political decision to be made in the context of climate change is how much effort to expend on countering it versus adapting to it. Such a decision will have to be informed and based on detailed and periodic assessments of the risks associated with climate change. As we go ahead, and the nature of international climate agreements become far more stringent and binding, India needs to think about whether it can achieve a compliance mechanism domestically, between the state governments and the union government. For all those contributing to climate policy, both state and non-state actors, it is important for them to analyze and assess the operational aspects of such a mechanism, ensuring domestic climate co-operation without disturbing the political stability within the country.

Achieving an agreement domestically would go a long way in sending a strong signal internationally and strengthen India's call for foreign finance and partnerships. Nonetheless, to see effective domestic efforts towards tackling climate change, India will need appropriate legal frameworks at the national and state level, an institution with the necessary executive powers, and a transparent and democratic mechanism for stakeholder engagement.

If India as a nation has to achieve the targets it has set for itself as part of its INDC communiqué and, more importantly, act for its own welfare, there is an urgent need to forge a mechanism that builds accountability at all three levels of governance in India – national, state and local. The role of leadership will be critical and the right approach with a true sense of commitment can ensure that what may today seem a distant dream can turn into a reality.

27 www.thehindu.com/news/national/ndc-to-be-scrapped-niti-aayog-council-likely-to-get-its-powers/article8051108.ece

15 Comparing the US and India on climate change
How the tables turned

Armin Rosencranz and Rajnish Wadehra

Introduction

The fortunes have reversed: the US's promising lead was knocked out of place by the courts even before the victory of the Republicans led by Donald Trump, and India's stumbling start picked up considerable speed with successful e-auctions for solar and wind installations. India is now well on its way to meeting some of its climate goals, while the US has decided to withdraw from the 2015 Paris Agreement.

In the first section of this chapter the authors study the legal challenges which blocked the Obama administration's Clean Power Plan, President Trump's announcement of withdrawal from the Paris Agreement, and several concerted actions the departments of energy and the interior have taken to repeal restrictive air, water, mining and fracking laws.

The US had committed in its nationally determined contributions (NDCs) to reduce greenhouse gas emissions by 26–28% over 2005 levels by 2025, and the economy seems to be well on the way to achieving these targets with the momentum already created. It remains to be seen whether coal-fired power plants draw substantial investments in the absence of any drastically new clean coal technologies, but coal will be mined and can be exported. To repeal unpopular rules and still achieve reductions in emissions would serve the administration well, if the resultant surge in economic activity is maintained.

The second section studies India's power availability following its surprisingly quick success in starting up several green-field solar and wind farms. But a critical element is the rising coal production. India has not committed to any absolute reductions of emissions in its NDCs, but it has promised merely to bring down its share of dependence on fossil fuels to 60% and to reduce the emissions intensity of its Gross Domestic Product (GDP) by 33–35% by 2030.

The country's per capita power consumption is amongst the lowest in the world and it needs cheap coal power to fuel its development. India is going ahead with plans to increase its coal production to 1.5 billion tonnes annually by 2020.

Burning so much coal will undoubtedly emit more greenhouse gases, but the government seems to be counting on faster GDP growth, as the emissions intensity could still improve if the economy grows faster than emissions do. In absolute values emissions would not be contained at all – by some estimates they could rise threefold – notwithstanding the fact that Article 4.1 of the Paris Agreement obliges signatories to limit their peaks as early as possible. But the huge growth that looked promising in 2015 has slowed down due to structural factors. Manufacturing growth is poor and the economy is burdened further by demonetization and the imposition of a general goods and services tax, and employment generation is slack for India's teeming young millions. The slowing down of the growth rate, while emissions increase, could have a devastating effect on India's compliance.

US and Indian sovereign priorities have similarities in their optics. Both are led by strong-willed leaders who excel in dramatic populist rhetoric, both love coal and both play to cheers of their vote banks. But while the US looks inward to contain its advantages, and creates walls around itself, India is opening out to the world and globalizing. In the past decade, as the US became self-sufficient in oil and gas, it lost interest in the Middle East and pivoted towards Asia. The US and India have become close military and energy allies and this alliance should have a substantial impact on the future of renewable energy technology and investments in India.

The story of the US shift

Game theory

In a game theory framework analysis, Professor Rohit Prasad of the Management Development Institute, Gurgaon, credits the Trump administration with a great victory[1]. The federal government can do popular things, take popular decisions, and save its money, while still enjoying already declining greenhouse gas emissions. The rest of the world has not rescinded the Paris Agreement. As other nations contribute towards achieving their commitments, the US stands to benefit as a free rider while the global climate improves.

An unintended outcome of President Trump's withdrawal from the Paris Agreement has been that several US states and cities have strengthened their own plans to reduce their emissions. Governors of several states have declared their resolve to continue their efforts to abate climate disruption. The federal government is thus unburdened and is free to pursue its own agenda, while the states and municipalities take it upon themselves to achieve their climate goals.

This is not the first time that the US has exhibited that its calculated strategic sovereign priorities outweigh its international commitments. The US back-out from the Koyoto Protocol is still recent in human memory, and it wasn't so long ago that the US backed out of the League of Nations. In fact, there are some who view the United Nations as an organization that has been reduced to

---

1 www.livemint.com/Opinion/1adtSRXp9HP9GHJwalimuI/Donald-Trumps-Paris-Agreement-pullout-Masterstroke-death.html (accessed 27 September 2017)

a debating platform. Its teeth seem to have been taken away by successive years of squeezing its funding. Instead of sending UN peacekeeping forces to quell the troubles in Iraq and Afghanistan, the US seized the opportunity to make its own manoeuvres, merely informing the United Nations Headquarters about its actions at the last possible moment.

This is a very rational outcome of strategic thinking, as it embodies purely utilitarian strategizing to gain advantage, oblivious to the consequences of one's actions on others. The fundamentals of rational conservative economic thinking which leads to Pareto optimality based on Cartesian logic stand at odds with liberal tolerance. Pareto celebrates if one can get better without making anyone else worse off. But some nations can't see beyond their borders, or don't care that others are getting worse off, and act oblivious to the harm their actions might unknowingly cause others. Liberal thought, on the other hand, embraces differences, finds common grounds, and celebrates diversity. Even while it works to earn profits and get rich, it consciously refrains from doing so at a cost to others.

The problem of climate change is a 'tragedy of the commons'[2]. The environment is a shared resource, and looking at it from one's narrow point of view, which could be very logical and rational, no solution can be found. Rising above one's territory, above one's vista point, seeing the larger universe for what it is, is an exercise which can resolve this, and the Paris Agreement embodies humanity's first baby steps towards this goal.

With the fragmentation of the former Soviet Union, a sort of oligarchy emerged, and with the opening up of private enterprise in China, a new heavy-handed, single-party, state-sponsored capitalism has emerged. The US has reigned supreme as the world's largest free market for decades. In his famous treatise *The End of History and The Last Man*, Francis Fukuyama[3] explained how the centralized planning failed and the world came to acknowledge the democratic free enterprise market system as the best. The US was established as its architect at the pinnacle of its glory, and leader of the world in thought and action.

But the US's promising start on forestalling climate change impacts fell short. President Trump and his EPA Administrator Scott Pruitt emasculated Obama's plans, and have proceeded to dismantle several regulations. The Clean Power Plan never had a chance to take off even before Donald Trump's election. Several states, corporations and fossil fuel industry groups sued the US Environment Protection Agency (EPA), citing administrative overreach and pleaded to effectively stall the carefully crafted regulations until the administration changed. The Clean Power Plan was eventually repealed by the EPA in October 2017, relieving power stations and businesses from being constrained to clean their emissions.

---

2 Hardin, G. (1968), "The Tragedy of the Commons", *Science*, New Series, 162, (3859), American Association for the Advancement of Science (13 December), pp. 1243–1248; www.jstor.org/stable/1724745 (accessed October 2010, cross-ref http://pages.mtu.edu/~asmayer/rural_sustain/governance/Hardin%201968.pdf, accessed 24 November 2017)

3 Fukuyama, F. (1992, 2006), *The End of History and the Last Man*, Free Press, New York, NY

The events leading to the US announcement of withdrawal from the Paris Agreement and the reactions to this from the world community, as well as from several US states and cities, reveal a larger public consciousness. The re-iteration of their clean power goals by several states, led by New York and California, and by several city administrations reveals new lessons in federalism, but the divide is sharply segregated. Coal, oil and shale oil and gas-bearing states have sided with the Trump administration, while several other states challenged the repeals of climate regulations.

The Clean Power Plan challenged

The Obama administration had drafted detailed regulations to ensure the delivery of its climate goals well ahead of the Paris Agreement. Guided by Energy Secretary Ernest Muniz, EPA Administrator Gina McCarthy had drafted rules in 2014 under which states would need to submit plans to the federal government committing to reduce emissions, and restrictions were planned for industries not adopting specified technologies.

The Clean Air Act was interpreted by the US Supreme Court to provide the authority to the EPA to issue such regulations[4]. The Clean Power Plan was a set of regulations which required states to commit to reductions in carbon dioxide emissions, on a formula based on their energy mix. States could choose between rate-based goals (lb per MWh of power generation) or mass-based goals (carbon dioxide emissions measured in tons of CO_2 emitted)[5]. The regulations required all existing power plants to file initial plans of emissions by 2016–2018, which would come into effect in 2022.

New power plants using fossil fuels would be prohibited if they did not use carbon-capture technology, and existing power plants were required to reduce their carbon emissions. Restrictions were also placed on transportation and buildings. The health and well-being effects were researched and documented by the EPA. Cost savings of $20 billion and health benefits of $14–34 billion were evaluated and these benefits far outweighed the costs[6].

Plans would be federally enforceable and source-specific. State measures such as renewable energy requirements would be state enforceable. The EPA maintained authority to approve or reject each state's proposed plans and to administer a federal plan in states that did not choose to participate and report updates[7].

Intended nationally determined contributions proposed by the US at the United Nations Framework Convention for Climate Change's (UNFCCC) 21st

---

4 www.ucsusa.org/global_warming/solutions/reduce-emissions/the-clean-air-act.html#.WhpJilWWZ0w

5 www.ago.wv.gov/publicresources/epa/Pages/default.aspx; Office of Attorney General Patrick Morrisey (last accessed 1 June 2017)

6 www.ago.wv.gov/publicresources/epa/Pages/Existing-Coal-Fired-Power-Plants.aspx (last accessed 1 June 2017)

7 http://wspp.org/filestorage/oc_2015_fall_mtg_training_clean_power_plan_kushner.pdf

meeting of the Committee of Parties at Paris, of a reduction in emissions by 26–28% over 2005 levels, would be thus achieved. These were proudly announced at a joint event with China a year ahead of the Paris Agreement, to showcase that the US was ready to lead the world in tackling climate change.

Even before final regulations were published by the EPA, the draft plan which was put up for public comments was challenged by Virginia and ten other states, and by fossil fuel industry interests, at the designated US Circuit Court of Appeals for the DC Circuit. It is well known that over 90% of West Virginia's economy runs on fossil fuel and related fossil fuel businesses in coal-mining towns were losing jobs. These petitions were dismissed in August 2015 on the grounds that the proposed rule had not yet been finally published in the US Register. Being premature, the petitioners could file a request again once the rule was published[8].

Final rules, with slightly lower targets, were published in the Federal Register in August 2015 to come into effect from October, with revised provisions for states to engage with vulnerable low income, minority and tribal people, as well as with workers and their representatives[9]. Further actions were a $4 billion private sector commitment to scale up innovation in clean energy and the launching of a new Clean Energy Impact Center at the Department of Energy[10]. Incentives were granted to appease the interests of energy lobbies in the form of five-year tax credits for solar and wind power plants. Enacted separately in December 2015, the Clean Energy Incentive Scheme was notified in June 2016.

The EPA cited 'a moral obligation to leave our children a planet that's not polluted or damaged', a doubling of the incidence of asthma, rise in sea levels, terrestrial warming, extreme droughts, wildfires and heat waves. It claimed that it would lead to 90,000 fewer asthma attacks, 'will reduce premature deaths due to power plant emissions by 90%' and 'save consumers $155 billion between 2020 and 2030'.

The Department of Energy set a goal of reducing carbon emissions by 3 billion tons by 2030 by issuing standards for 29 categories of equipment and appliances and a code for commercial buildings. These measures would bring down emissions to 17% below 2005 levels by 2020 and 26–28% by 2025[11].

Led by West Virginia, 14 states filed an extraordinary petition seeking emergency relief on 13 August 2015, just ten days after the final plan was adopted. Several industry groups and associations including the US Chamber of Commerce filed suits and became interceptors or *amici*. Relief was denied by the DC Circuit Court of Appeals[12].

---

8 http://wspp.org/filestorage/oc_2015_fall_mtg_training_clean_power_plan_kushner.pdf
9 www.ago.wv.gov/publicresources/epa/Documents/Final%20Section%20111(b)%20Rule.pdf (last accessed 1 June 2017)
10 https://19january2017snapshot.epa.gov/cleanpowerplan/clean-power-plan-final-rule-table-contents_.html
11 www.obamawhitehouse.archives.gov/climate-change> for more about the President's Clean Power Plan (last accessed 1 June 2017); https://obamawhitehouse.archives.gov/blog/2015/08/03/clean-power-plan-myths-and-facts (last accessed 1 June 2017)
12 *State of West Virginia, et al. v. EPA*, US Court of Appeals for the DC Circuit, No. 15–1277

The panel consisted of Judge Judith Rogers, a Clinton appointee; Judge Karen Henderson, a Bush appointee; and Judge Sri Srinivasan, an Obama appointee. In their brief order, the judges wrote that the parties 'have not satisfied the stringent requirements for a stay pending court review'[13]. The judges, however, allowed fast-track hearings and scheduled oral arguments to be concluded in June 2016, to decide the legality of the rules quickly[14].

Writ of certiorari at the US Supreme Court

As the federal government started getting ready to implement the plan, more than 50% of US states, led again by West Virginia, with industry chambers of commerce and lobby groups, filed a plea to the US Supreme Court for a writ of *certiorari*.

The pleaders argued that US EPA had overstepped its authority under the Clean Air Act. Power plants were being forced to buy clean coal technology, or perish. Eight writs for prohibition or review were filed at the US Supreme Court, challenging the EPA's legal powers to order such drastic rules. The coalition included 27 states and, among others, the West Virginia Peabody Energy Corporation, the State of Virginia, State of North Dakota and Murray Energy.

Since the Clean Power Plan mandated huge investments which could not later be undone, they argued that the rule would hurt them in irreversible ways and asked for a stay[15]. They contended that the EPA had no authority to regulate emissions from power plants under sections 111 (d) of the Clean Air Act. Section 112 already regulates emissions and the act debars the EPA from regulating the same item under two sections of the law[16].

In an *amicus* brief, the US Chamber of Commerce submitted that costs of compliance would amount to at least $7.3 billion by 2030. It argued that Clean Air Act Section 111(d)1 'precludes EPA from directing states to establish standards of performance from any source of air pollution that is already regulated under Clean Air Acts clause 112 and multiple regulations burdening owners and operators of power plants would dramatically increase electricity costs while making electric service less reliable'[17].

The US Supreme Court ruled in February 2016 that the Clean Power Plan be stayed, and the matter be heard at the Federal Court of Appeals for the District of Columbia. The unprecedented order halted implementation of the regulations, and sent the case back to the DC Circuit Court of Appeals to arrive at a verdict.

---

13 "DC Circuit Court denies stay on EPA Clean Power Plan", Gavin Bade@GavinBade, 21 January 2016; www.utilitydive.com/news/dc-circuit-court-denies-stay-on-epa-clean-power-plan/412514/
14 DC Circuit Opinion (6-9-2015) (last accessed 11 June 2017)
15 www.usatoday.com/story/news/2016/01/21/clean-power-climate-change-obama-appeals-court/79134944 (last accessed 1 June 2017)
16 www.utilitydive.com/news/dc-circuit-court-denies-stay-on-epa-clean-power-plan/412514/ (last accessed 11 June 2017)
17 *State of West Virginia, et al. v. EPA*, US Supreme Court, No. 15A773

Justice Scalia's deciding vote stayed the Clean Power Plan 5: 4, supported by Chief Justice Roberts and Justices Kennedy, Thomas and Alito. Justices Ginsburg, Breyer, Sotomayor and Kagan opposed the stay.

The White House issued a strong statement disagreeing with this stay and hoping that the merits of the plan would eventually be evident, that it would continue to develop the details of the plan in this hope[18].

'En banc' review at Circuit Court of Appeals

Back at the District of Columbia Circuit Court of Appeals, the fast-track hearings promised in the earlier order of the three-judge bench didn't really take place as intended. The run-up to the presidential election saw a dragging on of dates and decisions.

Though the case had been heard earlier by a three-judge bench, and a hearing date of 2 June 2017 had been announced, the court decided in May 2017 to allow more time to the petitioners, so that any more aggrieved parties may join the litigation. In a convoluted explanation for delaying the hearing, the court stated: 'Because Respondents' proposal would unfairly prejudice as-yet-unknown parties that may challenge the reconsideration denial, Petitioners oppose Respondents' cross-motion to establish a modified briefing schedule. Moreover, because no party sought a stay of the Rule and the Rule remains in effect, Respondents will not be harmed by waiting a few additional weeks so that all parties may have input on a new proposed briefing schedule.'

This overruled the earlier order of the three-judge bench which, while denying the stay, had required that hearings be done before a three-judge panel on 2 June 2016, on two consecutive days if necessary, so that a judgment could be arrived at soon. The DC Circuit Court actually assumed the consent of parties in extending the date for the final arguments. The order of 16 May 2016 cancelling the hearings scheduled for that date, and scheduling *'en banc'* arguments to be held on 27 September 2016, pushed the date to a year since the Clean Power Plan was instituted and a just few weeks to go for the presidential election[19].

Chief Judge Merrick Garland and Judge Julia Pillard recused themselves and did not participate in this circuit court's proceedings. Garland was nominated to the Supreme Court to replace the late Justice Antonin Scalia (the US Senate refused later to act on this nomination citing the impending presidential election). This left a bench of eight, with four judges appointed by democratic presidents[20].

---

18 https://obamawhitehouse.archives.gov/the-press-office/2016/02/09/press-secretary-josh-earnest-supreme-courts-decision-stay-clean-power (last accessed 13 June 2017)
19 www.power-eng.com/articles/2016/05/full-d-c-circuit-to-hear-clean-power-plan-in-september.html (last accessed 1 June 2017)
20 www.usnews.com/news/politics/articles/2016-09-27/dc-appeals-court-set-to-hear-clean-power-plan-case

Arguments were led by West Virginia's Solicitor General Elbert Lin, who stated that by deploying a little-used provision of section 111(d), the EPA has sought to compel coal power-plant owners to either invest or shut down, as the targets set for power plants would have to be met if they must continue to generate power. These standards were impossible to meet. To survive, plants must buy emission credits from their competitors, thereby subsidizing their business rivals. Calling this 'generation shifting', he argued that the act allows the EPA to mandate improvements in performance, or 'best system of emission reductions', but not the power to force plants to invest in different technologies. EPA's lawyers cited Massachusetts v EPA (2007) to state that the EPA was actually ordered by the Supreme Court to start regulating carbon emissions besides automobile exhausts.

David Doniger, a senior counsel for Natural Resources Defense Council, concluded that the power of EPA to use of section 111(d) to regulate power plant emissions is settled in law, 'provided that it has been determined that pollution endangers public health and welfare'[21].

Justice, though, was still withheld, for reasons best known to the court, as no decision or ruling was announced and the next date was set for January 2017, past the lame duck period and past the date when the new administration would be sworn in. No one could tell who would win, but the timing seems more than coincidental, as the original ruling had allowed for fast-track hearings and, had the court considered it expedient, it could have chosen to actually adhere to a fast-track.

Enter the Trump administration

The Trump administration took charge in January and requested for time needed to get the government going, which the court granted, and the January hearing was adjourned to March.

Energy Secretary Rick Perry and EPA Administrator Scott Pruitt hit the ground running. They weren't waiting for the court's extension, and went about their task with zeal. Administrator Pruitt appears to have focused on easing air, water, environment and climate regulations, and Secretary Perry on reviving coal mines and creating jobs.

They secured President Trump's approval for an executive order in February 2017 to unwind the Waters of the US rules which had been issued by the Obama administration's EPA Administrator Gina McCarthy under the Clean Water Act of 1972, and in March to start rescinding the Clean Power Plan.

'C'mon, fellas. You know what this is? You know what this says?' President Trump said to the miners. 'You're going back to work.'[22] This Executive Order

---

21 www.nrdc.org/experts/david-doniger/whats-next-clean-power-plan (last accessed 1 June 2017)
22 www.nytimes.com/2017/03/28/climate/trump-executive-order-climate-change.html?_r=0 (last accessed 12 November 2017)

initiated a 180-day review to identify all regulations that impede energy production, lifted restrictions on hydraulic fracturing for gas and oil, lifted the moratorium on federal leasing of coal mines, and eliminated Obama-era guidelines that mandated all agencies to consider climate change impacts before taking any decisions.

'My administration is putting an end to the war on coal – going to have clean coal, really clean coal,' Trump declared, and also said the order would 'reverse government intrusion' and 'cancel job-killing regulations'[23].

EPA further defers the court's decision

The EPA requested the court in March to defer the final court hearing for another two months. Administrator Pruitt mentioned on the EPA's site that this gave time to the new administration to initiate procedures for evaluating and reassessing (and in all probability winding up) the Clean Power Plan.

Meanwhile, President Trump announced the US exit from the Paris Agreement from the Rose Garden at the White House and issued an Executive Order on 1 June. Thereafter, the complete rollback of the Clean Power Plan took place on 25 September 2017, rendering further decisions by the courts redundant.

The entire exercise frustrated the rules that may have made a marked difference in the air humans breathe. One can draw conclusions also about the waste of public time and money in the judicial process, which was undoubtedly used to stymie the course of a landmark regulation.

Meanwhile, the following actions completely changed the scenario of restrictions on industry driving the Trump administration's reform momentum: 1) A rule which required energy companies to collect data on emissions from oil and gas wells was withdrawn in March 2017, on the recommendations of Attorneys General of 11 states; 2) A ban on using the pesticide Chlorpyrifos, which EPA scientists had concluded during the Obama regime caused damage to children's development, was withdrawn on 30 March 2017 and its evaluation deferred to 2022.[24]; and 3) Undoing Waters of the US Rules, which the President had mentioned to be one of the worst regulations in US history; in June 2017 the EPA filed a detailed 42-page proposal to rescind this rule, and in September went ahead with repealing it. This was an Obama era rule which aimed at curbing pollution in the nation's waterways[25].

In June 2017 the EPA filed a proposal to delay the implementation of accidental release prevention of pollution requirements and risk management programme regulations by two years. These rules had been designed to oversee

---

23 www.cbsnews.com/news/trump-signs-executive-order-dismantling-obama-environmental-regulations/ (last accessed 27 November 2017)
24 www.nytimes.com/2017/05/15/health/pesticides-epa-chlorpyrifos-scott-pruitt.html (lLast accessed 28 November 2017)
25 www.nytimes.com/2017/06/27/climate/epa-rescind-water-pollution-regulation.html

methane leaks from drilling rigs, and obliged fossil fuel companies to adopt preventive measures[26].

The Department of the Interior rescinded 2015 regulations of the Bureau of Land Management which required disclosure of chemicals used in fracking fluids that found their way into waste water. Hydraulic fracturing pushes chemical fluids and water at high force on to shale rock to crack out oil and gas from its seams, and the techniques for drill wells explore vertically as well as horizontally underneath the surface.

These measures were designed to regulate well construction and the environmental impact of fracking on public lands. The repeal was justified citing savings of compliance costs of $14–34 million per year and also that existing state federal regulations 'would not leave hydraulic fracturing operations unregulated'[27]. Existing federal regulations, however, are more than 35 years old, when fracking was not known, and were made to safeguard normal oil well drilling operations.

The Obama administration's stream protection rule that disallowed coal mining near streams and rivulets was revoked by Congress in February 2014. Steven Gardner, who had opposed the rule and supported surface mining on mountain tops, had been Director at the Office of Surface Mining, Reclamation and Enforcement.

New standards to improve vehicular fuel efficiency were suspended. The Obama era practice of declaring 'interim status', 'standard review' or 'unreasonable risk of injury' while potentially harmful chemicals were being studied was declared to be confusing[28].

The EPA's overall budget was cut to $5.7 billion (from $8.3 billion in 2016). This cut down research in health and climate and would necessitate firing 20% of the EPA's staff of 15,000. National Oceanic and Atmospheric Administration's (NOAA) Sea Grant Program was completely cut; this programme had supported research in 30 US colleges and universities and had predicted a 3–7 feet rise in sea levels between 2050 and 2100. Such a rise might drown vast reaches of the US coastline. The US Geological Survey's budgets were cut by 15%. An 11% cut was imposed in the budget for the National Science Foundation, which organizes climate and scientific research. The EPA's research budgets have been cut to $250 million (from $483 million in 2016). The superfund sites budget for cleaning hazardous wastes was also cut by 25%. This was unexpected, as a task force report had mandated further supervision[29].

26 www.cbsnews.com/news/epas-pruitt-moves-to-roll-back-over-30-environmental-regulations-in-record-time/
27 www.washingtonpost.com/news/energy-environment/wp/2017/12/29/to-round-out-a-year-of-rollbacks-the-trump-administration-just-repealed-key-regulations-on-fracking/?utm_term=.465a38afff09&wpisrc=nl_energy202&wpmm=1 (last accessed 3 January 2018)
28 www.sciencemag.org/news/2018/01/trump-s-epa-once-public-chemical-safety-reviews-go-dark (last accessed 14 January 2018)
29 www.epa.gov/sites/production/files/2017-07/documents/superfund_task_force_report.pdf

US exit from the Paris Agreement

In a grand fulfilment of his election promise, President Trump announced the US exit from the Paris Agreement from the Rose Garden at the White House on 1 June 2017. It was denounced as a bad deal for the US to pay for countries like China and India to finance their renewable energy projects. He would try to renegotiate the terms, if possible, to win the US a better deal, and was stopping US payments to the Green Climate Fund. He said that he had an obligation to the people who had voted for him – 'I was elected to represent the citizens of Pittsburgh, not Paris' – and would therefore work to revive US coal mines and try to get them their lost jobs back.

He did not say that climate change was a hoax or perpetuated by China, as he had during his election campaign. But he brought up a research paper by Massachusetts Institute of Technology which stated that, with all the nationally determined contributions, the world's temperatures would come down only by a tiny fraction of a per cent, and that these commitments were not binding on nations, and might not be fulfilled.

'Even if the Paris Agreement were implemented in full, with total compliance from all nations, it is estimated it would only produce a two-tenths of one degree Celsius reduction in global temperature by the year 2100...[30]'

The obligation to implement its INDCs, which had already been ratified by the Obama administration and taken the status of Nationally Determined Contributions, thus became obsolete[31].

Studies that denied climate change were rather weak – a NASA report claimed these were a mere 3% of the published papers – and suffered from insufficiently compelling arguments. That the US had already achieved a trajectory of reduction in greenhouse gases was an argument that had some merit, as this had been achieved without stringent regulations, and could continue without international commitments.

The 'Make America Great again' slogan had now morphed into 'America First'. This appears to be a re-statement, presumably because the earlier slogan implied implicitly that America wasn't great anymore. This withdrawal was construed as the will of the people and its timing was politically impeccable. While several analysts saw it as an irresponsible and uncaring populist move, its nationalistic stance was nowhere in doubt. It was followed by a strategic plan of energy dominance in the world.

The Paris Agreement is not a treaty

The UNFCCC had framed the commitments at Paris not as a treaty, but as an accord, or an agreement, which would not be binding on elected parliaments of

---

30 www.reuters.com/article/us-usa-climatechange-trump-mit/trump-misunderstood-mit-climate-research-university-officials-say-idUSKBN18S6L0
31 http://unfccc.int/resource/docs/2015/cop21/eng/l09r01.pdf

the signatory nations. In the case of the US, it would not require the Senate's ratification. President Obama had formally ratified the agreement by an executive order. The failed Kyoto Protocol had been a treaty which was not approved by Congress, and had led to much embarrassment for the US, thus this was the method employed to keep commitments non-binding, as intended contributions, with no recourse for default.

But on the flip-side, this made the agreement equally easy to withdraw from. President Trump was advised that it was perfectly within his powers to withdraw unilaterally without needing the consent of the Senate or the Congress[32]. Hence, all that was required to announce the US's withdrawal was simply another presidential executive order, and it was issued by President Trump without seeking public opinion of concerned citizens or stakeholders on its ramifications.

The intention to renegotiate the agreement appears to be vague, as the UNFCCC does not have any provisions to renegotiate. The US could withdraw, but withdrawal would be possible only in 2020. Under the Paris Agreement, the earliest a party can give notice of withdrawal is 4 November 2019, and the earliest it would take effect would be from 4 November 2020.

Article 28 of the Paris Agreement states in Clause 1: 'At any time after three years from the date on which this agreement has entered into force for a party, that party may withdraw from this agreement by giving written notification to the Depositary'. Clause 2 states that any such withdrawal shall take effect on expiry of one year from the date of receipt of the notice with the Depositary[33]. Christina Figueres, head of the UNFCCC, said that the US could ask to be reinstated in 2020 or thereafter, after having once withdrawn, but quashed the idea of any renegotiation, as it is a multilateral agreement and any one country could not possibly renegotiate the terms.

The letter officially submitted to the UNFCCC on 4 August 2017 by the United States stated that the US was withdrawing from the Paris Agreement. It was unclear until this was done whether the US would withdraw from the UNFCCC or merely from the Paris Agreement. It mentioned the US intention to renegotiate the terms at the appropriate time, which was more like a reminder to itself, and until then it would continue to participate in the 23rd Committee of Parties at Bonn and activities of the Convention. UN Secretary General António Guterres expressed his disappointment and conveyed that he saw it as 'a major disappointment for global efforts to reduce greenhouse gases and promote global security'.

MIT professors deny Trump's charge

Massachusetts Institute of Technology officials denied the President's conclusion about their work, stating that President Trump had grossly misunderstood their

---

32 https://doi.org/10.21552/cclr/2017/3/16 https://cclr.lexxion.eu/issue/CCLR/2017/3 cross ref

33 https://unfccc.int/files/meetings/paris_nov_2015/application/pdf/paris_agreement_english_.pdf

research conclusions. Emily Putt, reporting for Reuters from New York, wrote in June 2017:

> That claim was attributed to research conducted by MIT, according to White House documents seen by Reuters. The Cambridge, Massachusetts university published a study in April 2016 titled 'How much of a difference will the Paris Agreement make?' showing that if countries abided by their pledges in the deal, global warming would slow by between 0.6 degree and 1.1 degrees Celsius by 2100.'
>
> 'We certainly do not support the withdrawal of the US from the Paris agreement,' said Erwan Monier, a lead researcher at the MIT Joint Program on the Science and Policy of Global Change, and one of the study's authors. 'If we don't do anything, we might shoot over 5 degrees or more and that would be catastrophic,' said John Reilly, the co-director of the program, adding that MIT's scientists had had no contact with the White House and were not offered a chance to explain their work.

Some researchers have estimated that business-as-usual or baseline policies would lead global temperatures to rise by 4.1–4.8° Celsius. Current trends until 2015, with policies as they were before Paris Agreement, would lead to a temperature rise of 3.3– 3.9°. Pledges of nations have been in several non-uniform formats and estimation of the impact of the Agreement could vary. Some studies have placed temperature rise post-Paris at 2.4–2.7°, according to Professor Swapna Pathak, Assistant Professor of Environment Studies at Oberlin College, Ohio. MIT's might be the most pessimistic amongst such studies, but most, including the Intergovernmental Panel on Climate Change in its report of October 2017, agree that Paris would not deliver the targeted 2° drop.

The UNFCCC faces the unenviable task of pushing nations to ratchet up their ambitions, and there is no denying that the world actually faces a less-than-desired reduction in global emissions. A diplomatic cable leak revealed that the administration has no intentions of renegotiating the Paris Agreement, in a briefing note from Secretary of State Rex Tillerson to the President dated 4 August 2017[34]. Secretary Tillerson was relieved of his position as Secretary of State in early 2018, ostensibly over differences of opinion over the agreement with Iran, but the strain seems to have started months before over the US withdrawal from the Paris Agreement.

The US continues to improve its emissions

In a follow-through of the momentum already generated, despite the absence of new climate regulations, the US has continued to achieve its intended climate goals. Greenhouse gas emissions have continued to decline, and the sentiment

---

34 www.climatechangenews.com/2017/08/09/diplomatic-cable-us-no-plans-renegotiate-paris-agreement/ (last accessed 17 November 2017)

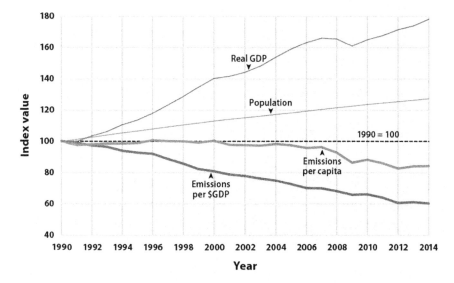

Figure 15.1 US greenhouse gas emissions per capita and per dollar of GDP, 1990–2014
www.epa.gov/climate-indicators/climate-change-indicators-us-greenhouse-gas-emissions

generated by the environmental movement has led to a conscious shift towards cleaner energy.

Prices of renewable energy equipment have dipped sharply, with homegrown technology and economies of scale. Falling prices encourage more people to switch over to renewables and as demand for renewables increases, it leads to economies of scale which enable lower costs. Businesses have taken advantage of this opportunity and blue solar-panelled rooftops are visible in several parts of the country. Technologies to develop newer and better renewable energies are also developing fast in the areas of concentrated heat and power, thin solar films, solar tiles and storage[35].

US emissions have been on a declining trajectory in the past decade, despite the fact that many regulations never came into effect. The consciousness of climate change in the industry seems to have paid off. US emissions have come down substantially and consistently since 2006, and data from the EPA provides some more details[36].

In 2014, US greenhouse gas emissions totalled 6,870 million metric tons (15.1 trillion pounds) of carbon dioxide equivalents. This total represents a 7% increase since 1990 but a 7% decrease since 2005[37].

---

35 See "US emissions trajectory till 2025 with and without Clean Power Plan" in *The New York Times*, 28 March 2017: www.nytimes.com/climate
36 www.epa.gov/climate-indicators/climate-change-indicators-us-greenhouse-gas-emissions
37 www.epa.gov/sites/production/files/2016-08/documents/print_us-ghg-emissions-2016.pdf

The Trump administration's reversals would have the desired effect if coal mines start opening up again and power plants go back to burning more coal, but it remains to be seen how far the market forces pick up on the openings provided by the administration. The impact of opening up the coal-mining industry might turn out to be negligible if actual investment in new coal power plants is slow and excess mined coal is exported.

Renewable energy installations have surged and greenhouse gas emissions have declined despite the absence of these rules. Coal powers only 30–35% of the US economy and employs fewer than 100,000 persons. The long-term trend away from coal would in fact benefit the Trump administration, as the US could claim to have done its bit and also saved itself from spending money to pay for energy supplies of developing nations.

The question many are asking is whether the momentum will last.

Perhaps the tipping point has already been reached, and the transformation could go ahead on its own steam. Consciousness has been raised in the world, starting with Stockholm 1980, Rio 1992, the flopped Kyoto Protocol, and then Rio+20 2012 followed by the Paris Agreement, with research done by several scientists and non-governmental bodies, led by the IPCC. This belief or hope underlies the Trump administration's conservative policies. It may well turn out, however, that even with several incentives for coal, investment might not flow in, and public sentiment might well triumph over the political.

Repeal of the Clean Power Plan

The doomed Clean Power Plan was finally repealed in October 2017. The uneven burden it had imposed on fossil fuel-producing and burning states had led to a clear divide between states and industries supporting their states. States which had already adopted renewables on a larger scale benefited from the plan, and this contentious divide was withdrawn by the complete repeal of the Clean Power Plan.

The plan had been an enabling regulation to achieve the US NDC goals in a planned 'wholesome' approach, geared to deliver results. A conventional tactic could have been used to tax polluters under the polluter pays principle, but the Obama administration had consciously chosen this route, which never actually managed to see the light of day[38].

States led by California and New York, which did not oppose the plan, have set up enabling regulations to continue with their erstwhile goals. Coalitions of cities have sprung up and representations were made at the COP 23 at Bonn to recognize participation of cities and municipalities as stakeholders in this mission.

President Obama's EPA Administrator Gina McCarthy mentioned in an interview that she hoped the core values of the EPA would stand to fulfil its goals. She stated that the administration would have to say either that 'the facts were wrong

---

38 www.bloomberg.com/news/articles/2017-10-04/trump-is-said-to-begin-repeal-of-obama-power-plant-emission-cuts (last accessed 7 October 2017)

or the science was wrong, or that it didn't get the law right', and would need to invite public comment, and if that is done the public health benefits would be seen to outweigh the savings in costs[39].

Waters of the United States Rule, 2015

The Clean Water Act was an amendment of the originally established Federal Water Pollution Control Act in 1972. 'The Waters of United States' or the Clean Water Rule of 2015 under this Act has also been nixed, citing overreach of government powers to inspect every ditch and puddle on private lands. The rule intended to protect the quality of drinking water that people draw from rivers and ponds. Gina McCarthy said she had invited extensive public comments over seven months, and studied the issues and concerns raised in nearly a million comments received. The EPA had defined river beds, lakes, ponds, ditches, etc., adequately and provided exemptions to farmers and ranchers, but not to builders.

Donald Trump had campaigned against the rule, which was challenged in courts and blocked as soon as it was enacted in 2015. Scott Pruitt had led the attack in courts as Attorney General of Oklahoma and removal of these rules was imminent. The EPA invited public comments, announcing its intention to repeal these rules, but provided a very short period of 21 days in December 2017 in which 39,000 comments were received, and the rules were repealed soon after. Twenty states in the US have their own laws which are more stringent than the Clean Water rule, while all others would enjoy the benefit of unchecked pollutants on removal of this federal legislation[40].

Coal ash rules and the Clean Air Act

Coal ash increases the risk of cancer of the lungs, bladder and colon, and more than 100 million tonnes of this waste is produced by the 400 US coal-fired power plants[41]. Gina McCarthy signed the Coal Combustion under Improvement of Regulation Final rules in 2014, which were published in the Federal Register in April 2015 and promptly challenged in seven different lawsuits filed by industry groups and municipalities. The EPA sought the court's permission to remand certain portions of these rules and shortly thereafter President Obama signed the Water Infrastructure for Improvements to the Nation Act of 2016 which authorized regulation to prevent toxic leaks of coal ash's harmful chemicals[42].

39 https://stateimpact.npr.org/pennsylvania/2017/10/04/qa-former-epa-administrator-gina-mccarthy-discusses-pruitt-wotus-and-clean-power-plan/ (last accessed 10 November 2017)
40 https://stateimpact.npr.org/pennsylvania/2017/02/27/trump-plans-to-ditch-obamas-protection-for-small-wetlands-and-waterways/ (last accessed 26 November 2017)
41 https://apnews.com/f3c75ec69bd7485590d467d76d766dfc/APNewsBreak:-US-utilities-find-water-pollution-at-ash-sites (last accessed 31 March 2018)
42 Environment Law at Harvard: http://environment.law.harvard.edu/2017/12/coal-ash-rule/ (last accessed 30 March 2018)

In Sept 2017 the EPA under Administrator Pruitt requested the court's permission to remand certain provisions of the rules, and proposed amendments which would lower costs of generating power by $100 million[43]. States would now have the privilege of making their own rules in certain respects. These are open for comment for 45 days and a public hearing is scheduled in Washington DC on 24 April 2018.

Groundwater monitoring reports, published by utilities under the same rules in March 2018, however, show higher levels of arsenic and radium at 70 sites across the US. Impacts are severe for less privileged communities who live near coal ash ponds and landfills at the fringes of such plants.

Pleas from several state administrations on ground-level ozone or smog which crosses over state borders have been delayed or rejected. A petition under the Clean Air Act's 'good neighbor' conditions from Connecticut to curtail smog on its east coast from a coal-fired power plant in Pennsylvania has been held up by the courts. Maryland has sued the EPA for missing the six-month deadline to respond to its petitions against 36 polluting power plants in Ohio, Kentcky, Indiana and West Virginia. The EPA has also been sued by health groups and cancer associations in 16 states for not implementing Obama-era ozone rules. The EPA has sought time until April 2018 to frame fresh rules[44].

The Government Performance and Results Modernisation Act of 2011 mandates a strategic plan which the EPA announced in February 2018 to run through until 2022. With a core mission of ensuring clean air, land, water and chemical safety, it emphasizes the rule of law and process, and the autonomy of federal and state regulators. It also stresses co-operative federalism to rebalance power with the states. The plan provides until 2022 to reduce the average period from identification of an environmental violation to its redressal, and baseline thresholds for this will be determined during 2018[45].

The EPA under Scott Pruitt

The New York Times interviewed 20 members of the 15,000 staff at the EPA, and arrived at the conclusion that Pruitt has outsourced critical work to law firms and his old allies, known to him since his days as the head of the Republican Attorney Generals Association, and that this association had collected since 2013 $4.2 million from Murray Energy, Exxon Mobil, Koch Industries, and others in the fossil fuel sector[46].

---

43 https://insideclimatenews.org/news/15032018/epa-regulations-coal-power-plant-waste-ash-ponds-toxins-health-jeremy-orr-water-testing (last accessed 31 March 2018)
44 https://insideclimatenews.org/news/21032018/trump-epa-smog-rules-pruitt-air-pollution-clean-power-plan-air-quality-connecticut-pennsyvlania-power-plants (last accessed 30 March 2018)
45 www.eenews.net/eenewspm/stories/1060073637 (last accessed 14 February 2018)
46 www.nytimes.com/2017/07/01/us/politics/trump-epa-chief-pruitt-regulations-climate-change.html (last accessed 27 November 2017)

It has been reported that the anti-climate policy is decided at the top, and staff are asked merely to execute the agenda. Career staff members of the EPA have voiced their protests against being left out of consultations. Pruitt, in his defence, voiced an opinion on the 'sue and settle' policy of the previous administration which would settle lawsuits from citizens' environmental groups by paying thousands of dollars, and in a way help these groups earn some money.

Xavier Becarra, Attorney General of California, has filed a public records lawsuit against Administrator Pruitt for failing to hand over documents relating to his ethics arrangements. Scott Pruitt had agreed to stay away from lawsuits over issues he had challenged the EPA over in court, as he should be recused or disqualified from participating in certain matters which evidence a conflict of interest as per federal ethics regulations. Pruitt maintains that he is not restrained from rolling back the rules, such as the Clean Power Plan and the Waters of United States or the Clean Water rules[47].

The career staff has been reduced by 700 employees as of March 2018, and 59 new hires had been registered lobbyists or lawyers for fossil fuel or chemical and other industries[48].

Energy Secretary Rick Perry's push for coal

Coal-fired power plants generated 30–35% of US energy in 2015–16, down from 57% in 1978. Coal employed about 100,000 persons, down from 250,000 in 1979 and 1 million in 1920. The case for creating jobs and opening closed coal mines drives the Trump administration's political agenda. Most of the nation's coal power plants are more than 40 years old and new clean coal technology is largely absent[49].

The Trump administration's attempts to save dying coal-fired power plants and revive jobs in a sagging economy are well known. While closed coal mines have been a cause of concern prompted by thousands of lost jobs and lost profits, the emergence of several successful businesses in renewable energy has filled up some of the gaps. Rooftops lined with solar panels and the countryside lined with wind and solar farms are now a familiar sight in the US.

New technologies in combined heat and power, carbon capture and concentrated solar troughs have been successfully implemented. Better batteries have led to the availability of storage suitable for renewable power and also for electric vehicles. But this is not the vote bank that voted President Trump into power.

One of the new technologies that Energy Secretary Rick Perry has proudly showcased is the coal-fired Petra Nova project, a joint venture of NRG Energy and

---

47 www.politico.com/tipsheets/morning-energy/2017/08/14/pruitts-commitment-to-transparency-questioned-221852
48 https://insideclimatenews.org/news/21032018/trump-epa-smog-rules-pruitt-air-pollution-clean-power-plan-air-quality-connecticut-pennsyvlania-power-plants (last accessed 30 March 2018)
49 https://energy.economictimes.indiatimes.com/news/coal/opinion-trump-tries-to-save-coal-but-probably-in-vain-kemp/61045996 (last accessed 23 October 2017)

JX Nippon Oil and Gas Exploration Corp. This a 240 MW unit at Thompsons, Texas, which captures up to 90% of the carbon dioxide emitted while burning coal; the captured CO_2 is pumped at high pressure in supercritical form into a nearby oilfield to push out additional oil, resulting in enhanced oil recovery[50].

Secretary Perry was quoted in *The Washington Post* saying: 'This is a tremendous example of how investments in clean technology can also lead to increased development of conventional sources.'

The DOE has offered $50 million in subsidies to design, build and operate two large-scale pilots for transformational clean coal-power generation. The budget maintained 30% tax write-offs for both wind and solar, but provided extra incentives for coal power plants which maintain extra inventory of coal.

The benefits of stable baseline coal power are being rewarded, while environmental restraints and restrictions contained in the Clean Power Plan have been repealed. Data for the past few years shows a surge in renewable energy installations and a healthy decline in emissions despite the absence of these rules. Coal plants have been closing across the country, and that could be for economic reasons rather than political, and it remains to be seen if this swerve in policies would actually serve to revive closed coal mines.

The plants which closed were small, old or inefficient. Efficiencies in old coal fired plants vary around 30%, whereas newer ultra-supercritical technology plants achieve 45% and combined-cycle plants which use gas and coal achieve over 65%. Gas is also known to be more versatile for grid efficiency, especially to balance what is being called the 'California Duck Curve', the dips and peaks of daytime and nighttime use and the absence of solar power at night.

The Department of Energy's views on grid resilience

The Department of Energy (DOE)'s Staff Report to the Secretary for Energy on grid reliability has expressed concern about baseload power and the resilience of the grid. Although whether forecasting is improving, grid discipline is hampered by varying dips and spurts in renewable sources of energy which cause 'creep fatigue interaction'. Gas-generated power with new combined-cycle technology is favoured for the baseload to ramp up and down efficiently, as it provides resilience to the grid, a layer over and above baseload coal and nuclear power.

The DOE claims that the EPA's 2016 New Resource Review hampers growth and raises costs. Even in the absence of the Clean Power rules, coal power plants which do not retrofit their carbon and sulphur dioxide emission control equipment are subject to the uncertain future of the NSR rules, and some who have installed these are retiring anyway.

One of the factors providing security to nuclear and coal-generated power is their availablity of raw materials on site. Perry has proposed to the Federal Energy

---

50 www.washingtonpost.com/news/energy-environment/wp/2017/04/12/obama-funded-this-clean-coal-plant-now-rick-perry-is-celebrating-its-opening/?sw_bypass=true&utm_term=.d342cadb429b

Regulatory Commission (FERC) incentives for coal and nuclear power plants to stock up with 90 days' raw material on site. This negates the work the FERC has done over the past few decades to bring market forces into play in the field of renewable energy[51].

Secretary Perry's proposal would boost the demand for coal and nuclear materials, increase power prices and bring on a spurt in spending on raw materials, with multiplier effects on growth of the economy. But it would only be temporary, as plants that carry a couple of weeks' worth of inventory would increase their purchase of coal and nuclear raw materials and thus provide some short-term benefit to mines opening back after withdrawal of the clean power rules.

Jody Freeman, Professor of Law at Harvard, and Joseph Goffman, a former official of the EPA, reported in *The New York Times*, lamented the lack of strong justification and public comment before proposing such a momentous change[52].

The FERC being an autonomous body can hear a request from the administration but must act on its own independent judgment. In a unanimous decision, the five-member commission, of whom four happen to be Trump appointees, rejected Secretary Perry's proposal in January 2018. It asked, however, utilities to come up with details on how they would ensure grid resilience, which was the aim of the proposal. The ruling stated that there was not enough evidence to make such a change and cited a cost of $1.19 billion, which consumers would have to pay in increased electricity tariffs if the proposal were to be implemented.

US energy consumption slows down

There is no denying the fact that growth of energy demand has slowed down from levels of 2.5% annually to 1% or less in recent years. Some of this could be due to better efficiencies in transmission and distribution, and it acts as a deterrent for coal-power producers to get back into production despite the incentives offered[53].

A report by the Energy Group, commissioned by the Advanced Energy Economy and American Wind Energy Association, concludes that the slowdown in coal is not a result of renewables, but is due to lower demand, cheap shale gas availability and efficient gas-fired power plants generating top-class power at low costs. So does the DOE's staff report.[54,55]

---

51 www.reuters.com/article/us-usa-powergrid-perry/u-s-energy-head-seeks-help-for-coal-nuclear-power-plants-idUSKCN1C42G0 (last accessed 12 November 2017)
52 www.nytimes.com/2017/10/25/opinion/rick-perry-coal-antimarket.html?rref=collection%2Ftimestopic%2FPerry%2C%20Rick&action=click&contentCollection=timestopics&region=stream&module=stream_unit&version=latest&contentPlacement=2&pgtype=collection (last accessed 12 November 2017)
53 https://energy.economictimes.indiatimes.com/news/coal/opinion-trump-tries-to-save
54 www.energy.gov/staff-report-secretary-electricity-markets-and-reliability (last accessed 31 March 2018)
55 See Bloomberg Quoting US Energy Information Administration: www.bloomberg.com/news/articles/2017-10-03/perry-s-coal-proposal-seen-unlikely-to-reverse-historic-decline

The future of American coal

The Trump administration has been vocal in blaming the Obama administration for waging a war on coal and the closure of hundreds of coal-fired power plants. It is uncertain whether it can succeed in reversing the decline of coal, although the political imperative for reviving the industry seems to be greater than the economic, environmental and social reasons.

Between 2000 and 2016, more than 500 coal-fired power-generating units were closed, according to an analysis of data from the US Energy Information Administration, but most of these were old, small and inefficient plants, according to a survey by the US National Energy Technology Laboratory. Surviving coal-fired plants generate an efficiency of about 34%, while many that closed had lower and uneconomic efficiencies[56].

Replacement of old worn-out coal plants with UMPPs (ultra mega power projects) suffers a cost disadvantage, as modern combined-cycle gas plants are much cheaper to run and maintain. These have the added advantage of being able to start in a couple of minutes and reach efficient operation in 15 minutes against many hours required by the best coal-fired technologies.

The administration's hope of reopening closed coal power plants therefore seems to be unfounded. Those that are nearing 40 years of age now will also no doubt close and be replaced by combined-cycle gas-fired plants. Pushing the coal power agenda thus seems to be doomed, and the way out to support miners would be to mine and export coal. This is what the administration tried to do at the 23rd Committee of Parties at Bonn, as US delegations were accompanied by coal producers who set up booths at the conference offering their coal for export. They were booed down by environment-conscious Europeans, but there are still many energy-hungry countries in the world which import large quantities of coal, including India and China.

The global impact can be felt in changes of ownerships of coal mines; some mines in the US are already owned by Indians and 2 million tonnes of American coal was exported in 2016–17 to Indian power plants. The freight costs of transporting American coal are prohibitive to Asia and Africa, and generating plants tend to be built to specific standards to take coal of specific range of calorific value and ash content. So the export agenda might take a few years to pan out until plants adapt and, until then, the administration would have to keep subsidising coal miners to make them compete with shale gas.

If clean coal technologies come up, such as converting coal to diesel down in the mine, this game can be played. But so far there aren't any signs of new technologies on the horizon. Subsidizing research in this field might produce results, but in the very long term. Until then, subsidies on mining and coal-power generation would be the only way to save this industry. The conservative nature of this

---

56 https://energy.economictimes.indiatimes.com/news/coal/opinion-trump-tries-to-save-coal-but-probably-in-vain-kemp/61045996 (last accessed 23 October 2017)

agenda is an expensive bet in the long run, played with tax payers' money. There is a tide of change which the administration seeks to stem and its success seems to be a very remote possibility.

The past two years have been tumultuous and, as is evident, environmental legislations of the democratic Obama administration have been repealed layer by layer. The phase of dismantling legislations seems to be coming to an end and now the phase of taking focused new forward action should begin for the administration. To foster growth by its conservative Republican beliefs, tax laws been amended, overheads have been reduced, legislations rescinded, budgets trimmed. Now, after the destruction, if Joseph Schumpeter was right, the phase of construction and creation should begin, as the administration hopes that investments will flow back to the United States and create jobs in conventional coal and energy industries.

The Trump administration has unleashed a conservative attack on climate legislation and the impacts of these concerted actions could be severe. Several actions appear to have been taken without sufficient scientific evidence. Some have compared the US with Saudi Arabia and Russia, and even with China, which are ruled with a single-minded determination not usually found in liberal pluralistic democracies.

His focus on US interests has been considered selfish by some and his personal preference for rich and influential people in his administration was also evident when he said on 21 June 2017, as reported by *The Hill*, that he would prefer not to have a 'poor person' in a top economic role in his administration[57].

It may suffice to conclude that it is a corporate culture to ruthlessly maximize profits, and this is certainly being achieved at the White House led by a billionaire. How long the White House stays white, or whether it succumbs to turning off-white with pollution, like the Taj Mahal in India, only time will tell.

India's mixed success story

The government invited private bidders to invest and commit prices at which power would be sold to the grid. With many of these already up and running, the energy mix of the country is being transformed. The slow process of equipping the largely state-owned grid to carry renewable power is limiting the transmission, and while several checks have been put in place to save the producers of variable renewable power from suffering, it remains a work in progress. Rooftop solar is lagging far behind its targets and the momentum in households and industries is missing. But at the same time, India continues to push its coal production to 1.5 billion tonnes by 2020. It has vast coal reserves and has lagged behind in ramping up its coal output in the past decades. In this section the authors study the two-faced nature of India's stance.

57 http://a.msn.com/r/2/BBD1Ncr?a=1&m=en-us http://thehill.com/homenews/administration/338912-trump-i-just-dont-want-a-poor-person-in-top-economic-roles (last accessed 27 February 2018)

Ratification of the Paris Agreement

India followed the same route for ratification as the Obama administration. Instead of taking the agreement to parliament, it was put up as a note for a meeting of the cabinet of ministers, headed by Prime Minister Narendra Modi, and approved. India's withdrawal could be equally quick and painless, should the need to do so occur at any future time, as it would not need to go to the parliament. This is not to imply that there are any signs of this need showing up, as precautionary safety valves have been put in place by India in its NDCs.

India has not committed to reduce any emissions, and has not even stated when its emissions might reach a peak and begin a trajectory of decline. All that India has committed to is in the form of a reduction in its dependence on fossil fuels and the carbon intensity of its GDP. India's NDCs state an endeavour to achieve 40% of its energy needs using non-fossil, including nuclear, sources of energy. This means that fossil fuels would continue to generate 60% of the nation's power. India has also committed to a 33–35% reduction in the carbon intensity of its GDP, which means the proportion of its GDP contributed by dependence on GHG-emitting generation processes[58].

All that India has committed in actual numbers is the creation of a carbon sink or afforestation covering 2.5 million km by 2035. India's bold plan to create renewable power generation capacity of 175 GW was stated at the Paris conference, but this was kept out of its INDCs, and offered just as an internal plan to achieve its goals.

Proportionate contributions allow room for carbon intensity and emissions to grow. It is obvious that the nation can grow a lot by 2030, as at a healthy 7.5% growth the economy will more than double its size every ten years. In fact, its emissions could grow three-fold by 2035 and could surpass US emissions to make India the second-largest emitter in the world after China.

India's NDCs are contingent upon receiving money to make this transition. The developed Annex 1 nations will need to pay the entire cost for this transition and it is obvious that the $100 billion committed will not be enough.

For emerging economies of non-Annex 1 nations, an estimate puts the total cost at $8 trillion. India has not stated exactly how much it needs for itself, but had pleaded the case for a total contribution of $2.5 trillion for G77 nations at the failed COP 19 at Copenhagen in 2010. India, the world's third-largest emitter at 2008 million tonnes carbon dioxide equivalent[59], had argued that its per capita emissions were much less, as when divided by its huge population, it emits much less per person than the top polluters in the world[60].

---

58 www.eia.gov/beta/international/analysis.cfm?iso=IND
59 BP Statistical Review 2015, www.bp.com/en/global/corporate/about-bp/energy-economics/statistical-review-of-world-energy/statistical-review-downloads.html
60 BP Statistical Review 2015, www.bp.com/en/global/corporate/about-bp/energy-economics/statistical-review-of-world-energy/statistical-review-downloads.html

India's policy dichotomy

India's internal policies reveal a critical dichotomy.

The National Energy Policy of 2017 lays out ambitious objectives, such as universal and uninterrupted access to electricity, and an expected three-fold rise in its per capita annual electricity consumption from 1075 KWh in 2015–16 to over 2900 KWh in 2040, taking into account the expected rise in population from 1.2 billion to 1.6 billion by 2040.

The target of creating 175 GW capacity of renewable energy generation is being implemented with fervour. State governments, in collaboration with the Solar Power Corporation of India, have awarded projects for more than 39 GW of renewable capacity and another 100 GW have attracted the interest of several private investors. Several overseas investors have set up businesses in India to get a piece of the action. As equipment prices dipped, prices bid for solar and wind power have correspondingly fallen, and aggressive rates have consistently been bid. Price parity might soon be achieved with coal power.

Targets for mini-hydro and bio-gas projects for the year 2020 are being met, and large farms of solar and wind are coming online at a fast pace. Rooftop photovoltaic panels, however, lag behind. The 40 GW sub-target for solar rooftop generation capacity by 2020 still seems beyond reach. India has refrained from taking China to the dispute resolution mechanism at the WTO for dumping despite frequent instances which the Indian government has consciously decided to ignore.

But its goal of producing 1.5 billion tonnes of coal is a huge, three-fold increase over 2014 levels, which negates the push for renewables. The National Electricity Plan has capped setting up new coal power generation capacity at the 50 GW already in construction. To add to the confusion, India continues to import nearly 200 million tonnes of coal annually, and several power plants, including large ultra-mega power plants, have been designed for imported varieties of coal.

Table 15.1 shows renewable capacity rising quickly. The biggest shift is in solar capacity, which has jumped from 5.5 to 14.7 GW, so that now 4.4% of the total capacity of the country is solar, while wind power has grown from 9% to nearly 10% of the country's total generation capacity. The entire growth in renewables has happened in the private sector, which shows the shifting nature of the momentum of India's growth, from public sector capital to private capital investment.

Parliament was informed by the Minister of State responsible for power, R.K. Singh, in November 2017 that India has crossed 62 GW of renewable energy capacity. The Central Electricity Authority (CEA) produced a load-balancing report that claimed that India would be a power surplus nation in 2017–18, and also that the target of 175 GW of renewable energy would be exceeded by 2022[61].

---

61 https://energy.economictimes.indiatimes.com/news/renewable/india-can-reach-17k-mw-renewable-energy-capacity-by-2022-government/62339920

Table 15.1 An analysis of capacities for power generation in India

Power generation capacity in India	Fossil fuels			Non-fossil		Renewables – up to 30 September 2017			
Owned by	Coal	Gas	Diesel	Nuclr	Hydro	S.Hdro	Wind	Bio	Solar
State govt Oct 2017	63.8	7.0	0.4	0	29.8	1	0	2	0
Cntr govt Oct 2017	55.2	7.5	0	6.8	11.7	0	0	0	0
Private Oct 2017	74.5	10.6	0.5	0	3.3	3.4	32.7	7.3	14.7
Total Oct 2017 Breakup of 331 GW	193.4	25.1	0.8	6.8	44.8	4.4	32.7	9.3	14.7%
Percent of Total Oct 2017	**58.4%**	**7.6%**	**0.2%**	**2%**	**13.5%**	**1.4%**	**9.9%**	**3%**	**4.4%**
Total Mar 2016 Breakup of 288 GW	175	25	0.8	5	42	4	25	5	5.5
Percent of Total Mar 2016	**60.5%**	**9%**	**0.3%**	**2%**	**14.5%**	**1.5%**	**9%**	**2%**	**2%**

Source: Central Electricity Authority, Govt. of India and Ministry of New and Renewable Energy, Govt. of India websites, Feb 2016, updated Oct 2017. Renewables data is as obtained by CEA from MNRE and relates to 30 Sept 2017 and is quoted by the CEA in their Executive Summary for 21 Oct 2017 and 31 Mar 2016

www.cea.nic.in/reports/monthly/executivesummary/2017/exe_summary-10.pdf

This rise in capacities has been accompanied by a slowdown in power demand. Thermal power plants in the country have been running at 56–58% of their installed capacities. The CEA concluded that India is close to achieving self-sufficiency and the growth in generation capacities should now be tapered off. The National Electricity Plan is based on the CEA's forecasts of economic growth and this has led to stringent curbs on new coal power plants. The plan decides not to allow thermal power plants to come up beyond the 50 GW capacity currently under construction.

Urban households have surpassed the power consumption of industry for the first time, and this shows rampant urbanization and a weak industrial demand. The dip in industrial demand could be attributed to the slowdown in the growth of the economy, challenged by the twin impacts of demonetization and imposition of the Goods and Services Tax in 2016–17.

Several coal power plants came online in the last decade and there has been a glut in the offtake of power. Spot power prices in power exchanges have stayed low all year, except during summer peaks, and several plants who do not have long-term power-purchase agreements with utilities, or whose captive coal blocks were cancelled in 2014, have become financially unviable, their loans non-performing assets, and many have been sold in the recent past.

Considering that India has to do a second 'global stocktake' in 2028 under the Paris Agreement to reveal new climate change initiatives, the over-ambitious policy recommendations and the heavy reliance on depleting fossil fuels are worrisome.

Actual generation compared to capacity

Actual generation data offers a very different picture to the capacities (Table 15.2).

All renewables together account for no more than 8% of the total power generated in India, which is an improvement from the past year, when it was merely 6.5%.

The share of thermal power or all fossil fuels together has shown a negligible decline from 78.29% in 2016 to 77.93% in 2017. Nuclear energy and small hydroelectric have also fallen.

It is still a long way to go for the 60% committed in India's NDCs.

Actual power output from renewable plants is beset with fluctuations and problems of grid management. Plant load factors are hampered by dust, wind, clouds and vagaries of the weather. The National Electricity Plan forecasts a 20% capacity utilization for renewable plants and a 70% utilisation for fossil fuel plants. At these levels there is a nice balance, but these levels are far from being achieved, and India could grow faster than the 6.5% growth rate envisaged by the CEA. With a stronger demand pull at any time, it would certainly be possible for fossil fuel plants, with available excess capacity, to ramp up their generation. The planned trebling of coal output could certainly provide for a substantial share of fossil fuels in India's total energy mix. The plan therefore seems to carry a latent leeway to accommodate growth in demand if it occurs beyond the CEA's forecasts.

Table 15.2 Actual generation of power in India

Power generation in India 2016/17**	Fossilfuels (Thermal power) Apr - Oct	Non-fossil Apr-Oct		Renewables Apr to Sept figures**			
Billion units	Coal + Gas + Diesel	Nuclr	Hydro	S.Hdro	Wind	Bio	Solar
Apr-Oct 2017 Total 56.3+ 709.43= 765.73 BU**	596.78 77.93%	20.19 2.63%	92.47 12.76%	4.58 0.60%	37.06 4.84%	3.81 0.50%	10.80 1.41%
Apr-Oct 2016 Total 47.26+ 679.39 = 726.65 BU**	568.89 78.29%	22.14 3.04%	88.35 12.16%	5.01 0.69%	31.60 4.35%	4.83 0.66%	5.73 0.79%

Renewables figures are from April–Sept 2016 and Apr–Sept 2017 ** hence these are not addable to. Thermal figures are for April–Oct 2016 and Apr–Oct 2017; this has been done merely for comparison purposes, as data for Oct 2017 is awaited by the CEA from MNRE and will be available in early Dec

India generates 1.16 billion units or KWh of power. The CEA has revealed a deficit of only 0.7% and a peak load of about 164 GW. With most towns, cities and villages not receiving uninterrupted power and suffering frequent outages, overall losses in transmission and distribution aggregating more than 20%, and distribution companies in severe losses despite high retail power prices, this provides a deficient picture, and perhaps demand backed without purchasing power needs to be studied to arrive at the real needs.

Further, there is unrevealed demand in those households which still use wood, animal dung and charcoal fires to cook. India's per capita consumption is amongst the lowest in the world and self-sufficiency in power would certainly need a measure different from that of the Central Electricity Authority.

With a Plant Load Factor of only 56.5%, the plan envisages that 175 GW of renewable energy would constitute 20% of India's total installed capacity in 2022 but, keeping the lower plant load factors of renewables in mind, only about 7% of its power generation[62].

62 www.cea.nic.in/reports/monthly/executivesummary/2017/exe_summary-10.pdf

The coal story

Coal continues its growth trajectory and contributes 58% of India's power capacity and about 68% of its actual power consumption. Benjamin Sporton of the World Coal Association, writing for *The Financial Times*, concluded that the trajectory of emissions given by a doubling of India's coal consumption could be contained by better emission control with the aid of multinational partners and technology alliances[63]. The International Energy Agency's Clean Coal Centre concurs[64]. The similarity with the Trump administration's views is unmistakable, but it necessary to verify the antecedents and drivers of arguments pushed forward by those who depend on coal for their livelihood. India will need huge investments it cannot afford at this stage to clean up its emissions from burning coal, and in all likelihood doubling of coal output will result in a substantial growth in emissions.

The policy does not say what would be the fate of new allottees of coal mines who have bid aggressively and won rights to mine coal for captive power generation. What would they do with their coal if they cant set up power plants to generate power with it?

The anomaly is that India will need only 741 million tonnes of coal in 2022 and 876 million tonnes in 2027. But the Ministry of Coal continues to push its ambitious targets to raise coal production to 1.5 billion tonnes by 2020, of which 500 million are planned to be produced by private coal mines and about 1,000 million tonnes by the public sector.

Generation of power is licence-free under the Electricity Act of 2003, so all that private mine allottees need is a connection to the grid. Since the grid is state-owned, this gives leverage to the central government to defer or delay connections, and it leaves the industry with a tacit understanding that, while investments in coal production are encouraged, investments in coal-power generation are discouraged.

India's coal-mining growth is not commensurate with India's projected coal power plants' growth. The first steps to tackle this anomaly seem to have been taken, as Coal India has come up with a Coal Vision 2030 document which attempts to tackle the issue of whether there is a need for producing so much coal. This could lead to a revision in the 1.5 billion tonne target of 2020, which would be most desirable.

The emergence of the private sector in power generation is nascent, as it was allowed entry only after India liberalised barely two decades ago. In the past three years, with slow industrial growth, many are faced with reduced demand for their power and reduced offtakes, especially in the spot markets.

---

63 www.ft.com/content/be41db14-93e3-11e7-a9e6-11d2f0ebb7f0
64 www.iea-coal.org/ and www.iea-coal.org.uk/site/2010/news-section/news-items/dont-write-off-coal-amid-indias-solar-boom (accessed 17 October 2017)

Inadequacies in the National Energy Policy and Plan

The transmission and distribution sectors lose 21.5 % of the power they buy due to Aggregate Technical and Commercial Losses. The UDAY (Ujjwal Discom Assurance Yojana) scheme of 2015, which was launched even before the National Electricity Plan of 2016 was drafted, has been subscribed to by all state governments except West Bengal, Nagaland and Odisha, but utilities have not yet shown signs of combating the systemic problem of loss of power, and reducing their aggregate technical and commercial (AT&C) losses, which is really a misnomer camouflaging the tolerance of electricity theft. The National Electricity Plan presumes that reductions in technical and commercial losses and feeder segregation, etc., committed by states would be achieved.

The net effect of the UDAY scheme so far is a rise in indebtedness of the state governments, as they have taken over 75% of their distribution companies' debts, and funded them with bonds, thus passing a large part of their $68 billion accumulated losses to the public. This is not the first time that such debts have been written off – they were wiped clean in 2003 under the Electricity Act which enabled the creation of distribution, transmission and generation companies by unbundling erstwhile state-owned electricity boards. Participating states have committed to reduce AT&C losses, to raise tariffs, control further financial losses, and take over up to 50% of future financial losses. But actual reduction of AT&C losses is still very little, as the national figure has come down just a bit from about 24% to about 21.5%, and to presume that targets would be met may prove to be erroneous.

The National Energy Policy fails to highlight the gradual substitution of internal combustion engines with electric vehicles, which would possibly amount to the withering of use of such engines by 2030 or even earlier, and complete transition to electric vehicles by 2040. The stalled Electric Vehicle Plan offered by the Ministry of Shipping, Road Transport and Highways was another piecemeal attempt that later had to be withdrawn.

Several European nations have announced their missions to go for 100% electric vehicles in the next two decades. This transformation in the automobile sector could be accompanied by grid-level and consumer-level electricity storage at homes, offices and industrial establishments. Storage and electric vehicles are cursorily mentioned, but the policy does not create a fresh impetus in this crucial area.

The policy does not address the issue of import tariffs on solar panels when dumping comes to light. India had excused Chinese solar manufacturers in 2014 and decided not to levy countervailing duties in order to make the renewable transition easy. But dumping has come to light again, and while the US has levied duties, India has not.

The policy suggests that its implementation would be monitored by a committee of secretaries chaired by the Chief Executive Officer of Niti Aayog, and the process supervised by a steering committee comprising members of the cabinet headed by the Prime Minister. This seems inadequate, as there is no institutional platform for mediating the complex web of vested interests of stakeholders engaged with different aspects of the energy sector. In fact, the lobby of private coal miners who suffered cancellations of their earlier coal blocks and had to bid for fresh allotments by e-auctions under orders of the Supreme Court, has had a nervously anxious role to play. Their coveted newly allotted coal mines will be rendered useless if India switches off its reliance on coal.

Nuclear power, which has its own constraints, is under a different and completely secretive ministry. Inadequacies of the Russian technology at Kundankulam, membership of the nuclear club, etc., are not addressed. The nation has suffered a 40% dip in its nuclear power production since August 2017, and the reasons have not been disclosed.

The National Electricity Plan, which is mandated every five years by the Electricity Act, has traditionally covers only power generation, transmission and distribution, and could not have much of a say in the details of fuel, oil and gas, for vehicles, ships and railways, etc. But the National Energy Policy could have considered all related sectors, with a focus on the environment.

The policy needs to be overhauled considerably by environment, energy and mobility sector experts, environmentalists and strategists to examine the paradigm shifts occurring in storage and electric vehicles as well as data analytics at the cusp of electric and electronic or internet of things, to promote new technologies in renewable resources such as smart grids and smart homes, battery storage, concentrated solar heat and power, etc., all of which have deep impacts on the environment.

Conclusion: comparing the US and the Indian stance

Both the US and India are led by conservative leaders who support business and value enterprise, but resist transformative change in policies except if it is to conserve existing industries. Both have their ear to the ground for votes and excel in winning popular mandates.

Both the US and India favour coal, but the difference is that the US says so blatantly and India does so behind wraps. Emissions from coal plants fuel climate change along with posing serious threats to public health and the environment.

While the US has clearly stated its withdrawal from the Paris Agreement, India continues to state publicly that it stands committed to do what it takes to meet its commitments, but keeps its escape route open. The US openly blames India and China for seeking American money to make their climate transitions, but India does not blame the US for pulling out of contributing to the $100 billion promised fund.

Both India and the US mandated stringent restrictions on power plant emissions in 2015. In both countries these have been resisted by coal fired power plants as they add substantially to costs. While in the US these have been repealed, in India these have been partially stalled.

The Indian guidelines were issued in Dec 2015 by the Ministry of Environment and Forests (MoEF) under the Environment protection Act of 1986, with a two year period for implementation, and are now in force, except that the Central Electricity Authority which is under the Ministry of Power issued new guidelines for Flue Gas DeSulphurisation with a staggered implementation periods of 2019–20 to 2024. This regulation is more strict, but 294 coal fired power plants with an installed capacity of 122 GW of have benefited from the extended 2024 timeline, and only four plants have been imposed with a deadline of 2019.

The industry was clamouring for extension, the ministry of power supported their case, but the MoEF did not relent. In fact, the National Green Tribunal had mandated to the MoEF in an order that curbs be put on emissions by 2017. Eventually, the Central Electricity Authority's statutory power granted to it by the Electricity Act of 2003 was used to stall the rules. While the environment guidelines only curtailed the amounts of emissions, the CEA rules stipulate a certain technology to reduce sulphur content in emissions. This was the most contentious of the environment guidelines, and the most expensive, others regulating carbon particulate matter and nitrogen oxides continue to be in place and the MoEF has publicly announced that it would consider extensions on a case to case basis. This obviously is a roundabout method of using the bureaucracy to stall unpleasant rules. Litigation to halt industry-unfriendly policies has taken the shape of an art-form in the US, but not yet in India, where lobbying still does the job.

To conclude, both nations don't like to impose curbs on pollution but the difference is that while India is a budding power, building its image diplomatically in the world arena, the US leads the world, excels in its economic prowess, and is there fore in disdain of whether others like its stance or not.

16 Cities and the Paris Agreement

Kelsey Coolidge

Introduction

'I was elected to represent the citizens of Pittsburgh, not Paris.' The now-famous quote by US President Donald Trump upon announcing his administration's decision to remove the United States from the Paris Agreement, or the Paris climate accord, was immediately rebutted by no other than the mayor of Pittsburgh himself. In an official tweet, Mayor Bill Peduto wrote, 'As the Mayor of Pittsburgh, I can assure you that we will follow the guidelines of the Paris Agreement for our people, our economy & future[1].' The Pittsburgh mayor is not alone. As of September 2017, more than 350 US cities and municipalities have pledged to uphold the Paris Agreement. And while this phenomenon might not immediately pique the interest of scholars of global governance and international law, this chapter will demonstrate the ways in which cities are becoming significant actors in international law worthy of academic attention.

Some of the cities that pledged to uphold the Paris Agreement are also active participants in other international networks of 'global cities' dedicated to mitigating climate change. These include the C40 Cities Climate Leadership Group and ICLEI, or Local Governments for Sustainability. Beyond pithy tweets and contentious domestic politics, the involvement US cities had in these international networks long before the Trump presidency suggests that these responses are not necessarily motivated by disdain for the president but rather by a long-term policy priority. Trump's announcement provided a political opportunity for cities already engaged in addressing climate change to unite under a unique domestic platform.

While these international networks include cities around the world, the role of US cities in these networks, and more broadly in international agreements on climate change mitigation, is of particular interest given the capricious responses of the United States to international efforts to combat climate change. Trump's announcement follows the Bush administration's retreat from the Kyoto Protocol

---

1 Aleem, Z. (2017), "Trump: I Was Elected to Represent Pittsburgh, Not Paris. Pittsburgh: Uh, We're with Paris", *Vox*, June 1; www.vox.com/policy-and-politics/2017/6/1/15726656/pittsburgh-mayor-trump-paris

in 2001. Arguably more so than any other comparable country in the world, the US has consistently failed to sign on to or uphold international climate change treaties. Regardless of the US government's stance, and perhaps directly in spite of it, many US cities have actively engaged in treaties related to mitigating climate change.

This engagement presents a compelling and relatively new question for scholars of global governance and international law: what role do or should sub-state actors play in international agreements on climate change, like the Paris Agreement? The Paris Agreement is a unique international agreement in that it is built of voluntary commitments by member states. In contrast to previous international agreements, the Paris Agreement did not prescribe specific emissions targets or timetables countries need to abide by. Theoretically, agreements like this could create the opportunity for sub-state actors to submit specific plans to an agreement, even if a sub-state's own national government refuses to do so.

Yet there are undoubtedly serious implications to involving cities in international agreements, bringing forth an important question: what are these potential implications for sovereignty and international treaty-making? The international system is based on the concept of state sovereignty. Directly undermining the state by working with sub-state actors creates the potential for conflict and, worst case, defection from existing treaties. Broadly, sub-state involvement in international treaties risks hollowing out state governments and reducing their ability to govern across multiple issues. This proves problematic as it could make the international system more chaotic and anarchic, to borrow from a traditionalist realism theory of international relations.

To explore these questions, this chapter examines the role of US cities in international agreements to mitigate climate change. It focuses on US cities because of the unreliable position of the US government on international agreements on climate change, and because of the active participation of US cities in the efforts of various international networks working on climate change mitigation. The first section reviews the existing (albeit limited) scholarship on this question, examining the history of cities' actions on climate change mitigation and discussing the definition and qualifications of a 'global city' and the potential legal and political effects of city involvement. The second section delves into greater detail on the existing international networks for cities and their efforts to mitigate climate change, focusing on the C40 network and ICLEI. These networks were chosen because of their requirements for membership as well as the near-ubiquitous discussion of these networks in the existing literature. This will introduce the US cities featured as case studies; these cities include New York, Philadelphia, Los Angeles and Washington DC. These cities were chosen because of their involvement in both of the existing networks, their qualifying as global cities, and information on their climate change mitigation plans being available. The third section details the specific programmes and policies in each of the selected cities specific to their involvement in these networks and their pledges to uphold the Paris Agreement. The fourth and final section explores the potential implications for state sovereignty and international law.

Literature review

The political climate in the US at the time of its announcement that it would pull out of the Paris Agreement notwithstanding, the role of cities in climate change and their individual decisions to uphold the agreement presents an unresolved question spanning the literature of political science, international relations, public policy and urban studies. Existing literature on the role of cities in international law is sparse. Most is found in legal studies journals or law reviews, but few references appear in political science or international relations literature. This is unsurprising, as much of that literature tends to emphasize the unitary nature of the state and power relations between states[2]. It is important to clarify that 'the state' in international relations literature refers to sovereign countries, whereas within US-based political science or public policy, 'the state' refers to regional and geographic subdivisions of the country (New Jersey, Virginia, Ohio, etc.). In this chapter, 'the state' refers to the international relations definition: a sovereign country. As such, cities do not stand out as relevant actors in foreign affairs. They can be defined as sub-state actors, i.e., local or municipal governments, cities, or regional governments, and are generally assumed to be agents of the national government. However, literature on cities and climate change mitigation exists and can largely be found in urban studies and public policy journals. While this literature discusses in broad terms the various international climate change treaties that cities pledge to, it otherwise under-emphasizes legal or political implications relating to international relations. Rather, this literature tends to analyze the effectiveness of these efforts at mitigating climate change[3], the partnerships between city governments and other stakeholders[4] and the public policy limitations of operating at the city level[5].

The global city and transnational municipal networks

Why do cities matter? Cities are important as the physical space where global governance is performed. They have a convening power with their ease of access to travel, easy-to-use public transportation, translation services, world-class health and sanitation services and cultural attractions. Cities with these features are

---

2 Morgenthau, H.J., Thompson, K.W. and Clinton, D. (2005), *Politics Among Nations*, 7th edition, Boston: McGraw-Hill Education
3 Betsill, M.M. (2001), "Mitigating Climate Change in US Cities: Opportunities and Obstacles", *Local Environment*, 6 (4), pp. 393–406, doi.org/10.1080/13549830120091699
4 Bromley-Trujillo, R., Butler, J.S., Poe, J. and Davis, W. (2016), "The Spreading of Innovation: State Adoptions of Energy and Climate Change Policy", *Review of Policy Research*, 33 (5), pp. 544–65, doi.org/10.1111/ropr.12189; Bulkeley, H. and Castán Broto, V. (2013), "Government by Experiment? Global Cities and the Governing of Climate Change", *Transactions of the Institute of British Geographers*, 38 (3), pp. 361–75, doi.org/10.1111/j.1475-5661.2012.00535.x
5 Chawla, A. (2017), "Climate-Induced Migration and Instability: The Role of City Governments", OEF Research, June 6, http://oefresearch.org/publications/climate-induced-migration-and-instability-role-city-governments

highly globalized in almost all social, political, and economic spaces. They provide some sense of familiarity for travellers from around the world. In the world's largest cities, travellers can find their culture's food replicated, or an expatriate community that shares the same cultural identifiers. Cities are the pinnacle of globalization and the best example of what a global culture is. It follows that they bear some significance to global politics. This will become increasingly true as the world becomes more urbanized and as cities host a growing majority of the world's population[6]. The academic literature that has used the city as a level of analysis has converged on a few key themes, namely a discussion of the 'global city'.

One of the earlier scholars of the term defined a global city as a centre to the 'globally reconstructed international production system'[7]. Sassen provides a more detailed definition outlining two necessary characteristics; global cities are 'sites for (1) the production of specialized services needed by complex organizations for running a spatially dispersed network of factories, offices, and service outlets; and (2) the production of financial innovations and the making of markets, both central to the internationalization and expansion of the financial industry'[8]. Much of the subsequent literature has emphasized the role of the global city in the international economic division of labour[9]. In seeking a common definition, 'the global city literature has consolidated urban studies and international political economy by analyzing the role of cities in the globalized finance, production, and associated service sectors[10]'. However, others note that global cities are not only defined by their role in the international political economy; many scholars also point to their role in cultural and political diffusion[11].

To a certain extent, all cities are global cities, using this scholarly definition. Distinguishing between a global city and a non-global city is somewhat difficult. For example, Denver, Colorado, a city in the United States that is much smaller than New York or Los Angeles, is headquarters to ten international companies[12]. Its mayor, Michael B. Hancock, led a delegation of Colorado city and state officials to London for a series of meetings designed to increase economic

---

6 "World's Population Increasingly Urban with More than Half Living in Urban Areas", United Nations Department of Economic and Social Affairs, 10 July 2014, world-urbanization-prospects-2014.html
7 Friedmann, J. (1986), "The World City Hypothesis", *Development and Change*, 17 (1), pp. 69–83, doi.org/10.1111/j.1467-7660.1986.tb00231.x
8 Sassen, S. (2006), *Territory, Authority, Rights: From Medieval to Global Assemblages*, Princeton, N.J.: Princeton University Press
9 Lee, T. (2012), "Global Cities and Transnational Climate Change Networks", *Global Environmental Politics*, 13, no. 1, pp. 108–27, doi.org/10.1162/GLEP_a_00156.
10 Lee, "Global Cities and Transnational Climate Change Networks", p. 111
11 Krause, R.M. (2011), "Policy Innovation, Intergovernmental Relations, and the Adoption of Climate Protection Initiatives by USCities", *Journal of Urban Affairs*, 33 (1), pp. 45–60, doi.org/10.1111/j.1467-9906.2010.00510.x
12 Chuang, T. (2016), "Ten Big Colorado Firms Make Fortune 500 List", The *Denver Post* blog, June 6, www.denverpost.com/2016/06/06/arrow-electronics-is-colorados-top-fortune-500-firm-again/

development and tourism between the two cities[13]. All cities, regardless of size, are globalized. A loose category like 'global city' that fails to define a clear outlier may not be a useful category in the first place. That being said, the C40 network provides a better working definition for global cities. This is important to use because it helps to more narrowly define the concept of a global city. The C40 network categorizes cities included in its network as one of three types: a megacity, an innovator city, or an observer city. A megacity is defined by either its population of more than 3 million in the city or 10 million or more in the metropolitan area or by being ranked as one of the top 25 cities in the world for GDP output[14]. An innovator city falls below these requirements but has otherwise 'shown clear leadership in environmental and climate change work[15]'. Observer city is a short-term category for new cities in the network, whether they are applying as a megacity or innovator city[16].

The definition of a megacity set forth by the C40 network is made useful for further defining a global city by attaching measurable characteristics. This is appropriate since much of the literature on global cities emphasizes the economic role global cities play. This scale matters. For example, New York City is the largest city in the country by a significant margin; 1 in every 38 people in the US lives in New York City, it comprises over two-fifths of New York State's entire population, and it has a larger population than 40 of the 50 US states[17]. If global cities can agree on a policy platform, it could have significant ripple effects. On climate change, cities agree on a policy platform. Specifically, global cities are operating through transnational municipal networks (TMNs) to solidify policy positions, share knowledge and act as a collective force. These networks are unique, defined by autonomous and free membership, and are self-governing, but where 'decisions taken within the network are directly implanted by its members[18]'. This definition is helpful for distinguishing the ways in which global cities are directly involved in global governance outside the traditional confines of global nation-state-based politics. According to some scholars, TMNs are a

---

13 "Mayor Hancock Leads Trade Mission to London as Part of Norwegian's Inaugural Flight from London/Gatwick to Denver", Office of the Mayor of Denver, 13 September 2017, www.denvergov.org/content/denvergov/en/mayors-office/newsroom/2017/mayor-hancock-leads-trade-mission-to-london-as-part-of-norwegian.html

14 Marinello, M. (2012), "C40 Announces New Guidelines for Membership Categories", C40 Cities Climate Leadership Group press release, 3 October, http://c40-production-images.s3.amazonaws.com/press_releases/images/25_C40_20Guidelines_20FINAL_2011.14.12.original.pdf?1388095701

15 Marinello, "C40 Announces New Guidelines"

16 Ibid.

17 "New York City Population: Population Facts", New York City Department of City Planning, accessed 19 November 2017, www1.nyc.gov/site/planning/data-maps/nyc-population/population-facts.page

18 Kern, K. and Bulkeley, H. (2009), "Cities, Europeanization and Multi-Level Governance: Governing Climate Change through Transnational Municipal Networks", *Journal of Common Market Studies*, 47 (2), pp. 309–32, doi.org/10.1111/j.1468-5965.2008.00806.x

'widely' neglected form of informal governance that cities have participated in for decades, but that related research and policy have failed to incorporate[19].

The C40 network and ICLEI are two examples of TMNs. Because the C40 network is selective about what cities can join, it creates an exclusive category with easily identifiable outliers. This approach is helpful in thinking about how to confine the definition of a global city. However, other TMNs are not always exclusive. For example, ICLEI places few limits on the size of the city that can join its network, but insists its membership remain actively in support of climate change mitigation. Voluntary membership to TMNs and implementation of the various projects or policies circulated through these networks is one of the key characteristics of the global city. TMNs play an emerging role in what Boudreau has described as the 'the centrality of urban politics in a global era[20]'.

Boudreau argues that urban politics has become important to global politics because of four factors: '(1) decentralization and increased intergovernmental relations; (2) conventionally municipal policy interests moving to the national and global scales and conventionally national and global policy interests moving to the local scale; (3) the re-scaling of civil society activities; and (4) the continued territorialization of the policy-making process[21]'. These suggest that the world's global cities are executing their relative power, namely through population size and share of GDP, to engage directly with global politics. In this respect, many scholars discuss how global cities are 'norm entrepreneurs'[22]. Toly cites Finnemore and Sikkink in describing norm entrepreneurs as those who 'construct and mobilize support for "particular standards of appropriateness" and convince states, potential norm leaders, to adopt these standards'.[23] By employing their relative power and participation in TMNs, global cities are pushing the envelope in global politics, particularly in the area of climate change mitigation. Toly argues that the presence of city-based TMNs are alone evidence of the ways cities are norm entrepreneurs in climate change politics[24]. Kern and Bulkeley describe this as the 'Europeanization' of cities, wherein cities developed shared policy platforms in the same fashion that the European Union's policies have trumped the local policies of its members[25]. Cities do this for other policy areas as well, like the women's rights policy legislation implemented in Davao City, Philippines, with support from local civil society organizations[26]. Furthermore, many of

---

19 Kern and Bulkeley, "Cities, Europeanization, and Multi-Level Governance"
20 Boudreau, J.-A. (2007), "The Centrality of Urban Politics in a Global Era", paper presented at the annual meeting of the American Political Science Association, Chicago, 31 August–1 September
21 Boudreau, "The Centrality of Urban Politics in a Global Era"
22 Toly, N.J. (2008), "Transnational Municipal Networks in Climate Politics: From Global Governance to Global Politics", *Globalizations*, 5 (3), pp. 341–56, doi.org/10.1080/14747730802252479.
23 Ibid.; see also Finnemore, M. and Sikkink, K. (1998), "International Norm Dynamics and Political Change", *International Organization*, 52 (4), pp. 887–917.
24 Toly, "Transnational Municipal Networks in Climate Politics"
25 Kern and Bulkeley (note 18), "Cities, Europeanization and Multi-Level Governance"
26 Coolidge, K. (2017), "Advancing Women's Rights in Davao City, Philippines: The Role of Local Civil Society", OEF Research, June 7, http://oefresearch.org/publications/advancing-womens-rights-davao-city-philippines-role-local-civil-society

the potential solutions to climate change, specifically reducing greenhouse gas (GHG) emissions, are currently under way in cities. These programmes include innovative financing options, building requirements and public transportation innovations. Cities are testing the projects and programmes necessary to mitigate climate change and, as a result, are asserting themselves into the political space of global governance.

Cities and international law

But what to make of global cities and international law? There is a vast body of literature exploring when, why, and under what conditions states comply with international law[27]. Some scholars argue that most countries abide by most international laws most of the time[28]. In general, international law aims to regulate the behaviour of states with the expectation that sub-state actors fall in line. Domestic legal systems vary with respect to the separation of powers among different levels of government. In the United States legal system, this is true with an important caveat. The federal system of government provides a division of power between the federal (national) government, state government, and state-to-local governments[29]. In this system, the US Constitution provides that whatever powers are not enumerated as federal powers are powers of the state government, which then decides the power of the local government. Local government had no formal status in the US Constitution until the Dillon's Rule of 1903, which explicitly declared that local governments are 'mere political subdivisions of the state for the purpose of exercising a part of its power.[30]'. Additionally, the United States Supreme Court ruled in 1937 that where state or local law conflicts with national law, national law is supreme[31]. Yet, this only matters insomuch as the federal government decides to enforce its power. When it comes to cities passing international treaties as local ordinances, the federal government has largely ignored them. Cities making pledges to the Paris Agreement with no federal government response might be setting a new precedent.

Regardless of the legal and political space that cities inhabit, many US cities have asserted themselves in international law by pledging to uphold international climate change treaties in open defiance of the national government's position. This appears to be based on decades of local action on climate change mitigation beginning in the 1980s, but activity rapidly increased following the Kyoto Protocol in the mid-2000s[32]. Following the George W. Bush administration's

---

27 Hongju Koh, H. (1997), "Why Do Nations Obey International Law?", *The Yale Law Journal*, 106 (8), pp. 2599–2659, doi.org/10.2307/797228
28 Henkin, L. (1979), *How Nations Behave*, 2nd edition, New York: Columbia University Press
29 Schroeder, H. and Bulkeley, H. (2009), "Global Cities and the Governance of Climate Change: What is the Role of Law in Cities?", *Fordham Urban Law Journal*, 36 (2), p. 313
30 Schroeder and Bulkeley (note 29), "Global Cities and the Governance of Climate Change", p. 325.
31 Schroeder and Bulkeley (note 29), "Global Cities and the Governance of Climate Change"
32 Ibid.

decision to withdraw from the Kyoto Protocol in 2001, several US states reacted, including California, which pledged to uphold the protocol's emissions standards[33]. US cities got involved with the Kyoto Protocol in the mid-2000s when members of the United States Conference of Mayors pledged to 'meet or beat the Kyoto Protocol targets in their own communities[34]'. Specifically, the US Conference of Mayors urged its members to take three actions: 1) to meet or beat targets in the Kyoto Protocol in their communities, 2) urge state and federal government to pass policies and programmes intended to meet or beat targets in the Kyoto Protocol; and (3) urge the US Congress to pass climate change mitigation legislation and establish a national emissions trading system[35].

Yet international treaties also do not recognize cities as being party to treaties. From the perspectives of both national government and international law, cities have no role to play in international law. Traditionally, the international effort to combat global climate change is confined to the international level, where countries are pressured to comply with comprehensive international agreements. While these international agreements aim to get countries to agree on shared goals, they also implicitly reinforce the state-based system of international relations where state sovereignty is upheld. Reinforcing state sovereignty often poses a challenge to addressing global issues. However, Collier notes that city governments play an active role in the 'multi-policy framework' focused on climate change mitigation policies in the European Union[36]. Within the unique institutional set-up of the European Union, cities emerge as significant actors.

Climate change is a quintessential challenge of global governance and the by-product of the industrialization that fed globalization. Yet, as Sassen notes, globalization is characterized by 'the emergence of conditions that weaken the exclusive authority of national states and thereby facilitate the ascendance of sub- and transnational spaces and actors in politico-civic processes once confined to the national scale[37]'. In this space, cities are emerging as important players in multi-scalar governance – meaning at all levels of governance – on climate change mitigation[38]. Thinking of climate change mitigation through the lens of multi-scalar governance is common in the existing literature, with some arguing that this is 'not only conceptually necessary but also an empirical reality[39]'. As such, climate change mitigation is a unique policy area because it requires a multi-scalar response. Governments at all levels can pass and implement policies that decrease GHG emissions.

33 Ibid.
34 Ibid.
35 Ibid.
36 Collier, U. (1997), "Local Authorities and Climate Protection in the European Union: Putting Subsidiarity into Practice?", *Local Environment*, 2, no. 1, pp. 39–57, doi.org/10.1080/13549839708725511
37 Drainville, A.C. and Sassen, S. (2004), *Contesting Globalization: Space and Place in the World Economy*, London; New York: Routledge
38 Bulkeley, H. and Moser, S.C. (2007), "Responding to Climate Change: Governance and Social Action beyond Kyoto", *Global Environmental Politics*, 7 (2), pp. 1–10
39 Bulkeley and Moser, "Responding to Climate Change."

Cities have participated directly in international talks on climate change mitigation, even if they do not formally hold legal status in international law. Both the C40 network and ICLEI detail their participation in two particularly prominent international conferences on climate change and sustainability: the United Nations Framework Convention on Climate Change Conference of Parties (COP)[40] and the United Nations Conference on Housing and Sustainable Urban Development (Habitat)[41]. These networks report direct linkages between their campaigns and sections of the Paris Agreement that highlight local and municipal governments. Article 7, section 2 of the Paris Agreement[42] specifically calls on parties to the agreement:

> Parties recognize that adaptation is a global challenge faced by all with local, subnational, national, regional and international dimensions, and that it is a key component of and makes a contribution to the long-term global response to climate change to protect people, livelihoods and ecosystems, taking into account the urgent and immediate needs of those developing country Parties that are particularly vulnerable to the adverse effects of climate change.

Article 7, section 5[43] continues:

> Parties acknowledge that adaptation action should follow a country-driven, gender-responsive, participatory and fully transparent approach, taking into consideration vulnerable groups, communities and ecosystems, and should be based on and guided by the best available science and, as appropriate, traditional knowledge, knowledge of indigenous peoples and local knowledge systems, with a view to integrating adaptation into relevant socio-economic and environmental policies and actions, where appropriate.

The Paris Agreement calls on states to recognize the relevant work by sub-state actors, like cities, but does not treat cities as entities that can uphold the Paris Agreement. By making public announcements and pledges to uphold the Paris Agreement, US cities have directly defied the US president's decision to remove the US from the Paris Agreement. Yet, even under the Bush administration and the Kyoto Protocol, the federal government has decided not to legally pursue cities that openly defy it. This might be due to the global city's size and influence

---

40 The United Nations Framework Convention on Climate Change is an international environmental treaty adopted in May 1992 with the stated objective of reducing greenhouse gas emissions. Since 1995, the parties to the convention have met regularly through the Conference of Parties.
41 The United Nations Conference on Housing and Sustainable Urban Development, or Habitat, has met in three iterations (Habitat I in 1976, Habitat II in 1996, and Habitat III in 2016) to address global, sustainable urbanization and to implement "The New Urban Agenda"
42 United Nations Framework Convention on Climate Change, "Paris Agreement", 21st Conference of the Parties, 2015 (Paris: United Nations) available at https://unfccc.int/files/meetings/paris_nov_2015/application/pdf/paris_agreement_english_.pdf
43 Ibid.

in domestic politics, which in turn means that the governance of city affairs is increasingly concerned with global affairs. It may also be a fluke of the US federal system, where US states have a lot more power over cities than the federal government does, *per se*.

Scholars of global governance, international politics and international law would do well to consider more seriously the role of cities in this space. The fact that US cities are independently doing more than the federal government to combat climate change presents a compelling dilemma for international law and the treaty-making process, especially if more US cities submit their plans to the Paris Agreement. US cities are voluntarily abiding by the Paris Agreement even as the federal government fails to do so. This leads to a few questions: what are US cities doing to uphold the Paris Agreement? What impact will it have for the US as a whole on meeting the Paris commitments? What does that mean for the international system?

The answers can be gauged based on the specific activities and policies cities are implementing. The following sections delve into greater detail about the ways US global cities have participated in TMNs dedicated to mitigating climate change, particularly in the C40 network and ICLEI. The cities selected for this chapter are participants in both networks and have each made a pledge to uphold the Paris Agreement. Most of the information about cities' plans for climate change mitigation was sourced from reports published by the C40 network and ICLEI, reports published by municipal governments, and media sources. Also included is a review of the C40 network and ICLEI and a more detailed explanation of which US global cities are included.

Local Governments for Sustainability

Local Governments for Sustainability, or ICLEI, was founded in New York City in 1990 by 200 municipal governments. According to the organization's website, its membership now includes over 1,500 cities from around the world. Roughly 83 US cities that are members of ICLEI have pledged to independently uphold the Paris Agreement[44]. The organization's governance structure is set up democratically and comprises the ICLEI Council, Global Executive Committee, Regional Executive Committee and Management Committee. The ICLEI Council works as the top decision-making and oversight body composed of all voting members of the regional committees, with power to amend the organization's charter and elect members to its Executive Committee. The Global Executive Council represents ICLEI at the global level. The Regional Executive Committees are based in eight defined regions of the world (North America, Latin America and the Caribbean, East Asia, South Asia, Southeast Asia, Africa, Europe, and Oceania) and is composed of three to five members from each, voted in by regional members. The Management Committee oversees ICLEI,

---

44 The author compiled a list of US cities that were members of ICLEI as of September 2017

Cities and the Paris Agreement 273

is appointed by the Executive Committee, and serves as the formal arbitration body[45].

All of ICLEI's programming is based on its ten Urban Agendas, which were constructed to help cities and local governments address sustainability challenges. These ten agendas include the Sustainable City Agenda, Low-Carbon City Agenda, Resource-Efficient and Productive City Agenda, Resilient City Agenda, BiodiverCity Agenda, Smart City Agenda, EcoMobile City Agenda, Sustainable Local Economy and Procurement Agenda, Sustainable City-Region Cooperation Agenda, and the Happy, Healthy, and Inclusive Communities Agenda. Each of these agendas has corresponding programmes, tools and services. These agendas also encompass objectives of other international environmental frameworks like the UN's Sustainable Development Goals and the Paris Agreement. For example, the Low-Carbon City Agenda has three corresponding programmes: the Green Climate Cities programme, the Urban Low Emission Development Strategies and the Energy-Safe Cities Initiative. The tools and services provided in this agenda provide cities with 'standards and guidelines for accounting and reporting greenhouse gas emissions, and planning for future local climate actions.[46]'.

The C40 network

The C40 network of cities was formally established in 2006 following an initiative launched by former London mayor Ken Livingstone. As mentioned, membership in this network is restricted to the world's largest cities as measured by population or GDP share, but also includes smaller cities that have made significant contributions to climate change mitigation. According to its website, C40 membership includes 91 cities representing 25% of global GDP and 1 in 12 people worldwide[47]. There are ten US cities that are members of the C40 network; Los Angeles, San Francisco, Chicago, New York, Philadelphia, Houston, Seattle and Washington DC are megacities, and New Orleans, Portland and Austin are innovator cities in the C40 network. Boston is home to the network's steering committee. Its governance is organized by a chair of the C40 network, a board of directors and a steering committee. The chair is a current mayor of a member city and is elected to a three-year term. The board of directors oversees the management and day-to-day activities of the network, and the current president of its board is former C40 chair and former mayor of New York Michael Bloomberg. The steering committee provides strategic direction and governance for the C40 network and rotates among C40 mayors.

The C40 provides issue-specific networks and programmes to its members. Both are targeted towards encouraging peer-to-peer exchanges of information, policy advice and recommendations, and lessons learned. The six current

---

45 "Governance", ICLEI Global, www.iclei.org/about/governance.html (accessed 15 October 2017)
46 "Low-Carbon City", ICLEI Global, www.iclei.org/activities/agendas/low-carbon-city.html (accessed 15 October 2017)
47 "The Power of C40 Cities", C40 Cities, www.c40.org/cities (accessed 15 October 2017)

networks include: adaptation; business, data, and innovation; energy and buildings; transportation; urban planning and development; and waste and water. They are organized around a 'four-pronged approach' to connect, inspire, advise and influence member cities on specific policies and programmes. For example, one of the adaptation networks, the Cool Cities network, 'supports city efforts to reduce the impact of the urban heat-island effect'. It focuses on sharing policies focused on 'UHI data monitoring and measurement, heat health vulnerability, integrating heat into long-term planning, and green and cool solutions[48]'.

The Compact of Mayors

The Compact of Mayors is a joint initiative of the C40 network and ICLEI launched by former UN Secretary-General Ban Ki-moon and supported by UN Habitat[49]. Available to any city or town regardless of size, the compact requires members to detail progress along a three-year programme. The programme is segmented into four phases. The first is called 'Commitment'. In the first phase, cities submit a letter of intent and agree to complete the remaining three phases within a three-year timeline. The second phase is 'Inventory', in which a city must assess the current impact climate change has on it. This includes measuring its overall greenhouse gas (GHG) emissions through provided Compact standardization; identifying current climate risks like flood, drought or temperature extremes; and reporting the results through carbon questionnaires. The third phase is 'Target', where members are required to create GHG mitigation targets. This phase includes identifying and reporting on emissions by source, like buildings or transportation, and setting initial GHG reduction targets. The fourth and final phase is 'Plan', which must be completed in the third year of joining of the Compact. In this phase, a member city must detail how it plans to reduce GHG emissions. All of the cities mentioned here are members of both ICLEI and the C40 network and also members of this programme. Progress on the Compact is detailed in each city's profile and is useful for contextualizing the actual progress that has been made[50]. Through this programme, cities have measured their current GHG emissions and developed plans to reduce those emissions within timelines of between 15 and 30 years.

Methods: city selection

The US global cities selected were determined by overlaying a list of US cities that pledged to uphold the Paris Agreement with US cities involved in the C40

48 "Cool Cities", C40 Cities, www.c40.org/networks/cool_cities (accessed 15 October 2017)
49 "Compact of Mayors", Compact of Mayors, www.compactofmayors.org/ (accessed 20 October 2017)
50 "Compact of Mayors" (note 49)

and ICLEI. This process helped to narrow the list of US cities and to ensure that those included had demonstrated long-term commitment to climate change mitigation. Because US cities' involvement in the Paris Agreement is a relatively new phenomenon, most of the cities featured in this chapter have made little progress in submitting a plan to, or implementing, the Paris Agreement. Thus, the profiles of cities review their involvement in the existing TMNs and, if available, their pledges to the Paris Agreement. Each one details specific policies and programmes that the city has pursued in an effort to reduce their GHG emissions.

New York City

Of the US global cities, New York is the most prominent. It was one of the founding members of the C40 cities network and former mayor Michael Bloomberg served as chair of the network from 2010–2013[51]. It is the largest city in the US in terms of population, at approximately 8.55 million, and by GDP, at $778 billion. In 2015, the city's emissions level was 52.18 million metric tons of CO_2 (MMT CO_2). According to the C40, New York has finished all phases of the Compact of Mayors programme, meaning that it has established a formal plan. In September 2017, the mayor's office released the city's plan for aligning its climate change mitigation plan with the Paris Agreement. In this plan, Mayor Bill de Blasio specifically mentions the Trump administration's intent to leave the Paris Agreement as a factor motivating this scaling up of action on climate change[52].

New York City's plan targets buildings and the transportation sector in order to reduce emissions. It states that 'fossil fuels in buildings for heat and hot water are the biggest source of GHG emissions, accounting for 39 percent of the city-wide total[53]'. As a result, New York has announced a new proposal requiring building-owners to invest more in efficient heating and cooling systems, insulation and water heaters in buildings larger than 25,000 square feet[54]. Other building-targeted plans include ensuring new buildings are energy efficient, providing the capital and financial tools to help build efficient buildings, and retrofitting old buildings to become more energy efficient. In regard to transportation, New York's plan targets private vehicle transportation, which accounts for 90% of its transportation-related GHG emissions[55]. The city has already announced it

---

51 "Chair of the C40", C40 Cities, www.c40.org/leadership (accessed 22 November 2017)
52 "One New York: The Plan for a Strong and Just City", The City of New York and Mayor Bill de Blasio, 22 April 2015, www.nyc.gov/html/onenyc/downloads/pdf/publications/OneNYC.pdf
53 "One New York: The Plan for a Strong and Just City", p. 6
54 Dennis, B. and Epstein, K. (2017), "New York's Buildings Emit Most of Its Greenhouse Gases. The Mayor Has a Plan to Change That", *Washington Post*, 13 September, sec. Energy and Environment, www.washingtonpost.com/news/energy-environment/wp/2017/09/13/new-yorks-buildings-emit-most-of-its-greenhouse-gases-the-mayor-has-a-plan-to-change-that/
55 "One New York: The Plan for a Strong and Just City," (note 52), p. 7

will install 50 new charging hubs for electric vehicles by 2020[56]. If New York can successfully implement these plans, it anticipates a rapid reduction in its GHG emissions from 52.18 MMT CO_2 in 2015 to 12.15 million MMT CO_2 in 2050[57].

Los Angeles

Los Angeles is the second-largest city in the US with a population of approximately 4.01 million and an annual GDP of $860 billion. It has completed the first phase of the Compact of Mayors commitment but has not updated the Compact on its progress on stated climate change mitigation goals. In April 2015, Mayor Eric Garcetti released Los Angeles' Sustainability City Plan[58]. In this plan, emissions levels were listed at 36.2 MMT CO_2 in 1990 and at 29 MMT CO_2 in 2013. The city outlines a GHG reduction target of 45% by 2025. The majority of LA's emissions are from stationary sources, meaning buildings and commercial and industrial sources. The city has prioritized diversifying the electrical grid to incorporate 50% renewable energy, increasing solar energy, reducing transportation emissions through increasing use of electric vehicles and public transit, and improving recycling and organic-waste management[59]. Recently, Mayor Garcetti announced plans to plant 40,000 trees over the next two years, install heat-reflecting roofs on new homes, and cover city streets with a reflective material to reduce the amount of heat trapped by asphalt.[60] The city has installed 1,000 chargers for electric cars and more than half of the city-owned fleet is electric. It has also implemented a pilot programme on an electric car-sharing scheme in low- and middle-income areas of the city[61].

As the largest city in California, Los Angeles benefits from many of the progressive policies and programmes the state has pursued. California pledged to uphold the Kyoto Protocol following the Bush administration's decision to leave, and also created the California Climate Registry in 2011[62], which 'promoted and protected businesses' early action to manage and reduce their greenhouse gas emissions', and has since expanded through the US and Canada[63].

---

56 Goba, K. (2017), "De Blasio Pushes Electric Cars in Brooklyn", *Kings County Politics* blog, 20 September, www.kingscountypolitics.com/de-blasio-pushes-electric-cars-brooklyn/
57 "One New York: The Plan for a Strong and Just City" (note 52)
58 "Los Angeles Climate Action Report: Updated 1990 Baseline and 2013 Emissions Inventory Summary", Office of the Mayor of Los Angeles, www.lamayor.org/sites/g/files/wph446/f/landing_pages/files/pLAn%20Climate%20Action-final-highres.pdf (accessed 27 November 2017)
59 Ibid.
60 Zareva, T. (2017), "Asphalt Roads Exacerbate Climate Change, but Los Angeles Has a Solution", *Big Think*, 13 September, http://bigthink.com/design-for-good/la-is-painting-its-streets-white-to-cool-down-the-city
61 Bourke, I. (2017), "Here are 16 Cities Tackling Inequality through Climate Action Schemes", *CityMetric*, 28 September, www.citymetric.com/horizons/here-are-16-cities-tackling-inequality-through- climate- action-schemes-3362
62 Schroeder and Bulkeley (note 29), "Global Cities and the Governance of Climate Change"
63 "About Us", The Climate Registry, www.theclimateregistry.org/who-we-are/about-us/ (accessed 20 October 2017)

California also has a statewide cap-and-trade programme that was recently extended to 2030[64].

Philadelphia

Philadelphia is the fifth-largest city in the United States by population, with approximately 1.56 million citizens, and an annual GDP of $346 billion. It reports GHG emissions of 19.21 MMT CO_2 as of 2012. According to the C40 network, it has completed all four phases of the Compact of Mayors, targeting buildings, private transportation and community-scale development sectors in its efforts. In September 2017, Mayor Jim Kenney announced the city's new plan for reducing GHG emissions. This plan begins by targeting city-owned buildings, with the philosophy that the city must set an example for private owners to follow. This Municipal Energy Master Plan for the Built Environment is tailored to meet the goals of the Paris Agreement by reducing GHG emissions by 50% by 2030, reducing energy use by 20% by 2030, generating or purchasing all electricity from renewable sources by 2030, and reducing the overall cost of energy[65].

The first key project in this plan is to retrofit the Philadelphia Art Museum, which is reported to be the largest energy consumer among all city-owned buildings. It is anticipated that the costs of the project will be covered by savings from the energy improvements; costs for the improvements are estimated at $9 million, but the city currently spends an average of $3 million a year for climate control and lighting alone in the four-acre building[66]. Second, the Philadelphia Energy Authority is seeking proposals for an RFP for a renewable energy power purchase agreement. Through the agreement, the city will commit to purchasing a large-scale renewable energy utility[67]. However, these recent plans are built off years of previous policies and programming. In 2013, Philadelphia won a $1 million grant from Bloomberg Philanthropies for its Social Enterprise Partnership, which overhauled the city's procurement process to encourage innovation from social entrepreneurs for solutions to the city's environmental problems[68]. That same

---

64 Edelman, A. (2017), "Forget Trump. The US is Storming Ahead on Climate Action Like Never Before", *NBC News*, 17 September, www.nbcnews.com/politics/politics-news/forget-trump-u-s-storms-ahead-climate-change-never-n801826
65 "Municipal Energy Master Plan", City of Philadelphia Office of Sustainability, https://beta.phila.gov/posts/office-of-sustainability/2017-09-26-municipal-energy-master-plan/ (accessed 2 November 2017)
66 Kummer, F. (2017), "Philly Moves to Slash Energy Use, Starting with $9M Retrofit of Art Museum", *Philadelphia Inquirer*, 27 September, www.philly.com/philly/health/philadelphia-releases-ambitious-plan-to-slash-energy-and-greenhouse-gas-use-including-9m-retrofit-of-art-museum-paris-accord-20170927.html
67 "Municipal Energy Master Plan", City of Philadelphia Office of Sustainability (note 65)
68 Reyes, J. (2012), "Philadelphia Social Enterprise Partnership: Bloomberg Philanthropies Names City Incubator a Finalist in $5M Mayors Challenge", *Technical.ly Philly*, 5 November, https://technical.ly/philly/2012/11/05/philadelphia-social-enterprise-partnership-bloomberg-philanthropies-names-city-incubator-a-finalist-in-5m-mayors-challenge/

year, the city passed legislation requiring large commercial buildings to benchmark and report energy and water usage[69]. The city also released a sustainability plan for public transportation, particularly rail and metro lines serving suburbs of Philadelphia[70].

Washington DC

Washington DC is a much smaller city in comparison to New York, LA, and Philadelphia, with a city population of approximately 670,000 and an annual GDP of $122 billion. Nonetheless, the C40 network classifies it as megacity. The city has completed all four phases of the Compact of Mayors programme, identifying buildings, community-scale development and private transportation as its largest sources of emissions. It measured its emissions in 2013 at 7.746 MMT CO_2 and has created a target of reducing emissions to 5.05 MMT CO_2 by 2032. Remarkably, in 2017, DC became the first LEED Platinum-certified city in the world by 'sustainably manag[ing] their energy, water consumption, waste treatment, and public transportation for residents[71]'. Government buildings are run by 100% renewable energy, and over half of all residents bike, walk or take public transportation to work. The city has passed a law requiring all energy suppliers to source half of their energy from renewable sources[72].

The city has also pursued a number of different programmes with support from the C40 and ICLEI. For example, DC has used the Property Assessed Clean Energy (PACE) financing tool[73] from the C40 network to support the rehabilitation of homeless shelters[74]. The city is home to one of the world's largest thermal hydrolysis installations, which supplies one-third of the power to the wastewater facility[75]. It has a public bike-sharing programme, Capital Bikeshare, that has recently entered into areas of the city in need of affordable

---

69 "Energy Efficient Building Gains Ground in Philadelphia", C40 Cities, 28 March 2013, www.c40.org/blog_posts/energy-efficient-building-gains-ground-in-philadelphia
70 "Case Study: SEPTAinable Transit Sustainability Plan", C40 Cities, 10 January 2013, www.c40.org/case_studies/septainable-transit-sustainability-plan
71 Mafi, N. (2017), "Washington DC Becomes First LEED Platinum City in the World", *Architectural Digest*, 15 September, www.architecturaldigest.com/story/washington-dc-becomes-first-leed-platinum-city-in-the-world
72 Bourke, "Here Are 16 Cities Tackling Inequality through Climate Action Schemes" (note 61)
73 The Property Assessed Clean Energy (PACE) tool is a financial tool that facilitates efficiency updates to buildings. PACE will cover 100% of the project's costs and allows property-owners to repay for up to 20 years. Mike, "What Is PACE?," *PACENation* (blog), http://pacenation.us/what-is-pace/
74 "Cities100: Washington, DC.– Green Finance Advances Housing Affordability", C40 Cities Case Study, 15 November 2016, www.c40.org/case_studies/cities100-washington-d-c-green-finance- advances-housing-affordability
75 "Cities100: Washington, DC – World's Largest Thermal Hydrolysis Plant", C40 Cities Case Study, 15 November 2016, www.c40.org/case_studies/cities100-washington-d-c-world-s-largest-thermal- hydrolysis-plant

transportation[76]. Washington DC was also the first city in the US to pass legislation requiring green building certification for public and private buildings. By January 2016, the city had 119 million square feet of LEED-certified buildings and more than 650 LEED-certified projects in the works[77].

Cities, climate change and implications for the international system

This is only a selection of US cities that are tackling climate change; many more are doing similar work. These examples will fail to convey any optimism without some assessment of what the real impact of these policies, if met, could be on overall GHG emissions levels in the US. One way to gauge this is to take the proportion of the selected US cities' GHG emissions level out of the overall US emissions level, comparing the projected decreased GHG emissions levels to the overall US emissions level. For simplicity, this assumes no change in the overall US emissions total. According to recent EPA assessments, the US's GHG emissions level is at 6,587 MMT CO_2 per year and the combined GHG emissions level of the selected cities is 108.136 MMT CO_2. This suggests that New York, Los Angeles, Philadelphia and Washington DC account for roughly 0.16% of all US emissions. If these cities meet their emissions goals, that percentage would decrease to roughly 0.065% (or 42.755 out of 6,587 MMT CO_2).

While an aggregate assessment of a city's emissions goals reveals that it has some impact on overall US emissions levels, it does not help clarify the legal or political ramifications of the city's role in climate change mitigation acting independently from the US federal government. The Paris Agreement calls on state parties to recognize how local actors are making progress towards climate change, and it's clear that US cities are advancing towards their GHG reduction targets. Given their participation over time, US global cities have proven to be far more reliable partners in mitigating global climate change than the US federal government. It is undeniable that sub-state actors have a role to play, but that role has been limited to informal activities with support from traditional international organizations, like the United Nations and the Compact of Mayors programme. What role do sub-state actors play in complying with international agreements?

For international environmental treaties, there are clear drawbacks to recognizing city governments as formal parties to the treaty. First, it's a threat to state sovereignty. National governments have an interest in maintaining a unified front in international affairs, and states understand their role in the international system

---

76 "Cities100: Washington, DC – Low-Cost Bike-Share Memberships for Low-Income Earners", C40 Cities Case Study, 15 November 2016, www.c40.org/case_studies/cities100-washington-d-c-low-cost-bike-share-memberships-for-low-income-earners
77 "C40 Good Practice Guides: Washington, DC – Green Code and Energy Efficiency Certification", C40 Cities Case Study, 15 February 2016, www.c40.org/case_studies/c40-good-practice-guides-washington-dc-green-code-and-energy-efficiency-certification

managing their country's foreign affairs vis-à-vis the balance of power among other states in the international system[78]. By empowering sub-state actors, the state hollows out and becomes less powerful, exerting less influence on its foreign affairs. Relatedly, and in the worst-case scenario, formally including sub-state actors in international law could embolden secessionist groups or ethnic factions within a state who would prefer a separate legal entity altogether. The Catalonia crisis in 2017 and the long-term question of Kurdish nationalism in light of the Syrian crisis, among other undetermined questions of statehood, will motivate states to block any attempts to include sub-state actors as parties to any international agreement. Realistically, the chances are that it is unlikely cities will play any formal role in international treaties.

However, this fear of a lack of state sovereignty does not exclude sub-state actors from using the international framework. Non-state actors have used international instruments to achieve their political goals in a variety of ways, often by pressuring their national governments to change behaviour. Cities are acting the way the literature in international relations describes non-state actors as acting: operating through international networks, using those networks as a platform for knowledge and information-sharing, and acting to circumvent national governments[79]. Yet cities deserve a status different from that of non-state actors. In the US, leaders of cities are democratically elected and held accountable to voters in local districts. Cities are direct service providers – more so than national governments. They are also directly affected by the effects of climate change and are investing billions of dollars in climate change mitigation and resiliency efforts. They are uniquely situated to have access to both the newest technology and knowledge and the ability to test and innovate on different approaches to reducing GHG emissions.

Importantly, cities are voluntarily declaring to uphold the Paris Agreement and are submitting their own emissions reduction targets along with lists of local goals. By design, this is not very different from what states are submitting to the Paris Agreement. The Paris Agreement is built upon the voluntary commitments of state parties and avoids setting any specific emissions targets, unlike previous treaty attempts. This is partly what made the Paris Agreement innovative and ultimately successfully, in contrast to the now-failed Kyoto Protocol, which sought to pin countries to specific GHG targets and timelines. It is possible that this new structure could provide for more formal involvement for sub-state actors to participate in international treaties. Because the Paris Agreement is a non-binding treaty and the mechanism relies on voluntary targets, cities could submit their local proposals as part of or independently from national proposals, thus contributing to achieving their countries' GHG emissions reduction targets and the global goal of avoiding the anticipated

---

78 Morgenthau, Thompson and Clinton, *Politics Among Nations* (note 2)
79 Keck, M.E. and Sikkink, K. (1998), *Activists beyond Borders: Advocacy Networks in International Politics*, Ithaca, NY: Cornell University Press

global temperature increase. The more transparency attached to the policies and programmes that reduce GHG emissions, the better. Transparency around the impact at different levels of governance can better reveal the link between local action and global action, as well as inform the appropriate way to scale interventions in climate change. This might necessitate defining a different status for city governments in international treaties and precipitate a formal acknowledgement of city affairs as global affairs.

However, the good that could come from these approaches could fall apart if states decide to pull out of international agreements. If the national government and local governments are in conflict, it complicates the status of countries' commitment to international treaties. While the US legal system would insist that national law takes precedence over local law, the US national government has done little to punish cities for their involvement in TMNs on climate change or prevent them from making pledges to the Paris Agreement. This alone is puzzling, because the US national government has targeted local governments in the past for undercutting its policies. Most recently, various US executive branch departments targeted or threatened to target 'sanctuary cities', or cities that plan to ignore new federal mandates on immigration[80]. As mentioned, US state governments have more legal power than city governments. Cities making pledges to the Paris Agreement with no federal government response might be setting a new precedent. At the very least, it signals the importance of some global issues and administration priorities; this means that cities can make significant progress on global issues that the national government decides not to. More optimistically, it might suggest that US global cities are truly norm entrepreneurs on climate change mitigation, driving forward overall innovation by forcing small changes in the way the US national government reacts to their work.

Global trends have made urban politics central to global politics[81]. Global cities have grown in size, population, share of global GDP and prominence. Future trends indicate that the world is urbanizing and that by 2050, 66% of the world's population will reside in cities[82]. As city governments begin to represent the majority of the global population, it follows that they will be incorporated as formal actors in international affairs, recognized by either the United Nations and other international organizations, incorporated as formal actors in international law, or at least a worthwhile subject of research in political science and international relations scholarship. With or without formal recognition, global cities will continue to make progress towards their stated goals and operate through TMNs as informal global mechanisms.

---

80 Ballesteros, C. (2017), "Trump and Jeff Sessions are Going after More Sanctuary Cities", *Newsweek*, 15 November, www.newsweek.com/sanctuary-cities-trump-sessions-department-justice-712965
81 Boudreau, "The Centrality of Urban Politics in a Global Era" (note 20)
82 "World's Population Increasingly Urban with More than Half Living in Urban Areas", United Nations Department of Economic and Social Affairs, 10 July 2014, www.un.org/en/development/desa/news/population/world-urbanization-prospects-2014.html

Conclusion

For policy issues as important and pressing as climate change, defiance of norms by a country as powerful – and as high an emitter of greenhouse gases – as the United States is reckless. In these circumstances, it's time for these sub-state actors to send the message internationally: powerful factions within the United States are on your side. Climate change is an exceptional case; it is the quintessential challenge of global governance, the tragedy of the commons, sitting at the intersection of industrialization and urbanization. Existing literature has identified global cities and their operations through TMNs as a new and novel form of international organization wherein cities co-ordinate policy platforms, specifically on climate change mitigation. Cities bear no legal status in international law, but some international organizations and international instruments have acknowledged the role that cities are playing in mitigating climate change. In the US context, the federal government has refrained from prohibiting cities' involvement in international climate agreements and, relatedly, city governments are pledging to uphold the Paris Agreement, submitting detailed plans that contribute to GHG reduction targets. The broader implications for state sovereignty and international law are apparent, but can be mitigated by developing more flexible treaty mechanisms inclusive of sub-state actors.

Nimble is not a word usually used to describe the United Nations and the international political order. This examination produces more questions than it does answers. As global trends lean towards urbanization, it poses the broader but nonetheless related question of how the international system will change with increasing urbanization. If cities continue to exert influence on global affairs, does that serve to weaken the international system, or does it call for a more malleable international order that can readily address global challenges? In short, the city must be considered an important actor in global affairs.

17 Beyond COP 21
What does the Paris Agreement mean for European climate and energy policy?

Annika Bose Styczynski

This chapter is about assessing the implications of the second international agreement under the United Nations Framework Convention on Climate Change (UNFCCC) on European climate and energy policy. After the Kyoto Protocol of 1997, the Paris Agreement of December 2015 stands for continuity of European diplomatic efforts to provide leadership in the fight against man-made climatic change globally. It is a sign of Europe's intact sense of international responsibility, bearing in mind that, with the invention of the coal-powered steam engine and the petroleum-fuelled internal combustion engine, the disproportionate rise in global greenhouse gas (GHG) emissions emanated from within Europe. With the discovery and understanding of the greenhouse gas effect, two Europeans – Joseph Fourier (1768–1830) and Svante Arrhenius (1859–1927) – laid the foundations of modern climate science research[1]. And today the World Economic Forum (WEF) regularly assesses the risks associated with climate change for our societies and the globe *in toto*. For the first time since the first edition of the WEF's Global Risk Report in 2005, extreme weather events, natural disasters and the failure to mitigate and adapt to climate change are among the top five hazards in both risk assessment categories – likelihood and impact – suggesting growing concern about the importance of climate change-related phenomena[2].

Vigilance, scientific rigour and diplomatic skills characterize the European approach to climate change negotiations. But what are the actual achievements in rewiring the European economy and effectively curtailing GHG emissions? In other words, how consistent are Europe's pledges? And what does the Paris Agreement mean for the further development of European climate and energy policy at EU level and in selected national cases?

To answer these questions the first part of the chapter is divided into three sections. It starts with a brief outline of the comprehensive integrity framework as conceptual lens to the topic, followed by a summary and assessment of the

---

[1] Schellnhuber, Hans Joachim (2015), *Selbstverbrennung. Die fatale Dreiecksbeziehung zwischen Klima, Mensch und Kohlenstoff* (Self-immolation. The fatal three-way relationship between climate, people and carbon), C. Bertelsmann Verlag, München
[2] WEF (2018), Global Risks Report 2018, 13th Edition, World Economic Forum, Geneva

relevant parts of the Paris Agreement. This first section concludes by reflecting on the accomplishments, in view of currently applying climate targets and the implications of the Paris Agreement on future European climate and energy policy. In the second part of the paper, national examples from Germany, Norway and France will be examined to illustrate leading cases of the European climate protection movement.

Conceptual approach

Institutional integrity as defined by Breakey and Cadman is a multilayered phenomenon best captured in a comprehensive integrity framework[3]. The most fundamental question of institutional integrity is that of why the institution – here the UNFCCC and its multilateral agreements – exists. On this basis three more aspects of institutional integrity are subject to scrutiny. To be comprehensively integer the institution has to adhere to its values (coherence integrity), has to exist in an environment that supports these values (context integrity), and has to act in consistency with publicly made claims (consistency integrity). In other words, comprehensive integrity describes the congruency of personal, societal and institutional values and deeds over time.

The integrity framework will be applied here to evaluate major aspects of European climate and energy policy formulation and implementation. Corresponding institutions are drawing their legitimacy mainly from the precautionary principle of mandating actions to avert or prepare for the associate dangers. One can generally argue that these institutions are imposing stricter rules and attempt to redefine the various economic, social and political games being played. Whether they stand up to the scrutiny of an holistic understanding of integrity is a central concern of this contribution.

Essentials and assessment of the Paris Agreement

The Paris Agreement of 12 December 2015 has been celebrated widely as a landmark event in international climate policy formulation. As a multilateral treaty that symbolizes a next step in the long-standing process of overcoming the divide between developing and developed countries, the agreement focuses on climate mitigation and adaptation plans stipulated in the member countries rather than taking a top-down approach. The agreement, however, obliges the signatory states to go back to the negotiation table every five years, aiming to gradually increase ambitions both with regard to goal formulation and ultimately also goal achievement. In this sense, the Paris Agreement is an expression of a long-term commitment that emphasizes global transparency and aspires to hold the global

---

3 Breakey, Hugh and Cadman, Tim (2015), "A Comprehensive Framework for Evaluating the Integrity of the Climate Regime Complex", in Breakey, Hugh, Popovski, Vesselin and Maguire, Rowena (eds.) (2015), *Ethical Values and the Integrtiy of the Climate Change Regime*, Ashgate, pp. 17–27

temperature increase well below 2° Celsius while seeking to limit the temperature increase to 1.5° Celsius above pre-industrial levels[4,5].

According to the Potsdam Institute for Climate Impact Research (PIK), we are currently emitting 41 Gt of carbon dioxide (and equivalents) annually worldwide. To remain within the 1.5–2° Celsius range of temperature increase, the midpoint carbon space available would be 600 Gt. If we continue to emit at the current pace, we are left with 15 years, which means that our emissions have to go down sharply to zero by 2032 to achieve the Paris Agreement goal[6].

When looking at global temperature increase, a number of different sources have been identified as contributing to global greenhouse gas emissions and subsequent temperature increase. Accordingly, the Intergovernmental Panel on Climate Change (IPCC) has distinguished five main sectors to account for the technical sources of emissions. These sectors include: 1) energy and transport; 2) industrial processes and product use; 3) agriculture; 4) land use, land-use change and forestry (LULUCF); as well as 5) waste management. Here, primary attention will be given to case studies in the energy and transport sector. The importance of this sector is emphasized by Carbon Tracker, a London-based NGO, which argues that the threshold of 2° Celsius global temperature increase would necessitate the conservation of more than 80% of the currently remaining coal reserves, 50% of gas reserves, and more than 30% of oil[7]. This being said, Paris is not solely, but essentially, about facilitating and accelerating the transition away from fossil fuels and towards renewable sources of energy.

To examine the integrity of the European climate regime in this respect, the following sections will take an historical perspective which traces developments beginning with the European emissions reference year 1990.

Taking stock and providing outlook: EU climate targets since 2008

Giving special attention to the notion of contextual integrity, this chapter takes an introspective view on the EU climate and energy policy process.

At the political level, beginning with the first commitment period of the Kyoto Protocol (2008–2012), the first round of climate and energy policy formulation at EU level concluded at the end of 2008 with the adoption of the Triple 20 by 2020 goals. The final compromise agreement contained goals of: 1) a reduction of GHG emissions by 20% below 1990 levels; 2) a share of 20% renewable sources of energy in the power mix; and 3) an increase of energy efficiency of 20%.

---

4 UNFCCC (2016), Adoption of the Paris Agreement, UNFCCC/CP/2015/L.9/Rev.1, https://unfccc.int/resource/docs/2015/cop21/eng/l09r01.pdf (accessed 8 March 2016)
5 Bodle, Ralph, Donat, Lena and Duwe, Matthias (2016), "The Paris Agreement; Analysis, Assessment and Outlook", German Federal Environment Agency (UBA) Research Paper, Dessau-Roßlau
6 Figueres, Christiana *et al.* (2017), "Three years to safeguard our climate", *Nature*, 546, pp. 593–595
7 Carrington, Damian (2015), "Leave fossil fuels buried to prevent climate change, study urges", *The Guardian*, www.theguardian.com/environment/2015/jan/07/much-worlds-fossil-fuel-reserve-must-stay-buried-prevent-climate-change-study-says (accessed 11 February 2018)

According to Eurostat[8], the European Union had achieved its emissions reduction target for 2020 already in 2013, even though the European Emissions Trading System (ETS) – a declared major instrument to spurring the transition – had become more of a challenge than a solution. 'Greenhouse gas emissions in the EU-28 (including international aviation but excluding LULUCF), stood at 4,611 million tonnes of CO_2-equivalents in 2013. This figure marked an overall reduction of 19.8% when compared with 1990, or some 1138 million tonnes of CO_2-equivalents[9,10]'. Despite a sharp drop in GHG emissions in 2009 (375.4 million tonnes of CO_2-equivalents in just one year) related to reduced industrial activity during the probably hardest year of the global financial and economic crisis, the year 2013 marks the lowest overall EU emissions on record since the beginning of the time-series in 1990[11]. Eurostat documents further show that, throughout the 23-year period (1990–2013), the largest drops are reported mainly for Central and Eastern European member countries such as Lithuania, Latvia, Romania, Bulgaria, Estonia, Slovakia, Hungary and the Czech Republic, with reductions ranging between 34% and more than 58%. Compared with the EU's share in global GHG emissions, however, the combined share of these Central and Eastern European EU countries totals less than 10%. In other words, their reductions contributed relatively little to overall EU emissions and emissions reductions, respectively.

At the other end of the spectrum, with a share of almost 13% of overall EU emissions in 2013, six member countries (Cyprus, Malta, Spain, Portugal, Ireland and Austria) reported relatively significant increases in emissions as compared to 1990. Except for Austria, which does not comply with its 13% reduction commitment under the EU burden-sharing agreement (Annex II to Decision 2002/358/EC), all these countries legitimately increased their emissions. They are beneficiaries of the burden-sharing agreement which defines how emissions reduction efforts within the EU would be distributed in order to comply with the EU's overall reduction goal of 8% as stipulated under the Kyoto Protocol for the period of 2008–2012.

The heavyweights of GHG emissions in Europe are Germany, the UK and France, with more than 21%, more than 13%, and almost 11% share of total EU GHG emissions in 2013, respectively. The German and French emissions reduction and implementation strategies will be looked into in more detail in the case study section of this chapter.

A second perspective on EU emissions provides the sectoral lens according to which the energy sector (without transport-related fuels) is accountable for more

---

8 Data based on the annual greenhouse gas inventory report by the European Environment Agency (EEA) on behalf of the European Union (EU) to the United Nations under the United Nations Framework Convention on Climate Change
9 LULUCF stands for 'Land Use and Land-Use Change and Forestry'
10 http://ec.europa.eu/eurostat/statistics-explained/index.php/Greenhouse_gas_emission_statistics (accessed 3 June 2016)
11 Ibid.

than 57% of EU-28 GHG emissions in 2013. The transport sector, including international aviation, follows with more than 22%. Since 1990, the share of the energy sector has shrunk. In contrast, transport sector emissions have become so dominant that concerns were raised that the sector would undermine the reduction achievements of other sectors. The European Commission, however, explicitly persists in its viewpoint that '[c]urbing mobility is not an option[12]'. Other sectors, such as agriculture, industrial processes and product use, and the management of waste, account for approximately 10%, roughly 8% and more than 3%, respectively, of total GHG emissions in the EU-28 of 2013.

In early 2014, towards the end of Connie Hedegaard's term in office as first European Commissioner for Climate Action (2010–2014), the EU 2030 Climate and Energy Package was published. The 2030 legislation stipulates: 1) a binding cut in GHG emissions of at least 40% compared to 1990 levels; 2) a binding share of at least 27% of renewable energy consumption; and lastly 3) energy savings of 27% shall be achieved by 2030 against the business-as-usual (BAU) scenario. Both the 2020 and the 2030 legislation show that generic policy formulation at EU level precedes and corroborates the Paris Agreement, which – at the policy formulation level – is an expression of internal coherence.

In its report "The Road from Paris, the European Commission admits that the 1.5°C goal would require 'higher ambitions', however, '[a] clear understanding of the specific policy implications of a 1.5°C goal needs to be developed[13]'. The IPCC would be mandated to address the underlying questions in a special report until 2018. Consequently, a corresponding revision of the EU climate and energy policy would take place at the earliest in 2023, when the world community has to present its goals for the time after 2030[14].

On 5 October 2016, the provisions for enforcement of the Paris Agreement were officially fulfilled and the agreement could go into force on 4 November 2016 – 30 days after meeting the minimum requirements[15]. At the end of November, after the 22nd Conference of the Parties in Marrakech, Morocco, the EU Commission presented – significantly earlier than expected – a comprehensive Winter Package proposing revised EU climate and energy targets[16].

---

12 COM (2011), White Paper, "Roadmap to a Single European Transport Area – Towards a competitive and resource efficient transport system", European Commission, Brussels, 28 March, p. 5
13 EC (2016), "The Road from Paris: assessing the implications of the Paris Agreement and accompanying the proposal for a Council decision on the signing, on behalf of the European Union, of the Paris agreement adopted under the United Nations Framework Convention on Climate Change", Communication from the Commission to the European Parliament and the Council, Brussels, 2 March, COM (2016) 110 final, https://ec.europa.eu/transparency/regdoc/rep/1/2016/en/1-2016-110-en-f1-1.pdf (access 7 March 2016)
14 Bojanowski, Axel (2016), "EU sperrt sich gegen strengere Klimaziele" (EU balks at stricter climate goals), *Spiegel* Online. www.spiegel.de/wissenschaft/natur/eu-kommission-sperrt-sich-gegen-strengere-co2-ziele- a-1080279.html (accessed 2 March 2016)
15 The provisions stipulate that 55 countries comprising 55% of total greenhouse gas emissions must have ratified the agreement
16 http://europa.eu/rapid/press-release_IP-16-4009_en.htm (accessed 13 December 2016)

This proposal pursues three main goals: putting energy efficiency first with a 30% energy efficiency target by 2030, achieving global leadership in renewable energies, and giving a fair deal to consumers.

On the occasion of the publication of this proposal, incumbent EU Commissioner for Climate Action and Energy, Miguel Arias Cañete, stated: 'Our proposals provide a strong market pull for new technologies, set the right conditions for investors, empower consumers, make energy markets work better and help us meet our climate targets. I'm particularly proud of the binding 30% energy efficiency target, as it will reduce our dependency on energy imports, create jobs and cut more emissions. Europe is on the brink of a clean energy revolution. And just as we did in Paris, we can only get this right if we work together. With these proposals, the Commission has cleared the way to a more competitive, modern and cleaner energy system. Now we count on European Parliament and our Member States to make it a reality.[17]'

The proposal contains updates on a number of directives (Energy Efficiency, Energy Performance of Buildings, Energy Labelling and Renewable Energy). In the area of energy efficiency, for example, around 400,000 new jobs will be created and €70 billion saved through insulation measures in buildings and more efficient technical installations leading to reduced import of oil and gas. More specifically, the eco-design directive focuses on energy-efficient consumer goods with high saving potential. This directive and its amendment are expected to yield energy savings beyond 2020 in the dimension of Italy's and Sweden's annual energy consumption together[18]. Beyond that, the package also speaks of the modernization of structures of the electricity market, wants national grid operators to collaborate better across borders through regional operation centres, and aims to partly abolish priority access of renewable energies to the grid[19].

For the time being, a better assessment of the integrity and degree of implementation of the EU 2020 and 2030 targets can be achieved by delving into national contexts and implementation processes, which the next section is dedicated to.

European cases of commitment: Germany, Norway and France

The EC's Winter Package of November 2016 is struggling to strike a balance between the progressive and the defensive forces within the EU while addressing current challenges of the tranformation process. At the same time, the EU seems to be well under way to meet its own overall emissions reduction targets, while Germany as Europe's largest emitter is about to redefine energy dependence.

17 Ibid.
18 http://europa.eu/rapid/press-release_IP-16-4009_en.htm (accessed 30 November 2016)
19 Becker, Markus (2016), "EU-Kommission will Energiemarkt umkempeln", (EU Commission wants to change energy market), *Spiegel* Online, www.spiegel.de/wirtschaft/soziales/winterpaket-eu-kommission-will-mehr-energie-effizienz-a-1123782.html (acessed 30 November 2016)

Germany

The German narrative of an energy transition has attracted attention from all over the globe. Since 1990 the share of renewable energies in the electricity mix has increased from 18.000 GWh, mainly in hydropower, to more than ten times as much in 2015. Electricity generation in Germany today accounts for roughly 30% renewable sources of energy (13.3% wind; 7.7% biomass; 5.9% solar PV; 3% hydropower)[20] and is very likely to reach the 35% target even before 2020.

Since the early 1990s, the StrEG (Stromeinspeisungsgesetz, the Act on the Sale of Electricity to the Grid), and since 2000 the EEG (Gesetz für den Vorrang Erneuerbarer Energien, the Act on Granting Priority to Renewable Energy) authored by members of Parliament Michaele Hustedt and Hans-Josef Fell (Green Party), Hermann Scheer and Dietmar Schütz (SPD), have paved the road for these developments through encouraging and incentivizing bottom-up initiatives for renewable energy deployment.

At the same time, CO_2-equivalent emissions have decreased from roughly 1,250 million tones in 1990 to 900 million tones in 2014[21]. The main reasons for this achievement are the increase of renewable sources of energy through the Feed-In-Tariff of the EEG (roughly 20%) and the industrial transition of Eastern Germany (roughly 36%). This indicates that, already prior to the EU 2020 and 2030 climate and energy framework, Germany drove an agenda aligning with and exceeding the country's Kyoto Protocol obligations. The next important step in convergence with EU level ambitions was the decision of the German cabinet to adopt the Integrated Energy and Climate Program (IEKP) in August 2007, which fell in the midst of the 16th Federal Parliament (2005–2009) – the first grand coalition of Christian and Social Democrats under Chancellor Angela Merkel (CDU). The first formalized revision of a German Energy Program since 1991, which already clearly exceeds the EU 2020 and 2030 targets, however, came into being only in 2010 under a coalition of Christian Democrats and the Liberal Party.

In conjunction with the 2010 Energy Concept, the CDU/CSU (Christian Democratic Union–Christian Social Union) and FDP (Free Democratic Party) had extended the lifetime of nuclear reactors – a decision which experienced abrupt revision shortly later. The tremors of the Fukushima Daiichi nuclear power plant accident went deeply and persistently into the ruling political circles. The initial three-month moratorium, under which seven old nuclear power plants (built before 1980) were shut off temporarily, turned into a permanent decision in June 2011 with the remaining nine reactors to be gradually decomissioned by 2022. Since nuclear energy is considered a carbon-free source of energy, its removal from the German energy mix is not in line with the overall goal of emissions reductions. Given the goal of GHG emissions reductions of 40% by 2020 (down to 750 million tonnes of CO_2-equivalent), this makes the extensive

---

20 German Association of Energy and Water (BDEW)
21 UBA (German Environmental Agency), 2015

Table 17.1 German federal energy concept, 2010

	Climate change	Renewable energies		Efficiency			Energy productivity	Upgrading of buildings
	GHG versus 1990	Power	Primary energy balance	Primary energy	Power	Transport		
2020	−40%	35%	18%	−20%	−10%	−10%	+2,1%	Upgrading energy performance 1% → 2% by 2020 Reduction of heat require-ments by 20% By 2050 reduction of primary energy demands by 80%
2030	−55%	50%	30%	↓	↓	−40%		
2040	−70%	65%	45%	−50%	−25%			
2050	80–95%	80%	60%					

Source: Schafhausen, Franzjosef (2012), "Die Energiewende – Chancen und Risiken" ("The Energy Transitions – Chances and Risks), Arnsberger Energiedialog "Energieeffizienz in der Wirtschaft", Dortmund, 6 Dezember 2012. www.bezreg-arnsberg.nrw.de/themen/a/arnsberger_energie_dialoge/veranstaltungen/veranstaltung_12_12_06/schafhausen_12_12_06.pdf (accessed 14 January 2014)

deconstruction of the German coal sector even more necessary. According to Fraunhofer ISE energy charts, the share of coal (lignite and hard coal) in 2004 was 54.4 GW, already reduced to 47.38 GW in 2014. On top of that, the German Government decided in mid-2015 that over the next years, 2.6 GW, i.e., five larger lignite coal power plants, will be decommissioned. These plants will be considered a capacity reserve, but will not play a role in the future power market design[22]. The German coal phase-out will also be an important contribution to reducing EU-wide emissions, because cheap coal power is smutching the power mix in Germany's neighbouring countries[23].

These three examples – the increase in renewable energies, the phase-out of nuclear energy and a gradual reduction of coal in the German primary energy mix – clearly indicate a form of comprehensive integrity where democratic and institutional values of climate protection prevail and endure. The Climate Protection Plan 2050 of the German Federal Government presented on the occasion of COP 22 in Marrakech strongly carries the mission forward by pointing the way to an almost complete renunciation of GHG emissions, with an indispensable gradual phase-out of lignite and significant emissions reductions in the transport sector[24]. Despite all this, Germany will hardly be able to meet its domestic emissions reduction targets of 2020. However, the recent coalition agreement of Social Democrats and Christian Democrats is a strong commitment to the national, European and international climate targets. It states that the action gap has to be reduced until 2020 and a law on the compliance with the 2030 goals would be needed[25].

Today, when comparing the industrial base of Germany and EU associate Norway, it is striking that the export surplus of both countries is generated in heavily fossil fuel-based industries – Germany in premium car manufacturing and Norway in oil and gas. Triggered by the scandal of German car manufacturer VW under US vehicle emissions standards, investigations by the Institute of Environmental Physics at Ruprecht Karls University Heidelberg and others have shown that multiple German and European brands and vehicle types also do not comply with the Euro standards under which they were licensed and registered. Although this concerns primarily NOx values of diesel engines, it reveals the disruptive and disintegrating nature of ever more stringent pollution regulation.

---

22 *Der Spiegel* (2015). "Koalition beerdigt Klimaabgabe", (Coalition buries climate excise tax), www.spiegel.de/wirtschaft/soziales/energie-koalition-streicht-klimaabgabe-fuer-kohlekraftwerke-a-1041662.html (accessed 2 July 2015)
23 A map of the flows of European electricity across borders can be found here: http://electricitymap.tmrow.co/
24 www.spiegel.de/wissenschaft/natur/klimaschutzplan-2050-regierung-einigt-sich-nach-streit-a-1120863.html (accessed 11 November 2016)
25 Coalition Agreement (2018), "Ein neuer Aufbruch für Europa. Eine neue Dynamik für Deutschland. Ein neuer Zusammenhalt für unser Land. Koalitionsvertrag zwischen CDU, CSU und SPD. Berlin, 7. Februar 2018" (A new decampment for Europe. A new dynamic for Germany. A new solidarity for our country. Coalition agreement between CDU, CSU and SPD) www.cdu.de/koalitionsvertrag-2018 (accessed 8 February 2018)

This should make us more vigilant and can be a warning that emissions reductions on paper and in reality are clearly two different pairs of shoes.

Without speaking of fraud, the following section will describe similarly disruptive tensions in the Norwegian case.

Norway

Norway got lucky when it discovered its oil and gas reserves in the 1960s.

In light of the climate challenge, however, the country probably got even more lucky by the fact that Norway's total electricity production in 2011, according to Statistics Norway, was based on more than 96% renewables sources, mainly traditional hydropower[26]. Due to the high share of renewable sources of energy in the primary energy mix, the country's most emissions intense sector is not the energy but the transport sector. This very fact has been addressed by the most holistic and successful public policy package in support of electric vehicles globally.

The country's fossil fuel industry, however, is under similar stress as western car manufacturers, and puts the Norwegian Prime Minister Erna Solberg on the defensive. 'We will change some things in all parts of the Norwegian economy in order to meet the ambitious goals that we have set in the parliament. But this does not mean that there won't be a need for both oil and gas in the coming decades[27].' But, as stated earlier, the 1.5–2° Celsius range would require keeping a significant share of still-available fossil fuel resources untouched. In consequence, Bellona Foundation, an international environmental NGO based in Oslo, which made its name through laying the foundations for the country's electric vehicle policy, speaks of a 'locked door' for Norwegian oil after 2030[28]. To follow the Norwegian narrative into the future will provide both interesting if not game-changing insights.

On the other hand, well aware of James Hansen's speech in front of the US Congress in 1988, Norway engaged in policy formulation that would help to curb emissions from climate gases in 1989. Rommetvedt *et al.* (2001) call it the three 'milestones' that constitute Norway's national climate policy of the 1990s: first, the Storting decision to limit emissions for the year 2000 at 1989-level[29]; second, a policy concerned with the introduction of a CO_2 tax (including its exemptions); and third, a report of the Norwegian Government to the Storting decision on climate policy, which 'ran up against the most long-standing principle

---

26 Styczynski, Annika (2015), "The Gearbox of Sustainable Innovation. A Comparative Case Study of the Policy Process of Electric Mobility in Norway and Germany", Doctorate thesis, Freie Universität Berlin, 2 June 2015
27 EurActiv (2015), "Norway gives COP21 a cold shoulder", www.euractiv.com/section/climate-environment/news/norway-gives-cop21-a-cold-shoulder/ (accessed 12 March 2016)
28 Ibid.
29 Rommetvedt, H. *et al.* (2001), "Corporatism and Lobbyism in Norwegian Environmental Policy-Making", in Nagel, Stuart S., *Handbook of Global Technology Policy*, Marcel Dekker, New York, p. 438

of Norwegian environmental policy; that is, the principle that Norwegian policy should not put Norwegian industry and commerce in a disadvantageous position compared with foreign competitors[30]. On this, Oskar Grimstad, climate policy spokesperson of the ruling Progress Party, argued that '[i]f Norwegian petroleum activity is shut down, it will be replaced by increased production in other countries that have higher emissions[31]'.

The degree to which the country is concerned over the climate change issue is reflected in further government reports, analyses and increasingly stricter policies. In June 2007, the Norwegian Pollution Control Authority (Statens forurensningstilsyn, SFT) stepped forward with an analysis of mitigation options for the time-line up until 2020, which identified a total of 57 mitigation actions in different sectors, from CCS, buildings and road traffic, to oil and gas extraction, industry, agriculture, waste, and ships. The Ministry gives to consider that '[t]he analysis mainly considers technical mitigation measures. It does not to any great extent include options involving major social change, changes in production levels or changes in behaviour[32]'. This basically accords with the assumption that 'technology is nowadays widely considered the key solution to the dilemma of getting national governments to agree to ambitious carbon reductions while at the same time safeguarding economic development and welfare[33]'.

In Norway, the Low Emissions Commission established by the Government in March 2005 concluded that 'reducing Norwegian emissions by about two-thirds by 2050 is necessary, feasible and not prohibitively expensive[34]'. On the basis of the Norwegian five-year Climate Policy White Paper series, the 2008 Climate Consensus established goals for the reduction of greenhouse gases of 25%, i.e. 15–17 million tonnes of climate-related emissions reductions to be carried out in Norway until 2020. Interestingly, it is only in the 2012 Norwegian Climate Policy White Paper (*Klimameldingen*) that the Norwegian transport sector, which is accountable for roughly one-third of total domestic greenhouse gas emissions in 2010, receives due attention[35]. Then Minister for Transport and

---

30 Dryzek, J.S. et al. (2003), *Green States and Social Movements: Environmentalism in the United States, United Kingdom, Germany, and Norway*, Oxford University Press, Oxford, p. 172
31 EurActiv (2015), "Norway gives COP21 a cold shoulder", www.euractiv.com/section/climate-environment/news/norway-gives-cop21-a-cold-shoulder/ (accessed 12 March 2016)
32 MD (2007), "Norwegian Climate Policy Report No. 34 to the Storting (2006–2007), Recommendation from the Ministry of the Environment (MD), Oslo, June 22, 2007", p. 6; www.regjeringen.no/nb/dep/kld/dok/regpubl/stmeld/2006-2007/report-no-34- 2006-2007-to-the-storting.html?id=507152 (accessed 29 March 2014)
33 Nilsson, M. and Rickne, A. (2012), "Governing innovation for sustainable technology. Introduction and conceptual basis", in: Nilsson, M. et al. (eds.), *Paving the Road to Sustainable Transport. Governance and innovation in low-carbon vehicles*. Routledge Studies in Ecological Economics, London and New York, p. 3
34 MD (2007), see note 32, p. 28
35 MD (2012), "Norwegian Climate Policy Report No. 21 to the Storting (2011–2012) (white paper) Summary Recommendation from the Ministry of the Environment (MD), Oslo, April 25, 2012", p. 9; www.regjeringen.no/pages/38117723/PDFS/STM201120120021000EN_PDFS. pdf (accessed 10 April 2014)

Communication, Magnhild Meltveit Kleppa (Centre Party, 2009–12) pointed out that Norway's climate protection plan has a carbon dioxide emissions reduction target of eventually 30% by 2020. 'The electric car is a very important tool for that, knowing that 40 percent of our emissions come from the transport sector and 60 percent of those come from road transport[36].'

As a consequence that shows the comprehensive integrity with which Norwegians have addressed the sector, the Norwegian market for zero emission vehicles has become the most progressed market, starting to globally lead in per capita vehicles registrations in 2011–12. According to Norsk Elbilforening, electric vehicles sales reached the 40% share in 2017[37]. On the basis of this success, Norway wants to go purely electric now. The country's recently presented traffic plan for 2018–29 stipulates that by 2025 all newly registered passenger cars, light commercial vehicles and buses shall run on electricity, and that by 2030 even half of the truck fleet newly coming into the market shall be electrified[38]. The effects on the country's emissions development are foreseeable. However, the contributions of an ever-increasing share of emissions from aviation, an industry which has not been regulated under the Paris Agreement, are not to be underestimated within Norway, across Europe and the world over. According to the International Energy Agency, more than 11% of global oil consumption occurs in the aviation industry[39]. To address the transport sector carbon lock-in, Europe's biggest aircraft manufacturer and technology developer Airbus is working on promising fuel cell solutions to also decarbonize the aviation sector.

Not quite there yet, but ahead of their time, a delegation from Norway went to France in the 1990s to learn from the city of La Rochelle, on the French Atlantic coast, about advancing electric vehicles. A high share of nuclear power in the French electricity mix provided favourable conditions for climate-neutral electric vehicles development and implementation early on.

France

France is one of the three major emitters among EU countries. Similar to the Norwegian case, the highest-emitting sector in France is transport, while energy sector emissions are relatively low due to the very high share of low-carbon nuclear energy. This could explain why France, alongside Finland, did not have to reduce its emissions under the EU burden-sharing agreement for the Kyoto Protocol (2008–12). Nevertheless, the French Government can pride

---

36 Meltveit Kleppa in Canning, Paul (2012), "Norway Leads the Way on Electric Cars", Agence France-Presse (AFP), April; www.care2.com/causes/norway-leads-the-way-on-electric-cars-video.html (accessed 2 October 2012)
37 Norsk Elbiforening (2017), "Norwegian EV market", https://elbil.no/english/norwegian-ev-market/ (accessed 12 February 2018)
38 Electrive.net (accessed 10 March 2016)
39 IEA (2011), www.theicct.org/blogs/staff/a-world-of-thoughts-on-phase-2 (accessed 12 February 2018)

itself on emissions cuts of more than 11% (excluding LULUCF) between 1990 and 2013[40].

In 2013, then French President François Hollande announced an energy transition law, which was ratified in mid-2015 as *Projet de Loi relatif à la transition énergétique pour la croissance verte* (the Energy Transition Law for Green Growth). The law includes the goal of reducing the share of nuclear energy from 75% to 50% until 2025. At the same time and beyond, renewables in final energy consumption shall increase to 32% cent until 2030 and energy consumption shall reduce by 50% until 2050.

According to the 2017 World Nuclear Industry Status Report, from a peak of more than 78% of nuclear electricity in 2005, the country starts to slowly show reductions with 56 operating reactors having produced 384 TWh (72%) in 2016[41]. It is a common false conclusion that Germany's nuclear phase-out would make the country an importer of French nuclear power. Rather, 'France remains a net importer of power from Germany, by 5.9 TWh in 2014, and has been for a number of years, because German wholesale electricity generally undercuts French wholesale prices[42].' After signing the Paris Agreement in April 2016, President Hollande announced that Fessenheim, the country's oldest nuclear power plant, would be decommissioned in 2017[43]. In the same breath, Ségolène Royale, the country's environmental minister, spoke of a doubling of wind farms, a tripling of solar power generation and 50% of renewables in heat production, which is a remarkable tightening of the 2010 National Renewable Energy Action Plan[44]. On the implementation side, according to Eurostat, the country has in fact increased its consumption of renewables from 9.5% in 2004 to 16% in 2016. That is still seven percentage points away from the goal of a 23% share of renewables in the French energy mix by 2020[45]. However, the meaning of the Paris Agreement for the host country of the 21st Conference of the Parties clearly lies in a window of opportunity for the French energy transition, with a high level of international attention guaranteed.

The development of hydrogen fuel cell technologies for various applications is a case in point. A solid domestic (ADEME, ANR) and EU funding (Fuel Cell and Hydrogen Joint Undertaking) base has promising potential to accelerate development and deployment of hydrogen and fuel cell technologies, including in aviation. For the road transport sector, the French National Low Carbon Strategy

---

40 www.statistiques.developpement-durable.gouv.fr/fileadmin/documents/Produits_editoriaux/Publications/Reperes/2015/highlights-key-figures-climate-2016-edition.pdf (accessed 17 June 2016)
41 Schneider, Mycle and Froggat, Anthony (2017), *The World Nuclear Industry Status Report*, p. 44; www.worldnuclearreport.org/-2017-.html
42 Ibid.
43 http://phys.org/news/2016-04-france-renewable-energies.html (accessed 17 June 2016)
44 See the *Programmations pluriannuelles de l'énergie* (PPE) of October 2016 for further details
45 Eurostat (2018), "Share of renewables in energy consumption in the EU reached 17% in 2016", Press release, 25 January 2018, http://ec.europa.eu/eurostat/documents/2995521/8612324/8-25012018-AP-EN.pdf/9d28caef-1961-4dd1-a901-af18f121fb2d (accessed 12 February 2018)

speaks of 29% emissions reduction through improvements in the energy efficiency of vehicles, i.e., vehicles consuming 2 litres of fuel per 100 km or running on electricity. This being said, it seems that much of what has been outlined under French national policy still has to prove its case to meet the criteria of comprehensive integrity in the case of climate and energy policy formulation and implementation.

Conclusion

The central focus of this chapter was policy formulation in the form of the 2008, 2014 and 2016 European Union (EU) climate and energy policy packages in relation to the Paris Agreement (2016) and selected national approaches to the implementation of a low-carbon technology strategy. Despite all adversities, the country cases show positive and promising signs of alignment to the climate protection movement preserving overall integrity of the endeavour. For a close-up integrity analysis, domestic political-economic factors are probably best explaining the variation in compliance among states to national policy targets and international agreements whether in terms of coherence integrity, context integrity or consistency integrity. In any case, a comprehensive integrity framework also has to acknowledge that '[c]limate change is an issue that presents great scientific and economic complexities, some very deep uncertainties, [and] profound ethical issues', as Mike Toman, Research Manager at the World Bank has put it[46]. The research on climate sensitivity and vast uncertainties about timing and intensity of climate responsivness to emissions reductions are two cases in point. Climate change is and remains a wicked problem, no matter how committed we are to solve it.

---

46 Toman, Mike (2014), www.worldbank.org/en/news/feature/2014/09/30/a-wicked-problem-controlling-global-climate-change (accessed 10 February 2018)

18 Strengths, weaknesses, opportunities and threats to the implementation of the Paris Agreement in the Latin American region

Trishna Mohan Kripalani and Gargi Katikithala

The world has certainly come a long way from arguing plausible deniability about the veracity of climate change to the prospects of signing a binding agreement to strengthen the global response to its threat. The adoption of the Paris Agreement at the largely momentous event of COP 21 recognizes that climate change represents an urgent and potentially irreversible threat to human societies and the planet[1]. It acknowledges that climate change is a common concern of humankind; and that parties should, when taking action to address climate change, respect, promote and consider their respective obligations on human rights, the right to health, the rights of indigenous peoples, local communities, migrants, children, persons with disabilities and people in vulnerable situations, and the right to development, as well as gender equality, empowerment of women and intergenerational equity[2].

While the Agreement is not enough to solve the problems concerning climate change, and there remains much to be achieved in terms of implementation and realization of goals, it sets the global community on a path to achieve a workable solution. The Agreement brought together not only states but also cities, companies, civil society groups and others that complement this effort since the first international conference in 1992. Moreover, after the breakdown of talks in Copenhagen in 2009, the success of the Paris Agreement was much welcomed.

The Paris Agreement provides for:

- Long-term mitigation goals and adaptation;
- A commitment to return regularly to make climate action stronger;
- A response to the impact of extreme climate events on the most vulnerable;
- The transparency needed to ensure action takes place; and
- Finance, capacity-building and technology to enable change.

1 United Nations Framework Convention on Climate Change, "Adoption of the Paris Agreement", http://unfccc.int/resource/docs/2015/cop21/eng/l09r01.pdf
2 Ibid.

It was duly recognized during COP 21 that in order to achieve the goals set out, there must be co-operation from not only the developed countries, but *all* countries. It also acknowledged the role that civil society can play, as well as subnational actions.

The role that several Latin American countries played during the climate summit was indispensable in helping the negotiations advance; this is especially true for Brazil, which maintains diplomatic relations with almost every UN member state, and in the role of Christiana Figueres, the Costa-Rican diplomat who has led the UNFCCC since 2009.

Scope

The paper seeks to analyze the strengths, weaknesses, opportunities and threats to the implementation of the Paris Agreement in the Latin American region. As the region is vast and not unified in needs and means, the paper does not purport to cover all issues pertaining to every country. However, it endeavours to identify overarching issues using country examples. It will stress the aspect of lack of political will and integrity; the issues of climate action vis-à-vis state development and rights of the indigenous people. The paper will first discuss weaknesses and threats, which will then be followed by strengths and opportunities.

Weaknesses and threats

As has been evident in recent years, Latin American countries have increased their influence at the United Nations climate change negotiations and offered potential solutions on coping with global warming. They want to be part of the global solution. However, the Latin American countries have had a fragmented approach at the negotiations. Competing interests and struggles to balance climate action with development plague this region. Hence, the different strands of negotiations can be roughly grouped into three separate identities:

1) Brazil, which has tended to negotiate as part of the BASIC bloc along with China, India, and South Africa;
2) The Bolivarian Alliance for the Americas (ALBA), comprising eft-leaning countries, led by Bolivia, Cuba, Ecuador and Venezuela, has been very vocal about the injustice of global climate change, and whose economies remain highly dependent on the export of climate change-causing fossil fuels;
3) The market-liberal Independent Alliance of Latin America and the Caribbean (AILAC), led by Chile, Colombia and Peru, which has pledged significant emissions reductions on the strict condition that others follow their lead.[3]

---

3 Constatine, Giles, "Climate Change & COP21: What Do Latin American Nations Have To Offer?", *Pulsamerica: Impartial, Direct, Independent, The Impartial Latin American News Magazine*, www.pulsamerica.co.uk/2015/12/04/climate-change-cop21-what-do-latin-american-nations-have-to-offer/

This fragmented approach could pose problems when trying to reconcile national needs at the time of formulating the national policies and plans for implementation of the Agreement.

Lack of integrity and political will

The concept of integrity can be applied to different sorts of agents, including institutions and persons. Institutions can be either formal or informal, including government branches, as in the case of Brazil, which will be examined below. An agent has full integrity if its activities, values and ethics, internal constitution and external relations accord with its self-understanding of its values, i.e., with what it 'stands for'. This definition draws on the inter-related ideas that integrity involves acting in accordance with one's publicly asserted values ('consistency'), being integrated ('coherent'), and that the institution fits with its external environment ('context'): Full integrity requires coherence-integrity, context-integrity and consistency-integrity on an ongoing basis.

- Consistency-integrity refers to the institution's acts, and judges whether such acts are consistent with the institution's public proclamations.
- Coherence-integrity refers to internal organizational arrangements and members' values, and judges whether these institutional qualities cohere in ensuring that the institution's professed values are implemented.
- Context-integrity refers to the environment surrounding the institution and judges whether this context aligns with the institution's successful pursuit of its goals.

Though Latin America is responsible for only 10% of global emissions, it still has a significant role to play. Latin American countries' intended nationally determined contributions (INDCs) have been criticized as insufficiently ambitious. The majority are incompatible with holding the increase in the global average temperature to well below 2° or 1.5° Celsius[4]. Most countries are promising a 'below business-as-usual' reduction as opposed to an absolute reduction in carbon emissions. A study by Climate Action Tracker (CAT) concluded that pledges by Brazil, Argentina, Chile and Peru, which together account for 72% of the continent's greenhouse gas emissions, are 'too timid' and do not go far enough to hold global warming below 2°C[5]. The governments have largely stuck to their current policies, which are insufficient and not forward-thinking to

---

4 Araya, Monica, "Can Latin American Diplomacy At COP21 Spur Interest in the Paris Deal Back Home?", Nivela, www.nivela.org/articles/can-latin-american-diplomacy-at-cop21-spur-interest-in-the-paris-deal-back-home/en
5 "Latin America Needs Stronger Climate Pledges – Analysts", Climate Home – Climate Change News, www.climatechangenews.com/2015/10/29/latin-america-needs-stronger-climate-pledges-analysts/

meet the paradigm of climate change. Costa Rica's pledges and de-carbonization strategy are considered the best in Latin America[6].

The INDC submitted by Brazil, though considered as one of the better ones, has been met by criticism that its goal of reducing carbon emissions by 37% from 2005 levels by 2025 actually represents a step back. The country has already achieved similar emissions cuts over the past decade, largely due to a near 80% decrease in deforestation rates. Therefore, Brazil's pledge could be read as meaning minimal emissions cuts between now and 2025[7].

Further, this region has often been accused of having a lack of consistent goals and a clear path for implementation. The promises as well as policies change quickly, without achieving its slated goals.

For instance, President Dilma Rousseff of Brazil had committed to reforesting 12 million hectares of degraded land and restoring 15 million hectares of cattle pasture. It is noteworthy that Brazil's Forest Code includes a list of placeholders for positive economic incentives for reducing deforestation and restoring degraded areas, as well as a provision for a market in forest reserves (Cotas de Reserva Ambiental). These commitments, though welcome, are tasks that will require not only massive effort but also funds. While Brazil is in the midst of an acute economic and political crisis, one wonders how the monies for these policies will be raised. Brazil's National Climate Change Policy called for the creation of a Brazilian Emissions Reductions Market. The government has made no progress in regulating or funding mechanisms or programmes that could generate the necessary money[8]. Moreover, recent changes to Brazil's Forest Code will, according to numerous environmental experts and activists, make it harder for the country to maintain its progress in curbing deforestation[9]. On the one hand these goals are enunciated, while on the other the Forest Code is made unfavourable to its achievement, all the while without an implementation plan. The timidity of the INDCs and lack of implementation mechanisms for the goals stated with regard to land in Brazil show lack of political will.

While Brazil's INDC seems promising and has even had a significant role to play at the time of negotiations, on deeper analysis it seems to be minimalistic in its endeavour. Brazil purports to be part of the solution, but the public proclamations by the government-asserted goals seem antithetic to the changes in the law, that place hurdles in the way of curbing deforestation. Further, there is a lack of an implementation plan, without any route to realize the funds required to make the concerted effort. It may be argued that the context, i.e., the prevailing

---

6 Constatine, Giles, "Climate Change & COP21: What Do Latin American Nations Have To Offer?", see note 3

7 Ibid.

8 Schwartzman, Steve (2017), "Why and How Brazil Should Do More to Stop Deforestation and Climate Change", EDF Talks Global Climate, http://blogs.edf.org/climatetalks/2015/10/07/why-and-how-brazil-should-do-more-to-stop-deforestation-and-climate-change/

9 "Brazil's Demarcation Plans Put People and Planet at Risk – Climate News Network" (2017), Climate News Network, http://climatenewsnetwork.net/brazils-demarcation-plans-put-people-and-planet-at-risk

environment, is in favour of reducing emissions, and the goals with regard to INDCs and deforestation align with it; however, when the pursuit of goals is looked at, only time will tell whether this will be achieved. Further, a similar analysis can be undertaken of Brazil's energy sector. Although renewables are seen for electricity generation, a large part of the conversation with regard to transport has been left out.

Climate action vs. state development and rights of indigenous people

An example from the ALBA group is that of Ecuador. Ecuador has given rights to the environment in its constitution and recognizes what is called '*Pacha Mama*', however, it is a region poised and ready to move over the scale of development. Large-scale extractive industries are currently being established, and the trend is to exploit fossil fuels. Almost all of the Amazon region has been siphoned off in concessions. Ecuador's Yasuní-ITT Initiative seeks compensation for roughly half the estimated value of certain untapped oil deposits, in order to leave these resources untouched.

One such instance is the large-scale copper and gold-mining concessions granted by the Ecuadorian Government to the Canadian and Chinese mining companies, Lundine and ECSA, respectively, in the Cordillera del Condor mountain range. This is a range in southeastern Ecuador that lies within the territory of the Shuar ethnic group, the second-largest indigenous group in Ecuador. The mining operations in this region will utilize open-pit mining.

Accumulating evidence shows that mining has more often than not slowed development and caused substantial long-term damaging impacts. Moreover, communities that live near the extractive industries' projects bear the brunt of the negative environmental, social and cultural impacts, not to mention the overall impact on climate change.

Ecuador is the first country to establish in its Constitution the rights of nature, ensuring that people have the legal authority to enforce these rights on behalf of the ecosystems, i.e., the ecosystem itself can be named as the defendant. In 2007–08, the Ecuadorian Government rewrote its Constitution in which the rights of nature and indigenous people and their territorial rights are recognized. The new Constitution states that, 'Nature, or Pacha Mama, where life is reproduced and occurs, has the right to integral respect for its existence and for the maintenance and regeneration of its life cycles, structure, functions and evolutionary processes. All persons, communities, peoples and nations can call upon public authorities to enforce the rights of nature' (Ecuadorian Constitution, 2008[10]).

In addition, the rights of the indigenous people are stated 'as Indigenous communes, communities, peoples and nations are recognized and guaranteed... To participate in the use, usufruct, administration and conservation of natural

---

10 http://pdba.georgetown.edu/Constitutions/Ecuador/english08.html

renewable resources located on their lands'. Furthermore, 'To keep and promote their practices of managing biodiversity and their natural environment... The State shall establish and implement programs with the participation of the community to ensure the conservation and sustainable use of biodiversity' (Ecuadorian Constitution, 2008).

Following the ratification of the new Constitution, Ecuador produced a new national development plan based upon the indigenous concept of *sumak kawsay*, or Good Living (*buen vivir*). In sum, Good Living has been defined as 'covering needs, achieving a dignified quality of life and death; loving and being loved; the healthy flourishing of all individuals in peace and harmony with nature; and achieving an indefinite reproduction perpetuation of human cultures...'. The Ecuadorian Constitution, 2008, also defines this term as 'The right of the population to live in a healthy and ecologically balanced environment that guarantees sustainability and the good way of living...' (Ecuadorian Constitution, 2008).

However, in 2009, President Rafael Correa adopted a new Mining Law that seemingly contradicts the Constitution of 2008, the National Plan of Good Living and Ecuador's goal of implementing the right of *buen vivir*. The new law supports the exploitation of mining resources. When protest to this new law arose, he stated that 'everyone is against the destruction of nature but if our development depends on it... it will be exploited'. He also added that there will be 'zero tolerance for anyone who tries to call strikes or generate chaos' (Acción Ecológica, 2011).

The paradox the nation is embroiled in is that the concessions that are being given out to industries undermine the goal of *buen vivir*, however, this same task is being envisaged as the basis of achieving *buen vivir*. It's the same task but with two different outcomes. The expansion of extractive activities is undertaken to increase national revenues, at the cost of cultural, social, environmental factors, and to consequently invest these monies in infrastructures and research in preparation for a post-extractive industry-based economy. In this manner, it is stated that the country is meeting the national plan of *buen vivir*.

State development goals pose a clear threat to climate action. There is clear inconsistency in action, as well as on paper; this, too, is indicative of a lack of integrity.

Further, the extractive industries can be a double-edged sword. The state generates revenues in the form of taxes, fees and royalties from the industries that are said to be targeted for local, regional and national, social and infrastructure projects. The industries tout the benefit of job creation that spurs the demand for consumer goods and services, thereby leading to economic growth and development for the nation. Unfortunately, the economic benefits often flow to the national capital, foreign shareholders and corrupt elites. In addition, the extractive industries often leave a swath of long-term environmental, social and cultural damage that cannot be remediated. This impact is better known as the 'resource curse' Therefore, climate-related policies are undermined by the need to use natural resources for economic growth.

In addition, there is the concern of the indigenous peoples, due to which the existence of extractive industries is being challenged time and again. It is interesting to note that the voice of the indigenous peoples has reached even the Inter-American Commission on Human Rights. Further, research presented at the Global Landscape Forum, a side event at COP 21, indicates that 20% of the carbon in tropical forests lies on indigenous lands. Moreover, the Paris Agreement acknowledges that parties should respect, promote and consider their respective obligations on the rights of indigenous people. Although states often envisage the realization of the rights of indigenous peoples as a weakness by the state, we would argue that it can actually be a strength.

A balance is yet to be achieved on the goal of development vis-à-vis preservation of the environment. The Latin American region is rich in fossil fuel and home to the Amazon and its indigenous peoples. It also has some of the fastest-developing economies; foreseeably, its needs for energy and infrastructure will drastically increase in the coming years. Keeping in mind the right to development, the concerns while formulating any implementation plan for the Agreement will need to take into account not only the needs of the indigenous peoples, but also their opinions and proposed solutions. Reliance ought to be placed on proposed solutions presented by them for example with regard to deforestation and land rights.

The role that indigenous peoples can play in the form of strength is fortified by the following presence at COP 21:

> On December 6, a worldwide coalition of indigenous groups paddled a symbolic canoe up the Seine into central Paris. This was no mere publicity stunt. It signified the important role the indigenous peoples had to play with regard to climate change concerns. On December 8, indigenous groups from Belize, Bolivia, Brazil, Colombia, Guyana, and Honduras won the prestigious Equator Prize from the United Nations Development Programme (UNDP) for their environmental stewardship and advocacy on behalf of indigenous lands – sometimes in opposition to powerful government and private interests. In addition to the countries represented at the award ceremony, indigenous groups from Costa Rica, Ecuador, Guatemala, and Panama were all involved in COP 21, either in national consultations of emissions reduction plans or as direct participants in the conference itself. Many came to Paris dressed in their traditional clothing.[11]

States' goals of development pose a serious threat in the implementation of the Agreement. However, we would characterize the presence of the indigenous people as a strength, provided the governments start to use their know-how and

---

11 "From Zeros to Heroes: The Highs and Lows of Latin America and the Caribbean at COP 21" (2017), NACLA, https://nacla.org/news/2015/12/15/zeros-heroes-highs-and-lows-latin-america-and-caribbean-cop-21

facilitate their participation while developing the national plan. The paper will now move forward to analyse the strengths and opportunities in the region.

Strengths and opportunities

The Paris Agreement is considered a triumph for the Latin American region, perhaps because this region is considered one of the most vulnerable to bearing the brunt of climate change. Some of these effects have already emerged with regard to the droughts, floods, rising temperatures and the ever-prominent El Niño effect. If these increasing risks to the people and environment were not enough for the region to hop to action, further concern was raised by the research from the United Nation's Economic Commission for Latin America and the Caribbean (CEPAL) that suggested the annual economic costs of climate change in Latin America could reach around 2.5% of the region's total GDP.

Emergence of High Ambition and availability of renewables

The Paris Agreement saw small island states and middle-income countries, including many from Latin America, join forces in an unprecedented manner. On the final day of the negotiations, Brazil broke away from BASIC and joined the High Ambition Coalition. This new group of countries is not a negotiating bloc, but it has grown to encompass over 100 countries without distinctions of rich/ poor /large /small/developing /developed, etc. The High Ambition Coalition pushed for four major issues: a legally binding agreement, a long-term goal on global warming, a five-year review of countries' emissions, and a standardized method of tracking progress. As a major developing economy, Brazil's adherence to the coalition was seen as pivotal to ensure the Agreement's success. Further, the Climate Vulnerable Forum advocated for the inclusion of limiting global warming to 1.5°C in the agreement and called for 100% renewable energy by 2050. The forum includes 43 developing countries such as Costa Rica, the Dominican Republic, Guatemala and Honduras. This was a promising end to the COP 21. However, there is a need to maintain these pressures and momentum.

Further, the Agreement is to facilitate a transition to low-carbon economies by mobilizing new investments in key sectors, such as energy. In addition to public funding, calls from many companies, investors, financial institutions and other non-state actors have proved significant. Latin America not only has fossil fuel, but also an abundance of renewable resources.

Although the region will be required to almost double its installed power capacity to roughly 600 GW by 2030, the Inter-American Development Bank says Latin America can meet its future energy needs through renewable sources including solar and wind, which are sufficient to cover its projected 2050 electricity needs 22 times over.

According to Bloomberg New Energy Finance, Latin America has four of the top ten countries for clean energy investment: Brazil, Chile, Mexico and Uruguay. Overall, in 2014, these four countries saw a total of $23 billion in clean energy investments. An increase in these green investments means more jobs, public health improvements and savings in energy costs. For example, according to New Climate Institute, Chile could save $5.3 billion each year on fossil fuels, avoid 1,500 deaths in Santiago due to air pollution and create 11,000 green jobs if it gets on a trajectory toward 100% renewables[12].

Although investments in sustainable development in the region have been increasing, the climate policy across the region suffers from weak implementation. However, this period has also coincided with high prices in oil. With the recent fall in oil prices there is a threat posed to the availability of finance for renewables[13]. This could be turned into a market driven opportunity led by state policies on energy auctions, as adopted by Brazil and Chile.

The progress of the Paris Agreement was assessed at the UN Climate Change Conference, which took place in Bonn in November 2017. Latin American countries which are most vulnerable to climate change have been making palpable efforts to mitigate their vulnerabilities by becoming signatories to the Agreement and trying to achieve growth through cleaner and low-carbon forms of energy. The region as a whole has upped its investment in renewable sources of energies.

> Last month, Argentina held its second renewable energy auction, RenovAr Ronda 2, which drew offers totaling 9,403 Megawatts, eight times greater than the previous auction. Recently, IDB Invest, the private sector arm of the IDB, signed a $104 million financing package for a wind farm in Buenos Aires province, Argentina and in Mexico signed a $75 million loan to finance the Solem PV project, which will be the largest solar power plant in Latin America. In 2016, we financed a total of *$2.69 billion* in climate-related activities such as loans, grants and technical cooperation, bringing us closer to our goal of 30 percent of total approvals of this kind by 2020. We are making progress, but know that significantly more is needed to fully align financial flows towards low-carbon and resilient development.[14]

Sub-national actions

For the Latin American region this is a great opportunity, as there are several best-practice examples from the region that could be used by others.

---

12 http://switchboard.nrdc.org/blogs/cherrera/the_paris_agreement_explained.html; also see *Assessing the Achieved and Missed Benefits of Chile's Intended Nationally Determined Contribution (INDC)* (2015), ebook, New Climate Institute, https://newclimateinstitute.files.wordpress.com/2015/09/assessing-the-achieved-and-missed-benefits-of-chile.pdf

13 Reis, Ciro Marques (2017), *Will the Expansion of Wind and Solar Energy Sources Resist the Fall in Oil Prices? An Overview of Latin America and the Caribbean*, ebook, Konrad Adenauer Stiftung, www.kas.de/wf/doc/kas_43643-1522-2-30.pdf?160202161833

14 Amin, Amal-Le, "Latin America and the Caribbean steps up the implementation of the Paris Agreement", *Global Americans*, https://theglobalamericans.org/2017/11/latin-america-caribbean-steps-implementation-paris-agreement/

In particular, the Amazon states of Mato Grosso and Pará have been considerably praised for reduction in emissions. Further plans and proposals have been put on the table to progress further. Mato Grosso Governor Pedro Taques rolled out the state's 'Produce, Conserve, Include' strategy at an event in Paris for investors and companies. The goal is to leave intact the native vegetation (forest and savanna), while increasing agriculture on already cleared lands. Further, the plan proposes to eradicate illegal deforestation and compensate the owners, as well as restore 2.9 million hectares of degraded land by 2030. Also by 2030, 100% of family farmers are scheduled to get technical assistance, in order to ramp up their share of production of the food consumed in the state and increase household incomes. Overall, between reducing deforestation and restoring degraded lands, the strategy aims to deliver 6 billion tons of CO_2 reductions and removals by 2030[15].

For 23 out of the 33 Latin American countries to have signed the Paris Agreement in 2015 is unarguably a firm commitment of these countries to limit rise in mean global temperatures. In order to achieve this:

> Mexico made an unconditional target to reduce 25 percent of its greenhouse gases and short-lived climate pollutant emissions such as black carbon below "business-as-usual" projections for 2030. This commitment implies a 22 percent reduction of greenhouse gases and a reduction of 51 percent of black carbon. Mexico also set a conditional target: it will reduce its emissions and pollutants to 40 percent below business-as-usual in 2030 if certain conditions, such as a global carbon price, access to financial resources, and provisions for technology transfer, are met.
>
> Brazil pledged to reduce emissions by 37 percent by 2025 and 43 percent by 2030, compared to 2005 levels. It pledges to eliminate illegal deforestation, restore and reforest 12 million hectares, and recover 15 million hectares of degraded pastures and enhance 5 million hectares of integrated cropland-livestock-forestry systems by 2030.
>
> Chile made an unconditional target of a 30 percent reduction of CO2 emissions-intensity of GDP below 2007 levels by 2030 and to 45 percent with further international support.
>
> Costa Rica made one of the most ambitious pledges, setting an unconditional target to keep net emissions below 9.37 MtCO2e by 2030, with proposed emissions per capita of 1.73 net tons by 2030, 1.19 net tons per capita by 2050 and -0.27 net tons per capita by 2100. [16]

---

15 Schwartzman, Steve, "Amazon States, Global Leaders in Emissions Reductions", EDF Talks Global Climate, http://blogs.edf.org/climatetalks/2015/12/11/amazon-states-global-leaders-in-emissions-reductions/#more-4801

16 Edwards, Guy, "How President Trump's Threat to Torpedo US Climate Policies Puts Latin America at Risk", *Global Americans,* https://theglobalamericans.org/2017/02/president-trumps-threat-torpedo-u-s-climate-policies-puts-latin-america-risk/

Reducing greenhouse gas emissions and using carbon-pricing mechanisms to restrict average global temperature rise to 1.5° Celsius as envisaged by the Paris Agreement is high on the agenda of the Latin American countries.

> According to the International Finance Corporation, Latin America and the Caribbean are likely to see USD 1 trillion of clean energy investment opportunities by 2040, of which USD 600bn are expected to materialize by 2030.[17]

The Latin American policy-makers have also been demonstrating a willingness to foster advancement of Article 6 of the Paris Agreement.

> Article 6 of the Paris Agreement sets out three economic instruments: transferring mitigation outcomes, essentially emissions trading schemes; designing a new Sustainable Development Mechanism, which would incentivize the private sector to develop emissions reduction and development projects; and setting a framework for non-market approaches, such as green bonds and carbon taxes.[18]

The Latin American and Caribbean Carbon forum meeting in October 2017 also indicates the region's commitment to the Paris Agreement by outlining its agenda for the meeting to discuss:

- Implementing Nationally Determined Contributions;
- Leveraging public and private finance for climate action;
- Carbon-pricing mechanisms and carbon markets;
- Sustainable development and transformational change;
- Public-private partnerships;
- Innovative business models to fight climate change. [19]

Conclusions

President Trump's decision to withdraw from the climate pact could have a devastating effect on Latin American efforts to combat climate change, and bring all their efforts to advance the Paris Agreement to naught. If the efforts of Latin American countries to combat climate change are not supported by the international community, and especially their powerful northern neighbour, then the consequences could be serious not just for the vulnerable nations themselves,

---

17 "Countries in Latin American and Caribbean Region Leading Climate Action", *Carbon Pricing Leadership*, www.carbonpricingleadership.org/news/2016/10/3/countries-in-latin-american-and-caribbean-region-leading-climate-action
18 Ibid.
19 "Latin American and Caribbean Carbon Forum (LACCF)" (2017), Latincarbon.com, http://latincarbon.com

but for the northern countries as well. There could be replay of the European refugee crises, but this time the refugees would not be from man-made conflict, but instead from natural disasters. Therefore, helping Latin America in its efforts to foster the Paris Agreement is as much in the interest of the international community as it is of Latin America itself.

The countries of Latin America ought to place their bet on renewables rather than investing in infrastructure and energy systems of the last century. The Latin American region wants to be a source of climate solutions, as the negotiations show. However, it is up to each country how they translate the strengths, weaknesses, opportunities and threats into proactive solutions that are a practical response to the region's sustainable development needs.

Index

23rd Committee of Parties at Bonn 26, 243, 246, 252

accountability: benefits of 'soft' law top-down approach 41; commitment 26, 29; facilitative mechanism of implementation 39; goals 30; mechanisms 178; recognition of shortcomings 31; support package in the Paris Agreement 35
Ad Hoc Working Group on the Durban Platform for Enhanced Action (ADP) 117, 170–71
Ad Hoc Working Group on the Paris Agreement (APA) 106–7, 112, 117, 124
adaptation 27–37, 39, 40, 131, 133–34, 150, 165, 172; *see also* Article 7 of the Paris Agreement
Adaptation Fund 15, 194–211, 230
Aichi Biodiversity Targets 88, 90, 93
Alliance of Latin America and the Caribbean (AILAC) 298
annex and non-annex countries 129; Paris Agreement 169
Arrhenius S. 283
Article 2 of the Paris Agremeent 23, 165, 170, 180; *see also* Common-but-Differentiated Responsibilities; public institutional justification
Article 4 of the Paris Agreement 26, 170–72, 176, 233; *see also* bottom-up; hard law; Nationally Determined Contributions
Article 6 of the Paris agreement 26, 170, 307; *see also* bottom-up; soft law; Sustainable Development Mechanism

Article 7 of the Paris agreement 27, 133, 271 *see also* Cancun Adaptation Framework
assessment: Article 8 of the Paris Agreement 29; collaborative efforts between the private sector and governments 39; country introspection 33; cycle duration 36; facilitation mechanisms 40; information quality 30; investment portfolios of financial institutions 28; mitigation commitments 34

Bali Action Plan 7, 66, 130, 189
Basel Convention 58, 68, 74, 77; Article 14 Basel Convention 73 *see also* COP 12; Regional and Coordinating Centres 73
Bharat VI 225
bilateral investment greements 191
biodiversity: conservation 82, 84, 98; goal 88; intellectual property rights of 84; loss 80, 85, 97; sustainable use of 82; target 87
black carbon 306
Bloomberg, M. 273, 275, 277, 304
Bolivarian Alliance for the Americas (ALBA) 298, 301
bottom-up approach 20, 23, 25, 26, 39, 40, 41, 42–47
Brazil 127, 129, 170, 182, 184, 189, 298–301, 303–6
Brazil's Forest Code 300
Breakey and Cadman 172, 175, 176
burden sharing agreement 3–4, 176, 286, 294
de Blasio, B. 273, 275, 277

310 *Index*

Cancun Adaptation Framework 133
Cancun Agreements 133
capacity-building 131; Article 11 of the Paris Agreement 29; financial, institutional and educational 145, 172; technological support 171; transparency 173
capitalism 155, 156
carbon credits 149
carbon-capture technology 235
carbon dioxide 212
carbon emissions 22, 33, 52, 108, 168, 170, 192, 235–36, 239, 285, 294, 299–300
carbon market 15, 194
carbon sink 38, 228, 254
carbon trading 149
Cartagena Protocol on Biosafety 84
Central Electricity Authority (CEA) 257–58
Circuit Court of Appeals 236–38
cities a.k.a. sub-state actors 16, 50, 133, 215, 233, 235, 246, 256, 258, 264–65, 270–71, 279–80, 282, 297
Civil Society Equity Review 129–30
Clean Development Mechanism (CDM) 6, 25, 22, 34, 148, 185–86; Article 12 of the Kyoto Protocol 22
clean power plan 232, 234–35, 237–40, 246, 249–50
Clean Water Act/rules 239, 247
Clear Air Act 235, 237, 248
Clearing House Mechanism 93
Climate Action Tracker (CAT) 299
climate finance 130, 135, 149; Article 9 of the Paris Agreement 27; duty of developing countries 128; Global North and Global South 145, 175
Climate Investment Funds (CIFs) 204
coal ash 247–48
coal mines 239–40, 242, 246, 249–50, 252, 259, 261
coal reserves 45, 255
codification 1–2
Coherence integrity 165, 178
collective carbon budget 123
collective progress 109–10, 113; moral 118–19 *see also* equity
collective responsibilities 81–82

Commitment period 21, 31, 34–36, 47, 126, 168, 170, 285; Kyoto to Paris 44; under the Paris Agreement 170
commitment/s: capacity building during the cycle 29; Paris Agreement 37; Paris Agreement and Doha Amendment 20–23; party efforts 26; scope and frequency 38; 'soft' law 41; sustainable process in the equity framework of periods 30–35; synergy of approaches 40
Common-but-Differentiated-Responsibilities (CBDR) 21, 25, 33, 37, 43, 128, 132, 137, 164–65, 170, 172; Article 3 of the UNFCCC 166–167; Article 6–22 of the Convention on Biological Diversity 81–83; CBDR under Paris Agreement 164, 165, 172, 173, 175, 178; climate finance 145; developing and developed countries 165; UNFCCC, Kyoto Protocol and Paris Agreement 166–67
compliance: approaches 11; assistance to a well-funded financial mechanism 6; delayed 168, 171; effective regime 5; implementation of agreements and obligations 4; incentives and sanctions 10; institutionalization in the regime 9; national policy targets and international agreements 17; non-compliance when getting benefits 8; non-compliance in the biodiversity convention 12; state compliance 7;
Compulsory Licensing 14, 15, 179, 180, 186, 188, 190, 192
Conference of the Parties serving as the meeting of the Parties to the Paris Agreement (CMA) 105, 109, 112–13, 117
Convention on Biological Diversity (CBD): binding operational commitments 80; ecosystem services 97 *see also* biodiversity; national biodiversity strategy 90; national sovereignty 80; Natural Biodiversity and Action Plans (NBSAPS) 83, 86, 89–91, 93–96, 98–99, 101, 103; non-binding goals 84
Convention on International Trade in Endangered Species of Wild Fauna

and Flora (CITES) 58, 68, 72, 75–76; National Legislation Project 73, 75
Cool cities network *see* urban heat-island effect, green and cool solutions
COP 12 68
COP 16 68, 133, 222
COP 17 16, 128, 222
COP 18 16, 223
COP 19 16, 127, 129, 223, 255
COP 20 133, 223
COP 21 16, 37, 45, 113, 134, 138, 139, 154, 180, 189, 297–98, 303; comparison with Kyoto Protocol 154; ethics 155
COP 22 127, 291
COP 23 26, 28, 246
COP 30 1, 39, 65, 87, 118, 120, 172, 175, 271
COP 7 21, 35, 68, 133

decarbonization 154
decision-making 31, 60, 87, 99, 140–41, 143–44, 147, 149, 178, 199, 200, 213, 217, 228, 272
deliberative legitimacy 114–15, 123
demonetization 222, 257
Department of Energy (USA) 250–51
depositary 42, 45, 243
developed countries 3, 6, 10, 13, 15, 21, 25, 27, 30, 34, 36–37, 39, 43, 46, 49, 68, 81–83, 93, 100–4, 111, 116, 128, 130, 149, 179–81, 184, 190, 203, 223, 298, 304; finance 28; 'Global North' 142; historical contribution 127; Montreal Protocol 168; non-annex countries 149; Paris Agreement 169, 171
developing countries 3, 6, 8, 10, 13, 15, 22, 24–25, 27, 37, 39, 46, 67, 78, 81, 101, 111, 128, 129, 130, 133, 135, 145, 154, 172, 174, 179–82, 184–85; CBDR 166; as 'Global South' 142; as non-annex countries 149; Paris Agreement 169, 171; small developing island countries (SIDs) 145; *see also* power balance between the North and South 145
Development Dividend 179
dispute settlement 141; Article 14 of the UNFCCC 51; Article 24 of the Paris Agreement in UN system 145; of the UNFCCC regime 51, 141, 142, 149; Doha Declaration 187–88
Doha Amendment 21–22

Early Warning Systems 27, 35, 194, 292
Earth Negotiation Bulletin 189n41, 222n2, 223n3
economic diversification 13, 137, 195, 203
economic sustainability 154, 196, 199
economics and climate change 151, 153, 155
Ecuador 17, 298, 301–3
effective implementation 2, 29, 31, 36, 63, 88, 93, 101, 102, 138, 139, 216; improvement 74
effectiveness 2, 8, 9, 25, 40, 48, 61, 78, 96, 130, 138, 139; Agreement concerned finance 134, 135; Kyoto Protocol 5; stocktake 132, 136; trade-off with equity 130
electric car/vehicle 249, 260–61, 276, 292, 294
emissions: assigned amount of units (AAU) 22, 167–68; certified Emission 22, 185, 278, 279; Clean Power Plan 235; COP 21 154; emission allocation 99, 108, 111, 128–29; emission reduction 4, 9–10, 17, 35, 39, 44–46, 100, 124, 128, 130, 138, 167–69, 171, 180, 182, 209, 224, 232, 286, 291; Emission Reductions Market 300, 306; emissions entitlement 108; Green Climate Fund 215; greenhouse gas emissions 6, 21, 43–45, 97, 99, 150, 169–71, 179–82, 195, 273–77, 299, 306–7; Kyoto Protocol 21–23, 25; operationalizing equity 127–30; USA 236–39, 244, 254
'en banc' review 238
energy efficiency 194, 204, 209
energy policy 194, 205, 255, 260–61, 283–85, 287, 296
Environment Protection Agency (EPA) 234, 279
Environment Sound Technologies (EST) 15, 179, 184n23, 185, 186n31
Environmental Social Management Plans/ ESMP 210

312 *Index*

equitable burden sharing 176
equity: Articles 2 and 4 of the Paris Agreement 128; capacity-based 108–9; distribution 109; Equity and CBDR 166, 176; genetic resources 81; intergenerational 108; intra-generational 108; need-based 108–109; principles 114–15, 119–21, 123; procedural virtues 114–15; *see also* Article 14 of the Paris Agreement
ethical responsibility 14, 155, 157–58, 161
ethics 13, 111–12, 115–16, 151–63, 249
European Commission 126, 287
European Trading System (ETS) 286
European Union (EU) 16, 168, 170, 179, 196, 207, 268, 270, 283–96
exclusive rights 186, 190
Executive order (USA) 230, 239, 240, 243

facilitation implementation mechanisms 62–64, 67–70, 74–76, 78; Article 15 of the Paris Agreement 31–32, 136, 177
fairness 45, 79, 81, 111, 114, 124, 157, 159; duty or obligations 157, 159; fair care 157, 159, 163; 'fair share' of mitigation targets 129
Falk, R. 20, 55
Federal government (USA) 16, 52, 197, 202, 233, 235, 237, 269–71, 279, 281–82, 291
follow-up mechanism 87, 92–94, 100 *see also* accountability
forests 26, 38, 215, 223, 223–24, 226, 228–29; afforestation 38, 225, 254, 300–1, 303, 306
fossil fuels 22, 48, 52, 129, 228, 232, 234–36, 241, 246, 248–49, 254, 257–58, 275, 285, 291–92, 298, 301, 303–5
fracking 232, 241

game theory 233, 253
Green Climate Fund (GCF) 36, 131–32, 206, 208, 214–15, 228, 242
genetic resources 82; Article
George W. Bush administration 263, 270, 273, 276
Germany 16, 184, 191, 230, 284, 286, 288–89, 291, 295

Global Biodiversity Outlook (GBO) 86, 90, 92, 95 *see also* reporting systems
global city 265–67, 302; globalisation 266, 270; population 255, 266–68, 276, 278, 281
global climate 6, 11, 13, 181, 192, 196, 212, 214, 233, 270, 279, 298
Global Environment Facility (GEF) 10, 70, 71, 74, 204
Global Stocktake (GST) 12–3, 24, 31, 105–7, 126–136, 175, 224, 227, 257; accountability mechanism 175–76; Article 14 of the Paris Agreement 30, 104, 126, 128, 135, 176, 227; Article 14 of the Paris Agreement Ad-Hoc Working Group 106–7; collective stocktake 136; equity 126–37; NDCs 100
global transparency 284
governance 31, 33–4, 40, 62, 78, 138, 139, 194, 269, 282; and Cancun 150; global governance 21, 53, 57, 100, 173–174, 263–265, 267, 269, 272, 282; quality 139, 141; and UNFCCC 142, 144, 145, 146, 147, 148; structure 140, 198–99, 272; systems 139; value 140;
Green Climate Fund **131,** 132, 206, 208, 214, **215,** 228
Green Technology Transfer 182, 186, 190
Gross Domestic Product (GDP) 110, 120, 222, 225, 228, 232, *245,* 256, 267–68, 273, 275–76, 281, 304, 306
Gutteres, A. 14, 152, 156, 163, 243

hard law 19–21, 23, 25, 32, 40, 41, 115
Human-Environment Systems 194
human rights 1, 5, 13, 15, 24, 48, 52, 108, 115, 117, 137, 156, 189, 197, 199, 297, 303
hybrid approach 47: hybrid architecture 45.

implementation: analysis of the agreements 40; Article 15 of the Paris Agreement 39; capacity-building 63, 65–66, 72–73; commitments and state compliance related to agreements 11–18; compliance mechanisms 68; compliance through the Paris Agreement 32;

effectiveness and transparency 29; by the executive branch 1; gaps 62–63; implementation of the Paris agreement 3–4; innovation 31; insufficiency of international law 2; objectives and mechanisms of the Kyoto Protocol 21–27; problems of the Kyoto Protocol 20; related to monitoring and reviewing 6–8; Responsibility to Protect 41; strategies 36; transparency 30; treaty 19; work of committees 37

indigenous people 17, 24, 27, 50, 297–98, 301–3

Integrated Energy and Climate Program (IEKP) 289

integrity: institutional, coherence, context, consistency, comprehensive 2, 17, 26, 34, 41, 118–19, 166, 178, 212, 218, 221, 283–85, 288, 291, 294, 296, 298–99, 302

Intellectual Property Rights 84, 180, 181, 186, 187, 189, 190

Intended Nationally Determined Contributions (INDC) 113, 127, 128, 129, 132–33, 154, 182–83, 223–27, 231, 242, 254, 299–301

Inter-American Development Bank 208, 304

intergenerational equity 48

Intergovernmental Panel on Climate Change (IPCC) 23, 34, 106, *107*, 113, 118, *124*, 127, 132–32, 181, 188, 246, 285, 287

Interim Inter-Ministerial Climate Change Secretariat (IIMCCS) 195, 208

international agreement 1, 6–7, 15–17, 53, 61–63, 92, 101, 128, 133, 139–40, 180, 187, 189, 192–93, 197, 201, 224, 263–64, 270, 280–81, 283, 296

International Assessment and Review (IAR) 101–2

International Consultation and Analysis (ICA) 101–2

International Development Banks (IDBs) 204

international networks 16, 263–64, 280

International Panel for Climate Change (IPCC) 23, 34, 106–8, 113, 118, 124, 127–29, 132, 133, 181–82, 188, 246, 285, 287

International Solar Alliance 49

intrusiveness 8, 18

Joint Implementation Mechanism 22–23, 25

Kierkegaard, S. 161–63
Ki-Moon, B. 156, 180, 189, 274
Kyoto Protocol: CBDR 164, 167; Compliance Committee 150; COP 21 154; emission standards 270; Enforcement Branch 175; European Union 283–94; Patents 184; top-down approach 44–46, 55, 169, 178

Latin American countries 17, 23, 272, 298, 305–7

Least Developed Countries (LDCs) 203, 223

less favourable yreatment 191

Livingstone K. 273

Lomborg, Bjorn 14, 153–56, 160–63

McCarthy, G. 235, 239, 246–47
McKibben, B. 152, 162–63
MCOP 94
Millenium Development Goals (MDG) 88
mining 232, 236, 241, 246, 253, 259, 264, 301–2
mitigation: Article 10 of the Kyoto Protocol 168; climate change mitigation 16, 49, 148, 150, 214; collective nature 171; conditional mitigation targets 132; 'hard' obligations 176; mitigation ambition 111, 128, 136; mitigation goals 28, 30, 32, 175; mitigation obligations 168; mitigation targets 25, 119, 126, 129, 131; substantive mitigation obligations under Paris Agreement 172; voluntary mitigation targets 132

Monitoring, Reporting and Verification (MRV) 30, 101, 229

monopoly: rights 183, 190; powers 189; profits 192

Montreal Protocol on Substances that Deplete the Ozone Layer 5–7, 43, 133, 171; delayed compliance 171; differentiation in substantive obligations 168

314 *Index*

multilateral consideration of progress 165, 175, 176
Multilateral Development Banks (MDB) 204
Multilateral Environmental Agreements (MEA) 130, 166, 173, 177

Nagoya Protocol on Access and Benefit Sharing 84
National Action Plan on Climate Change (NAPCC – India) 212, 214, **216**
National Adaptation Plans (NAPS) 36, 133, 203
National Adaptation Programs of Action (NAPA) 133, 197, 203
National electricity policy and plan 205, 255, 257–58, 260–61
National Implementing Entities (NIEs) 204
National Oceanic and Atmospheric Administration (NOAA) 241
Nationally determined contributions (NDCs): accountability 175; accountability and sustenance 3; Article 3 of the Paris Agreement 25; bottom-up 105, 111, 113–14; emission targets 47–48; flaws and insufficiencies 17; mandates, 'ratification fatigues' and equity-based considerations 11–13; nationally determined targets 46; party commitments 2; reduction objectives 44; required emission reductions 46; strategic long term approach 5–8; top-down 111, 114; transparency 172; voluntary commitments 46; voluntary efforts 46; *see also* intended nationally determined contributions (INDCs)
negotiations (including goal formulations) 4, 5, 7, 16, 31, 32, 39, 81, 91, 101–2, 130, 139, 149, 167, 186, 189–91, 196–97, 212, 283–84, 298, 300, 304, 308
NGOs: incorporation related to compliance 5; monitoring and benefits 8; negotiation on multilateral agreements and state assistance 7; reduced emissions 10; resources 17
non-compliance: Article 14 of the Paris Agreement 51; Article 41 of the UN Charter 52; Article 42 of the UN Charter 52; legal procedure 51; post-ratification 53
non-party stakeholders 49, 149
Norway 16, 52, 284, 288, 291–94

objective: Article 9 of the Paris Agreement 36; Article 11 of the Paris Agreement 29, 229; Article 15 of the Paris Agreement 28; binding obligations of the Kyoto Protocol 38; developed countries 24; five-year cycles 31; Kyoto Protocol 22; national sustainable development 27, reduction 34
obligation: approach of the Republican administration 23; CBDR of the Kyoto Protocol 33; comparison 20; 'hard' and 'soft' law 19; nature of obligations and the activities involved 38; OECD CPI Report on Climate Finance 134; Paris Agreement being a review mechanism of IPCC 34; role of parties 36; theme of finance 28
oil and gas: shale gas 45, 233, 235, 240–41, 250, 261, 288, 291–93
organizational responsibility 140, 141, 144, 146
oversight mechanisms 66, 102, 165, 173, 175

Paris Committee for Capacity Building 149
patent suppression 180, 183, 184, 186, 187, 190, 192
Permanent Court of International Justice 191
Perry, R. 239, 249–51
political will 3, 17, 94, 156, 185, 298–300
power: blue solar-panelled rooftops 245; hydropower 289, 292, 304–5; nuclear 1, 5, 289, 294–95; solar 194, 204, 208–9, 225, 250, 255; tiles and storage 295, 289; wind 289
precautionary principle 158, 163, 254
President Obama administration 232, 235, 239, 241, 242, 246, 252–54
Primary Energy 181, 290–92
Principles, Criteria and Indicators (PC&I) 140

Index 315

Pritchard, D.J. 83
productive deliberation 141, 144, 147
Pruit, Scott 234, 239–40, 247–49
public health 188, 191, 239, 247, 261, 305
public institutional justification (PIJ) 165, 178

Quantified Emission Limitation and Reduction Committee (QELRO) 167–68, 177; *see also* Article 2 and 3 of the Kyoto Protocol
quantitative emission limitation 44

Rajamani L. 47, 102
Ramsar Convention 58, 69, 72; Executive Secretariats 72, 73; *see also* reporting systems
Ratification Fatigue 21, 42, 44, 55, 179, 243, 254, 302
reciprocity process 92
regional centres 71, 73, 75
renewable energy 1, 15, 130, 181, 205–6, 209–233, 235, 242, 245–46, 249–51, 255, 257, 259, 276–78, 285, 287–89, 291–92, 295; investment 301–2, 304–5, 308; technology 15
reporting systems 65, 69, 72, 74
resilience centre 15, 213, 214, 221;
responsibility 13–4, 34, 95, 155, 157, 159, 166, 211, 283; historical responsibility 121, 130; organizational responsibility 140; responsibility to protect 55
Rio Declaration 64
Rio Earth Summit of 1992 21, 67, 79; Rio and CBDR 165–66; Rio and North-South discourses 167

Sassen S. 266, 270
Scaling Solar Programme 209
Section 111(d) and Section 112 of the Clean Air Act 237, 239
sector: private 27–28, 33, 39, 145, 197, 204, 206, 208, 255, 260, 305, 307; public 4, 82, 91, 257, 259, 304
self-differentiation 171
signatories 14, 44, 155, 233, 243, 284, 305
social justice 15, 81, 156–57, 159, 162–63, 195, 197, 201, 209–10;

environmental justice 198–99; socio-environmental justice 200
'soft' law 2, 8, 19, 20, 25, 26, 27, 32, 40, 41
sovereignty 11, 80–81, 102, 135, 191, 264, 270, 280
stakeholders 16, 20, 49–50, 65, 91, 96, 99, 138, 142–43, 145, 149–50, 200, 202, 205–6, 210, 216–17, 220, 231, 243, 246, 261, 265
state reporting of performance 65, 74, 90, 174
Stewart, R 174
Stockholm Convention 58, 68, 74, 77 *see also* COP 7
strategic patenting 183
Strategic Plan for Biodiversity 80, 85, 98, 100 *see also* capacity-building
Strategy on Resource Mobilisation(SRM) 89, 100
Subsidiary Body of Implementation (SBI) 93, 95–96, 100, 103
subsidy 225
sustainable development agenda 63
Sustainable Development Goals (SDGs) 13, 24, 137, 273
Sustainable Development Mechanism (SDM) 26, 298

target: in the agreement 2; compliance of the Paris Agreement 8–13; reduction targets 16
technical expert review 29, 30, 102, 174; technical expertise 34
technology transfer 6, 10, 12, 15, 29, 30, 35, 73, 81, 87, 93, 193, 306; Article 10 of the Kyoto Protocol 185; Article 10 of the Paris Agreement 28, 181; capacity building 66; elements of implementation of Paris Agreement 131; transparency 173
temperature goal 105, 123
Tillerson R. 244
top-down 19, 20, 25, 33, 40, 41
Totin, E. 215–16
Transitional municipal networks (TMNs) 267–68, 272, 275, 281–82: C40 263–64, 267–68, 271–79: cities 16 *see also* ICLEI or local governments for sustainability

transparency 2, 11, 21, 24, 26, 39, 141, 148, 173; Article 13 of the Paris Agreement 29–30, 174–75; internal and external transparency 172; framework 102
transport sector: aviation 133, 196, 285, 287, 291–95
triple 20 16, 285
Trump administration 233, 235, 239–40, 246, 249, 252–53, 259, 262, 275
Trump, Donald 16, 151, 163, 233, 234, 247, 263, 275, 306–7

United Nations Conference on Environment and Development (UNCED) 79, 80, 140 *see also* Earth Summit
United Nations Environment Programme (UNEP) 59, 65, 66; UNEP Gap Report 133
United Nations Framework Convention on Climate Change (UNFCCC): Cancun 203; CBDR 166–167; Comparison with Paris Agreement 197–103; differential climate action 167; government indicators 142, 143; Green compulsory licensing 180, 186, 192–193; managerial, facilitative, 'non-adversarial', 'non-punitive' 177; quality of governance on basis of region 144; quality of governance by sector 146
Universal Declaration of Human Rights (UDHR) 117, 123
urban heat-island effect, green and cool solutions 274
US Clean Power Plan (USCPP) 153
US Supreme court 235, 237, 269

Vienna Convention 43

White paper 287, 293
Winter package 287–88
World Bank 128–29, 204, 209, 296
World Economic Forum (WEF) 282–83
World Resources Institute 134
Writ of certiorari 237

Made in the USA
Coppell, TX
23 June 2021